Cases on Distance Delivery and Learning Outcomes:
Emerging Trends and Programs

Deb Gearhart
Troy University, USA

INFORMATION SCIENCE REFERENCE
Hershey · New York

Director of Editorial Content:	Kristin Klinger
Senior Managing Editor:	Jamie Snavely
Assistant Managing Editor:	Michael Brehm
Publishing Assistant:	Sean Woznicki
Typesetter:	Mike Killian, Sean Woznicki
Cover Design:	Lisa Tosheff
Printed at:	Yurchak Printing Inc.

Published in the United States of America by
Information Science Reference (an imprint of IGI Global)
701 E. Chocolate Avenue
Hershey PA 17033
Tel: 717-533-8845
Fax: 717-533-8661
E-mail: cust@igi-global.com
Web site: http://www.igi-global.com/reference

Library of Congress Cataloging-in-Publication Data

Cases on distance delivery and learning outcomes : emerging trends and programs / Deb Gearhart, editor.
p. cm.
Includes bibliographical references and index.

Summary: "This book provides practitioners in the field of distance education a case book that is a valuable reference of best practices in the field"--Provided by publisher.

ISBN 978-1-60566-870-3 (hardcover) -- ISBN 978-1-60566-871-0 (ebook) 1. Distance education--Planning. 2. Distance education--Research. I. Gearhart, Deborah L.
LC5800.P5C37 2010
371.35--dc22

2009040839

British Cataloguing in Publication Data
A Cataloguing in Publication record for this book is available from the British Library.

Table of Contents

Section 2
Cases on Distance Learning Programs

Section 3
International Cases

Detailed Table of Contents

With increased emphasis on transparency and learning outcomes, the NCA Higher Learning Commission has challenged institutions to demonstrate greater accountability in educational programming (2000). This article traces the steps taken to establish a set of guiding principles, or best practices, for endorsement by the American Association for Collegiate Independent Study (AACIS) membership. The AACIS membership is comprised of practitioners from institutions engaged in the provision of self-paced, independent educational opportunities for students who require flexible options, including year-round enrollment. While it was anticipated that agreement on all aspects of online delivery would be impossible, a consensus on core principles was seen as attainable. The purpose of the initiative is to lay the groundwork for further refinement of standards of practice by practitioners engaged in the delivery of online, distance, and independent learning programs.

This chapter addresses applied strategies for modularizing e-learning along both an academic university track and a commercial one for practitioners in the field. Professional certifications and academic qualifications have been seen as complementary in some ways, antithetical in others. Another way to visualize both is as one-in-the-same, albeit with versioning for the various differences. The chunking of a curriculum for both a formalist college setting and for a business one involves creative applications of the module format, particularly given the disparate needs and learning outcomes of the two different learner audiences. This case examines the differences between the learning needs of both demographics. This case then sets the dual-design scene from an instructional designer point-of-view.

This case study outlines problems with student conduct in an online undergraduate program and explains how a student code was applied to resolve the issues and institute procedures to reduce future incidents of academic dishonesty and incivility. The study describes several instances of student misconduct and explains how online program administrators responded by improving communication with both students and faculty and by modifying course design and development processes, as well as instructional practices. It also reports on how other administrators assisted in handling resolution and discipline. While technology itself may both complicate the maintenance of conduct standards and provide new ways to protect academic integrity, this study demonstrates that the introduction of technology should not change the rules.

This case study examines how achieving close alignment between course objectives and course assessments should be an essential goal in the course design/revision process regardless of what mode of delivery is involved. By examining the revision of two courses (Western Civilization I & II) offered at Thomas Edison State College, the authors demonstrate how the application of sound instructional design principles to achieve this alignment resulted in the measurable improvement of student learning outcomes. The major issue examined in this study is how to achieve a close correlation between what a course states that a student should be able to do after successful completion, what practice it offers them to achieve this proficiency, and how they are assessed to determine what level of proficiency they have attained.

This chapter will examine the organizational and technological challenges encountered by the highly successful Great Plains Interactive Distance Education Alliance (Great Plains IDEA) since its members began offering multi-university online academic programs in 2000. Members include the following universities: Colorado State, Iowa State, Kansas State, Texas Tech, Michigan State, Missouri, Montana State, Nebraska, North Dakota State, Oklahoma State, and South Dakota State. Inter-institutional online academic programs are a cost-effective means of rapidly increasing access and addressing emerging educational needs. The chapter explains how the Great Plains IDEA began, operates, and has evolved, as well as the mistakes made, lessons learned, and upcoming challenges. A major technological chal-

At a time when higher education is struggling to provide educational services to students on traditional campuses, many institutions are experiencing growth in the online degree programs that they provide. This case study looks at one institution's effort to deal with rapid growth in online education while maintaining academic integrity and quality in program delivery, managing program growth through the development and use of policy and procedures.

This chapter introduces the description of wiki appropriation. It considers the tool inclusion in an online innovation supported in student and group-centered learning approach, to improve the quality of the blended learning offered. It states that the university course of Educational Technology, through wiki application as an appropriated methodology–beyond its consideration as a Web 2.0 tool–gives the opportunity to enhance the student protagonism into the deepening of the conceptual filed of the subject by a collaborative knowledge construction. The evidence shows that the application worth to develop the learning strategies towards the student comprehension and its social skills in universities contexts. The results show that the transformation of reactive attitudes into creative ones is a long process of change mediated by emotions and meta-cognitive work. Both facilitate a change of students focus, perspectives and mentalities, understandable by the help of collective learning, among diverse variables.

This chapter aims at describing the networks within the Research and Development in Empirical Studies (REDES) Project, an intercultural enterprise aimed at promoting empirical studies of culture. Probably the first and only international project in the Humanities which prepares new researchers through online communication, the experience is believed to be scalable to other areas as it enables students to acquire the methodological techniques needed to undertake and carry out research in an environment totally different from the traditional iniversity classroom. The chapter traces the history of the project from its foundation in 2002, explains in detail how it was set up and evaluates the contributions of this joint

Preface

INTRODUCTION

As technology has been improving distance education course delivery, online education has been growing in leaps and bounds. The most recent Sloan-C report stated that in the U. S. alone there were almost 3.9 million students taking at least one online course during the fall of 2007 term (Allen & Seaman, 2008). Distance education is generally defined as any form of instructional activities in which the instructor and learner are separated from each other in physical distance (Wang & Gearhart, 2006). This broad definition of distance education has provided for the wide variety of cases in this book. The best practices are based on those established by most accrediting agencies:

1. Institutional Context and Commitment
2. Curriculum and Instruction
3. Faculty Support
4. Student Support
5. Evaluation and Assessment

Distance education is a field of constant change. The cases of this book will provide the reader with an excellent basis of best practices for distance course delivery and learning outcomes, program based cases, and cases from international authors. The field of distance education is often criticized for its lack of scholarly work to benefit the literature in the field. This case book will benefit the literature in a fast growing field.

OBJECTIVE

The overall objective of this case book is to provide practitioners in the field of distance education a case book that is a valuable reference of best practices in the field. Professional development is vital to practitioners of any field. In distance education, the field is constantly changing with the advances in technology and where best to provide professional development than through a reference case book for distance education practitioners.

The case book has several prospective audiences, among which are:

- Practitioners in the field of distance education
- Distance education program administrators

- A reference for libraries
- A potential textbook in distance education courses
- A resource for professional development in the training of distance education practitioners and faculty teaching at a distance.

THE ORGANIZATION OF THIS BOOK

The three sections of this casebook are designed to help provide the reader with cases on best practices of distance education, cases addressing emerging practices, cases on issues that developed as distance education grows, and provides an international perspective on distance education. The cases cover many of the critical topics in distance education today, such as cost and design strategies of distance course design, communication in distance courses, and dealing with student conduct in distance courses, developing distance programs to meet student needs and how to manage growth of distance programs, and use of new technology tools in distance courses.

Section 1 presents cases on the best practices of distance education and online learning, dealing with emerging practices and issues in the field.

In Chapter 1, Dr. Gera Burton sets the stage for the case book establishing best practices and guiding principles for distance education that have been in existence and refined for decades.

Chapter 2 is applicable to both academic and commercial use of distance education. Dr. Shalin addresses the dual-design strategies for the use of modules in e-learning.

Dr. Blakey addresses a developing best practice in Chapter 3, that of blended course delivery. Although many students like the flexibility of online learning, they are still more comfortable in the traditional classroom. This case discusses the issues related to developing blended course delivery.

Many institutions use learning (or course) management systems. In Chapter 4, Wang and Shao look at using quality assurance standards in a learning management system to provide objective-oriented assessment.

The costs pertaining to distance education programs are always a concern of institutions. Dr. Kendrick looks that the cost effectiveness of course redesign for distance education in Chapter 5. The case reviews both literature and application of cost effectiveness in distance education.

In Chapter 6, Drs. Wherry and Lundberg Windes look at student conduct in an undergraduate online program. In this case, technology should not the change the rules for student conduct. Student conduct in the online environment is often different than in the classroom and institutions want to protect academic integrity in distance courses. It is an important case for administrators.

A successful course delivery needs to closely align the course objectives with the course assessments. In Chapter 7, Lightfoot and Schwager demonstrate how application of instructional design principles help improve student learning outcomes.

In Section 2, the cases move to reviewing distance learning at the program level.

In Chapter 8, Moxley, Maes, and Anderson examine the challenges encountered in a distance education alliance, the Great Plains IDEA, a multi-institutional organization, from its formation and the lessons learned along the way.

Drs. Moore and Gimbert describe a case where distance education, through online learning communities, assist in the preparation of new teachers in Chapter 9. The e-tutorials provide electronic test preparation, virtual collaboration with peers, and e-coaching.

Distance education programs look to find their niche to provide the best learning experiences from their programs. In Chapter 10, Smith and Nagel investigate finding their niche in an external degree completion program in their case.

How quickly distance education is growing is the emphasis of the final two cases in Section 2. In Chapter 11, Dr. Blakey discusses the challenges faced by the institution and students when a distance education program grows. Dr. Gearhart discusses in Chapter 12, how a distance education program manages growth and maintains academic integrity through the development of policy and procedures.

The final section deals with distance learning from an international perspective with contributions from authors around the world.

In Chapter 13, Dr. Fainholc introduces wiki appropriation and the use of the Web 2.0 tool to assist in the development of learning strategies towards enhancing student comprehension in distance education courses.

A group of international researchers discuss, in Chapter 14, how the development of an intercultural project in the humanities using technology and based outside of the traditional classroom.

In the final case, Dr. Sharma discusses the emerging trends and best practices in student services of an international distance education program in India.

Deb Gearhart
Editor

REFERENCES

Allen, E. I., & Seaman, J. (2008). Staying the course: Online education in the United States. *The Sloan Consortium*. Retrieved on November 13, 2008, from http://www.sloan-c.org

Wang, H., & Gearhart, D. L. (2006). *Designing and developing Web-based instruction.* Upper Saddle River, NJ: Pearson Merrill Prentice Hall.

Section 1

Cases on Best Practices in Distance Delivery and Learning Outcomes

Chapter 1
Encapsulating the Basic Tenets: Best Practices in Independent, Distance, and Online Learning

Gera Burton
University of Missouri, USA

EXECUTIVE SUMMARY

With increased emphasis on transparency and learning outcomes, the NCA Higher Learning Commission has challenged institutions to demonstrate greater accountability in educational programming (2000). This chapter traces the steps taken to establish a set of guiding principles, or best practices, for endorsement by the American Association for Collegiate Independent Study (AACIS) membership. The AACIS membership is comprised of practitioners from institutions engaged in the provision of self-paced, independent educational opportunities for students who require flexible options, including year-round enrollment. While it was anticipated that agreement on all aspects of online delivery would be impossible, a consensus on core principles was seen as attainable. The purpose of the initiative was to lay the groundwork for further refinement of standards of practice by practitioners engaged in the delivery of online, distance, and independent learning programs.

BACKGROUND

Despite assurances from a variety of commercial sources, there is still no consensus on what would be considered best practices in distance or online learning. Purveyors of expensive learning management systems (LMSs, or software used for delivering, tracking, and managing training) execute impressive marketing campaigns by touting the benefits of their products to a largely uninformed public—and an only slightly better informed group of administrators—eager to share in the success of the online education boom. Surprisingly, seasoned independent study practitioners have been hesitant to set down the basic tenets upon which courses are designed, developed, and delivered. Yet, on university campuses, nationally and internationally, there exists a wealth of Independent Study expertise on every aspect of extending flexible, cost-effective educational programming to students at a distance.

DOI: 10.4018/978-1-60566-870-3.ch001

These professionals confirm, as their evidence suggests, that "students learn best when they act as independent, critical researchers within their disciplines" (Curran, 2008, p. 35).

As early as 1873, educational pioneer Anna Eliot Ticknor founded the Society to Encourage Studies at Home to provide learning opportunities for women by correspondence. With little fanfare, the Society reached more than seven thousand women around the United States over a 24-year period (Bergmann, 2001). In the wake of Ticknor's success, numerous societies sprang up to serve a dispersed population of men and women who were eager to learn. Since the turn of the 20th century, self-directed learners have continued to seek individualized, asynchronous options outside the classroom and beyond the confines of the traditional semester. The time-honored independent study methodology could not have been sustained for decades if a gap had not existed in higher education for self-directed learners. Ignored by most traditional educational institutions, millions of individuals—members of the military, missionaries, athletes, performers, and scores of others—outside the educational mainstream (collectively known as 'non-traditional students') empowered themselves by opting for self-paced, asynchronous, independent study arrangements.

Because educators, especially at the postsecondary level, have been reluctant to admit that learning may take place outside the traditional classroom, they have often blocked attempts to broaden their institutions' scope for non-traditional learners. Whether this was driven by a desire to protect their own turf or arose from a genuine sense of skepticism about the independent learning process, they have frequently held university-sponsored Independent Study programs to a higher standard than that demanded of on-campus programs. Some administrators even employed classroom-based quality standards to exclude programs that did not conform to the dominant model. This study reveals that the history of Independent Study is one of reassuring central administration of its capacity to 'measure up' to the standards of the classroom.

The 2008 Report of the National Survey of Student Engagement (NSSE), *Promoting Engagement for All Students: The Imperative to Look Within*, while focused on the resident student, reveals interesting findings about the independent learner. Relative to classroom-based learners, both first-time and senior online learners reported "more deep approaches to learning in their coursework" (p. 16). The Report proceeds to speculate on the reasons for this finding:

It may be that students who pursue online courses—such as older students for whom the flexibility and convenience of the medium may be particularly important, given work or family commitments—are those who embrace the spirit of independent, student-centered, intellectually engaging learning as captured by the deep learning measures. It may also be the case that professors who teach online courses make more intentional use of deep approaches to learning in their lesson plans. (p. 16)

The fact that online and independent learners outperformed their on-campus counterparts in reflective learning and higher-order thinking comes as no surprise to Independent Study practitioners. The NSSE Report concludes what they have long known: "Online courses seem to stimulate more intellectual challenge and educational gains" (p. 16). Because Independent Study students may enroll at a time of their choosing and take up to nine months to complete a course, they have the opportunity for reflection beyond the regular semester schedule. Distance tends to allow for more reflection, analysis, and surfacing of connections as students develop techniques to 'interrogate the text.' In practice, Independent Study is ideally suited for those courses designated 'writing intensive' or that demand more extensive writing assignments. Students have more time for those "deep approaches" (NSSE, p. 16). They have more time to read around the subject, develop a research topic, reflect on the writing process, revise, and avail of online writing tutorials to polish their writing styles.

Consistent with the movement to democratize knowledge, public universities and colleges—for the most part 'land-grant' institutions—recognized the need to offer distance programming for non-traditional students beyond the confines of the regular classroom. Due to the efforts of such visionaries as William Rainey Harper, founding president of the University of Chicago, William H. Lighty, Charles A. Wedemeyer at the University of Wisconsin-Madison, and Gayle Childs at the University of Nebraska, Correspondence Study—an asynchronous course delivery system—developed at land-grant institutions and prospered on the periphery. Correspondence Study became the first formally structured method for independent learners.

Traditionally, Independent Study member institutions responded to the needs of those excluded from what Wedemeyer called—using Jonathan Swift's satirical phrase for 18th-century ivory tower elitism—"the Great Gate of the Palace of Learning" (1981, p. 18). Recognizing the centrality of the individual to the learning process and the fact that when it comes to learning, 'one size does not fit all,' Independent Study practitioners developed a solid track record of providing asynchronous distance courses for so-called 'non-traditional students.' In the 1920s, 1930s, and 1940s, Extension Units integrated radio with correspondence study; in the 1950s and 1960s, television, film and tape recordings expanded its reach; and from the 1970s onward, computer and satellite technology brought independent study to another level.

In 1949, Gayle Childs, in his doctoral dissertation, looked at the effectiveness and reliability of correspondence study as a learning system. From then on, practitioners became eager to experiment with new technologies that promised to improve instruction and enhance learning. In 1951, on the strength of a Ford Foundation grant of $365,000 to investigate the application of television instruction in combination with Correspondence Study, Childs found no appreciable difference in the level of achievement of students taught in regular classrooms by means of television, or by a combination of Correspondence Study and television (Almeda, 1988). Childs pressed for additional research in the field of Correspondence Study methodology. Wedemeyer's publication of the *Brandenburg Memorial Essays on Correspondence Study, Volume I* in 1963 and *Volume II* in 1966, a collection of speeches and scholarly articles on correspondence instruction, represented a major contribution to the establishment of quality standards in the field.

Despite its appeal to millions of independent learners on all continents, the true dimensions of University-sponsored Correspondence Study were not publicized. Thus, its popularity was not fully appreciated by mainstream educators. This lack of recognition was due, in large part, to the so-called 'image problem' that arose as a result of the devious practices of unscrupulous proprietary schools. These were private, profit-maximizing organizations that recognized the thirst for non-traditional education on the part of a growing population fueled by successive waves of immigrants. The absence of a public relations response by universities to the bad publicity engendered by fraudsters—due, in some degree, to their ambivalence toward this non-conformist instructional model—threatened to destroy the university-sponsored correspondence movement in the United States (Wedemeyer, 1981).

Outside the United States, the Correspondence Study movement enjoyed even greater popularity, with varying degrees of government support. From 1970 onward, the integration of new media and technologies made a name change inevitable. Independent Study emerged as the preferred term for the field, with a renewed emphasis on the centrality of the learner, rather than the teaching format. Wedemeyer carefully crafted the definition that would gain widespread acceptance:

Independent Study consists of various forms of teaching-learning arrangements in which teacher and learner carry out their essential tasks and responsibilities apart from one another, communicating in a variety of ways, for the purposes of freeing internal learners from inappropriate class pacings or patterns, of providing external learners with opportunities to continue learning in their own environments, and developing in all learners the capacity to carry on self-directed learning, the ultimate maturity required of the educated person. Independent Study programs offer learners varying degrees of freedom in the self-determination of goals and activities, and in starting, stopping and pacing individualized learning programs which are carried on to the greatest extent possible at the convenience of the learners. (1977, p. 2114)

Beginning in 1964, with the support of a multi-million dollar grant from the Carnegie Foundation, Wedemeyer's ground-breaking Articulated Instructional Media (AIM) project to integrate media in independent study received much attention. Completed at the University of Wisconsin-Madison, the innovative project directly influenced the British Government's decision to launch the Open University in 1971, now known as the United Kingdom Open University (UKOU). The project revolved around a special sequence of interdisciplinary advanced courses in humanities, social studies, and the sciences, leading to a degree for adults. The courses combined Independent Study, telelectures, off-campus seminars, mobile laboratories, and libraries. The program became a catalyst for change in a university setting, encouraging professors to incorporate multimedia into their courses (Wedemeyer & Najem, 1969). Praising Wedemeyer's role, Michael G. Moore observed that the Open University was "an inheritor of [Wedemeyer's] inspiration, a beneficiary of his advice, and a learner from his wisdom" (2000, p. 1). The model was emulated by developing countries in Africa and Asia. In fact, throughout the 20th century, millions of learners in dozens of countries enthusiastically embraced correspondence/independent study in fulfillment of their educational goals.

It was not until the early 1990s that for-profit institutions recognized the void and adopted entrepreneurial strategies to capitalize on the burgeoning demand for flexible online educational alternatives on a grand scale. Recognizing the vast potential to harness the speed and efficiency of Internet communications for educational purposes, these institutions did not hesitate to exploit the opportunity to their advantage. Since then, technological innovation has produced a wide range of educational software with the potential to enhance the learning process and expand course offerings for a more diverse student body. However, despite rapid innovation and the growth in online availability, independent learners continued to find limited asynchronous university-sponsored resources online in pursuit of their educational goals.

By the close of the 1990s, the more successful Independent Study providers had already adapted to the online environment. This transition was often accomplished in the face of opposition from administrators and, more recently, from proponents of online 'learning communities' designed to replicate the dominant classroom model. Results of a recent study make the case for a more comprehensive approach to the provision of online learning. At Australia's Monash University, Murphy, Casey, and Fraser investigated the effectiveness of online learning. Contrary to the expectations of exponents of a classroom model, their data do not uphold the perception of online learning communities that mirror the social experience on campus. "Students want to be able to communicate with their teacher and receive communication in a timely manner but do not place a high value on the ability to establish or maintain online communication with each other"

(Murphy et al., 2007, p. 13). They also note that at the University of Adelaide, discussion boards have not proven to be a useful resource in learning (Murphy et al., 2007). This important work provides evidence that universities need to be more responsive to learner needs, especially those that call for more flexible systems that free students to proceed at their own pace. Despite the proliferation of online programs, a major concern remains. The independent learner continues to face biases similar to those faced by his or her counterparts of earlier times. Additional concerns relate to the limitations of LMSs, the paucity of evidence on their ability to enhance learning, and the need to provide comprehensive student services in a cost-effective manner.

With the proliferation of communication technologies in conventional higher education, the independent learner is again under threat. Due to the success of persuasive vendors, administrators have invested large sums in canned LMSs that promise to replicate the classroom experience without evidence or data to indicate whether or how learning takes place. In fact, the single greatest problem with distance education in the United States today may well be that it is "too tied to traditional models of higher education both administratively and pedagogically" (Berg, 2001, p. 6). Advocates of new technologies should be asked to demonstrate whether and how the technology is enhancing teaching and learning (Hawkins & Barone, 2003). Educause's 2007 Core Data Service Report cautions against misplacing emphasis on input measures to convince decision makers that 'more is better' when it comes to information technology (Hawkins & Rudy, p. ix). Instead, higher education administrators would be well served by focusing on measuring effectiveness, so that the public can ascertain which institutions are doing the best job with the fewest resources. Upon reflection, they might be wise to consider how resources are already being applied in the provision of custom-designed, cost-effective systems that have served a diversity of independent learners for more than a century. Undoubtedly, a great deal of research is required to produce sufficient evidence upon which to base decisions. One approach would be to conduct a review of the successful practices of long-established independent study programs. This article elucidates the process involved in formulating a set of best practices for Independent Study.

The study at hand traces the steps taken to solicit input from the membership of the American Association for Collegiate Independent Study (AACIS) to arrive at a set of agreed-upon standards and sets forth the results of the findings. The AACIS membership is comprised of practitioners from institutions engaged in the provision of self-paced, independent, educational programming for students who desire flexible options, including year-round enrollment. A majority of the association's members hold either faculty or professional appointments at regionally accredited higher educational institutions, some of which also house secondary level schools that provide online learning opportunities and high school diplomas. Members also represent colleges that combine educational levels, including high school, university, dual credit, and degree completion courses.

While agreement on all aspects of online delivery would likely be impossible, a consensus on core principles should be attainable. These findings are being developed to acknowledge their evolution as part of a continuum that includes the efforts of the pioneers of programming for "back door learners" (Wedemeyer, 1981). Independent Study practitioners have shown the determination and collective energy to define their own standards. Once agreement is reached, a draft set of these shared principles will be compiled for presentation to the AACIS Board of Directors for approval. Upon approval, they will be posted on the AACIS Web site for regular review and revision by a sub-committee appointed by the organization's president. Members will be encouraged to provide links from their home institutions'

Web sites to the *Best Practices in Independent, Distance and Online Learning* section of the AACIS Web site.

SETTING THE STAGE

A review of the history of the establishment of principles of best practices for institutions that facilitate independent study proves instructive. As early as 1915, representatives from 22 universities gathered in Madison, Wisconsin to found the National University Extension Association (NUEA). The creation of NUEA was, in part, a response to the tarnished image of the proprietary schools and diploma mills (Gerrity, 1976). From the outset, members displayed concern for high standards by excluding those institutions that were not 'university grade' schools (Edelson, 1991). A statement by University of Chicago representative Hervey Mallory provides insight into the prevailing concern: "I simply want to exclude those [schools] which are not of recognized standard" (1915, p. 218). In 1916, NUEA formed a committee on Nomenclature and another on Standardization of Credit Courses and charged them to provide recommendations regarding the maintenance of standards for university credit-bearing courses, including correspondence study.

A key year in the recognition of the equivalence of university extension courses, 1920 ushered in the adoption of policies that called for the approval of all courses and faculty by the appropriate academic departments. NUEA documents produced in 1924 and 1929 called for equal rigor, equivalent student and faculty qualifications, and appropriate testing in University Extension programs, allowing direct and close faculty oversight at all stages (Rohfeld, 1996). This meant that students who successfully completed extension courses would receive the same credit as resident students. In practice, institutions limited the level of correspondence credit that could be applied toward a degree from 25% to 50% of credit hours in an effort to tighten standards (Watkins, 1991). In 1931, the Committee established eight rules for good practice in Correspondence Study. However, as Pittman notes, "the NUEA presented its standards only as suggestions, hoping for voluntary compliance" (1998, p. 38). Practical steps to allay the fears of skeptics who questioned the levels of quality and rigor in non-traditional education, these policies have constituted the cornerstone of best practices employed by university-sponsored independent study programs to the present day.

For any discussion of standards in education, the oft-cited Chickering and Gamson's *Seven Standards of Good Practice for Undergraduate Education* merit attention. These Standards have been accepted by administrators as measures for judging the effectiveness of classroom teaching. With fundamentalist zeal, some administrators have endeavored to apply these principles to distance education. A review of each principle reveals that, in fact, Independent Study adheres to all but one of them and, that one principle, *by definition*, cannot be applicable to an autonomous, self-directed learning system.

- **Principle 1:** *The instructor encourages student-faculty contact and interaction.* Independent Study actively promotes this principle. Furthermore, instructors receive guidelines for the creation of quality feedback and assessment strategies for use with independent learners.
- **Principle 2:** *The instructor encourages student cooperation.* This principle is not applicable for independent learners because it assumes a classroom structure and calls for the creation of parameters that involve cooperative learning, group processing, team learning, and group projects. By its very definition, Independent Study allows the individual, self-directed learner to proceed at his or her own pace, unhindered by the schedules of other learners. The independent learner chooses an option

that does not require collaboration with an online community.

- **Principle 3:** *The instructor encourages active learning.* Independent Study demands that students take control of the learning experience, so active learning is not only encouraged, it is required. Independent learners are afforded the flexibility they require to read around a topic and conduct research at convenient times. They are not required to critique the work of other students.
- **Principle 4:** *The instructor gives prompt feedback.* Since research shows that prompt feedback promotes learning, Independent Study programs offer a range of incentives to ensure speedy—even immediate—instructor feedback. These incentives include preferential pay rates, close monitoring of turnaround times, and swift intervention, when necessary, to replace tardy instructors.
- **Principle 5:** *The instructor emphasizes time on task.* Independent Study courses provide performance objectives, grading and evaluation criteria, and grading rubrics. Courses require reflection, critical thinking, analysis, and reasoning.
- **Principle 6:** *The instructor communicates high expectations.* Independent Study courses address the same objectives as courses on campus. Assignments are equivalent in difficulty and depth to those required of on-campus sections. Student performance is assessed based on his/her own individual learning outcomes in accordance with standards set by the appropriate university department.
- **Principle 7:** *The instructor respects diverse talents and ways of learning.* By not limiting students to a semester timeframe, or requiring classroom attendance, Independent Study offers more flexibility for students to learn in a variety of settings. Students typically enroll year round, work at their own pace, and take up to nine months to complete their assignments. Courses require a range of assignments to allow for diverse talents and learning styles.

It is indeed noteworthy that, although constructed around the classroom model, six of the *Seven Standards of Good Practice for Undergraduate Education* are entirely compatible with Independent Study. Because *Principle 2* involves interaction with other learners in a synchronous learning environment, it has no application for Independent Study students. A factor often missed in any discussion of distance learning options is the fact that, whatever the methodology, the learner is required to demonstrate on *an individual basis*—ideally in a proctored exam setting—that he/she has achieved the learning outcomes.

With such an array of software options now on the market, how do Independent Study programs address the diverse needs of the distance learner? When it comes to curriculum planning, the provision of student services, and assessment and evaluation, upon what principles should administrators base their decisions? This study considers a range of such issues.

Despite noble efforts by practitioners to reassure academics of quality standards in Independent Study, most notably that of Thomas Russell's *No Significant Difference (NSD) Phenomenon*, whose fifth edition, published in 2001, represents a compilation of 355 research reports, summaries, and papers that document no significant differences in student outcomes with non-traditional courses, the skeptics remain. Even Gayle Childs' frequent assertion in his speeches and workshops that "Insofar as general educational methodology is concerned, the brain does not give a damn" (1973) fails to convince those most resistant to change.

CASE DESCRIPTION

Since to date, 'best practices' in the Independent Study field have not been clearly defined as such, a broad consensus on the field's basic tenets by practitioners is long overdue. This study endeavors to encapsulate a set of what are truly considered 'best practices,' or well-established principles, that enable autonomous, self-directed learners, described by an American Council on Education (ACE) Task Force as those who "instigate learning tasks through internal motivation and commitment" (1990, p. 24), to take control of their educational development. The study does not purport to be all encompassing; rather it represents an attempt to establish a foundation upon which to forge an established process.

Based on well-established principles that evolved over a century at the University of Missouri's Center for Distance and Independent Study (CDIS), with regular input from Independent Study colleagues at other institutions, the researcher compiled a draft set of basic guidelines as a departure point for the study. Focusing on the learner, the study began with a careful review of the literature to ascertain what formal steps, if any, had been taken to establish standards of best practice in Independent Study or in its previous incarnation, Correspondence Study, under the umbrella of the original professional organization, the NUEA, now the University Continuing Education Association (UCEA). Research on the history of establishing standards proved fruitful. As early as 1929, a recommendation emerged from the NUEA Commission on Standardization on such matters as course and credit equivalencies and instruction by regular faculty (Watkins, 1991). In 1931, the Committee established eight rules for good practice in Correspondence Study. These were set down as follows: Philosophy, Mission, Administration, Staff, Faculty, Instruction, Services, and Research and Evaluation.

In 1961, in order to establish quality standards and to differentiate university correspondence study from all proprietary offerings, Charles A. Wedemeyer, champion of open learning and then Director of Correspondence Study at the University of Wisconsin-Madison—who famously urged academics to extend learning opportunities to populations traditionally excluded from formal education—chaired a committee to review standards and practices. In 1963, the Committee produced a document setting down Independent Study's principles of best practices. These became known as *Criteria and Standards of the NUEA Division of Independent Study*. Eighty-six NUEA university-sponsored programs signed on to these guidelines that could be used for systematic evaluation, although as Pittman notes: "This extremely sketchy document dealt mainly with such mundane matters as prerequisites and examination security; it provided little, if any, help to anyone interested in serious program or course evaluation" (1992, p. 6). Burcaw noted that *Criteria and Standards* served as guidelines for assuring comparability with college credit courses offered in a face-to-face classroom setting (1981). The Committee on Criteria and Standards of the renamed NUCEA was charged with maintaining standards in line with current practices. In 1981 and 1989, the Committee completed major revisions to allow for the incorporation of available technologies with independent learning methodologies.

The 1963 *Criteria and Standards* document had shortcomings, however. Pittman called it "a defensive document" (1992, p. 7) that sought to reassure conventional educators and administrators that correspondence courses could reach the standards of the classroom. Anchored to the classroom model, members were reluctant to—even metaphorically—set the learner free of the constraints of the traditional classroom. In fact, the *Criteria and Standards* were voluntary guidelines—lacking real force or authority—for program evaluation. The eight areas identified were used as a basis for internal program reviews and licensing and as a point of departure by distance education administrators when drafting their own standards. For the purposes of this study, the researcher has refined the eight areas as follows: I. Mission; II. Curriculum Planning; III. Student

Services; IV. Research and Evaluation; and V. Marketing and Outreach.

For reasons beyond the scope of this article, members of the Independent Study Division of NUEA, later NUCEA, decided to create an independent association of members focused on the promotion of all aspects of Independent Study. In effect, however, a significant number of professionals—those whose universities held institutional membership in the older association—continued to maintain a relationship with NUCEA (renamed UCEA in 1996) in addition to their membership in a new association devoted to Independent Study. In 1992, a group of Independent Study practitioners formed a Steering Committee to give birth to the new professional organization dedicated to the primacy of the autonomous learner. Known as the American Association for Collegiate Independent Study, or AACIS, and governed by a nine-member Board of Directors, with a membership drawn from more than 50 universities, the group has held a professional meeting and workshop each year since its 1993 inaugural conference at the University of Missouri, Columbia, with the election of its first President, Roger G. Young. The Association launched a listserv for AACIS members that same year and introduced a Web site shortly thereafter.

Due to the strong bonds engendered by the circumstances that led to the birth of AACIS, on a cooperative basis, Independent Study practitioners adopted an informal approach to the establishment of standards, freely reviewing and revising quality standards by exchanging ideas with peers via the listserv. They received additional support from colleagues with the formation of committees and the activities of the annual conference and workshop. Significantly, members of AACIS established an awards process to evaluate courses from peer institutions, according to a set of criteria developed over many years. Named in honor of Helen Williams, an early leader in the field of non-traditional education at the University of Iowa, according to the AACIS Web site, "the Helen Williams Award recognizes both the author and the primary editor of the most outstanding independent study college level credit course initiated during the previous academic year" (2008, para. 1). The Monty McMahon Award recognizes "both the author and primary editor of the outstanding independent study pre-collegiate level (K-12) course initiated during the previous academic year" (2008, para 1). Monty McMahon, former director of the University of Nebraska's Department of Distance Education, created the first fully-accredited High School by Independent Study. Winners in each category—University, High School, and Non-Credit—are acknowledged at the annual conference and workshop.

While this organization of dedicated professionals has consistently engaged in peer review activities, it did not revive the Standards Committee that had been in existence under the old UCEA formulation until the early 1990s. Neither has the membership formalized the processes in the form of a standards document for dissemination. This Case Study documents the initial steps involved in the establishment and formalization of a set of Best Practices for Independent Study/Distance/Online Learning for endorsement by the membership of AACIS in 2009.

In early Fall 2008, the researcher contacted the AACIS President, John Skaare, Director of the North Dakota Center for Distance Education, with a proposal to formally update standards and guiding principles for the organization. Having read a summary of the rationale for such action, he immediately responded that it was indeed time to review principles of best practice for the organization. Furthermore, he invited the researcher to appear at the Annual Meeting of the AACIS Board of Directors held in St. Louis, Missouri in November of 2008 to make a presentation on the research proposal. On Thursday, November 6, the researcher presented a brief historical perspective on the development of the *Criteria and Standards* document drawn up by the Division of Independent Study, under the auspices of the organization then known as NUCEA. A discussion ensued on the subject that while the membership adhered to high standards that had evolved over time and routinely received input from AACIS colleagues, to that point, the process had not been subject to a formal review.

At the behest of the President, Board members built the case for a more formal approach to the process of establishing standards in the field of Independent Study. The proposal was favorably received by the members of the AACIS Board. Furthermore, the membership responded enthusiastically to the request to form a committee to review a draft set of principles. With full Board approval, four members volunteered to serve on the Best Practices Committee: Shannon Brown, Assistant Director, Independent Learning, University of Nevada, Reno; Dr. Sharon Walters, Assistant Director at the Office of Distance Education, Southern Illinois University-Carbondale; Kristyn Rose, Coordinator of Distance Education, Mesa State, Texas; and Yvette Wingfield, Senior Program Coordinator, University of Arizona. Later, Crystal Fey, a new member with Independent Learning at the University of Wisconsin-Madison, volunteered to serve. The researcher reminded the Board of the forthcoming presentation on the subject at the Annual Meeting.

During a panel session on the history of AACIS for the general membership on November 6, past-President and founder member Von Pittman, Director of the Center for Distance and Independent Study at the University of Missouri, spoke of the timeliness of the review of best practices and urged the membership to either attend the session or contact the newly formed Best Practices Committee with recommendations. At the conference session, the researcher presented a PowerPoint slideshow outlining a draft set of principles, based on established practices at the University of Missouri, eliciting confirmation of adherence by participants. There was a consensus that the principles should revolve around a mission for each institution. Participants agreed to provide input on each of the draft guidelines. It was decided that all responses would be reviewed, edited, refined, and compiled for inclusion in the final version. A summary for publication will make note of areas where no consensus was found. The final draft will be submitted to the AACIS Board of Directors for approval and dissemination via the organization's Web site. Since the establishment of quality standards will remain a work-in-progress, the Committee will conduct an annual review with proposals for revision, subject to Board approval.

Shortly after the conclusion of the conference, the researcher sent a copy of the PowerPoint presentation to each member of the committee with recommendations and a request for input. Bill Catchings, Assistant Director for Education and Technology at Louisiana State University's Office of Independent and Distance Learning, agreed to publish a draft set of Best Practices on the AACIS Web site. Dolores Shearon, Marketing Manager at the University of Missouri, offered to prepare a draft on guidelines for the establishment of ethical principles that apply to Marketing/Outreach efforts as they apply to University Extension programs. Because this section demands input from specialists in such areas as data mining and emerging Internet protocol, it was decided to tackle it last, to comprise Section V, at a later date, when the findings will be distributed to the committee for review prior to Board approval.

The purpose of this endeavor has been to stimulate input from the AACIS membership so that the basic principles will be further developed and refined to define the 'parameters of excellence' of the field of Independent Study. The intention is threefold: to open a dialogue with the membership on how to enhance quality; to educate potential students on a history of excellence in serving independent learners; and to inform accrediting bodies on adherence to standards. It is necessary to showcase how evolving practices in Independent Study not only adhere to, but set, high standards. The field is asserting its professionalism and distinction from the traditional, dominant educational model. Having researched the history of the establishment of best practices in the field and reviewed the range of guiding principles effective at this juncture, five critical—and frequently overlapping—areas are deemed to constitute a basis for elucidation on best practices in Independent, Distance, and Online Learning. These five broad areas have been selected to reflect the needs and special context of the independent learner:

I. Mission
II. Curriculum Planning
III. Student Services
IV. Assessment and Evaluation
V. Marketing Outreach

I. Mission

The mission statement is clear and consistent for print and electronic distribution. There is a shared understanding of the mission and all policies, procedures, and practices are consistent with it. The mission addresses access for learners outside the traditional classroom. The budget supports the mission.

The mission statement of the University of Missouri's Center for Distance and Independent Study serves as an example for its simplicity and straightforwardness:

To extend the instructional resources and programs of the University throughout the state of Missouri, the United States, and internationally, via an asynchronous distance delivery system.

While the Committee's approach was to recommend that AACIS members contribute to the development or revision of the mission statements at their home institutions, in the process of reviewing the concept of mission, it became apparent that the AACIS organization will need to revise its Web site to reflect its scope and purpose. This is especially important as the organization contemplates a name change in 2009.

II. Curriculum Planning

When addressing curriculum issues, some overarching questions need to be raised. What quality measures are in place to ensure that learning takes place? What strategies facilitate and reinforce learning? Does course content drive design? Do courses include a variety of materials/ textbooks to accommodate different learning styles? What factors influence the selection of software? Is there a process of regular course redesign or revision? If considering implementation of an LMS, does it offer enough flexibility to accommodate independent learners? Is customization taking place? Is a variety of assignments offered? Are assignments appropriate and consistent with learning objectives?

The following principles have been identified:

i. Since all courses begin with the academic units, there is a formal agreement with the Independent Study department on the development and delivery of courses. Annual approvals are obtained from the appropriate Chair for delivery of all courses in his or her department. Recommendations for new courses and on a revision cycle are sought from the academic department. An instructor of record is appointed for all courses offered.
ii. Faculty support takes place in the form of face-to-face orientations and workshops focused on learner outcomes with independent students. Electronic guidelines are provided for instructors. Faculty receives examples of quality feedback to promote high-quality responses in student assignments. An author

and instructor manual are provided in print and online formats. The institution's Web site includes guidelines for faculty on course development.

iii. Course proposals are required of faculty. Course proposals set out the course title, overview, textbook selection, materials, objectives, learning outcomes, lessons, assessments, and grading scales/rubrics.

iv. Course reviews are conducted in-house for courses in development, using a checklist of required elements for quality measurement. The consistency of lesson objectives is verified. Test items are reviewed to ensure measurement of learning outcomes.

v. Feedback is provided to faculty on proposed course design features and suggestions for improvement. Faculty members are encouraged to incorporate interactivity in courses for a variety of learning styles. This may take the form of student-to-material or student-to-instructor interaction. There should be a range of assignments that not only motivates and engages students, but allows opportunities for practice and reflection. There are incentives in place to promote prompt lesson feedback.

vi. Intellectual property issues are addressed in advance to provide clarification on ownership and rights. Copyrights are renewed as required, and all documents on ownership rights are kept current.

vii. A Course Development Team, consisting of coordinators, designers, and editors with strong editorial backgrounds is in place. The Team includes specialists in learning technologies. The Team adopts a 'house style' and adheres to academic standards, with scope for creativity. Courses comply with requirements of the academic department and include interactive elements to engage students and enhance learning.

viii. The Course Development Team maintains a Status Report of courses in development. The Report is updated to reflect changes such as newly commissioned courses, opened, and closed courses.

ix. The Web site provides a set of guidelines for students on strategies for successful course completion. These may include sample lessons to ascertain preparedness for independent study/online learning.

x. Course evaluations are distributed to students. Measures are taken to ensure a high rate of return. Evaluation items align with those of academic departments. Evaluations are reviewed regularly to identify specific areas in need of improvement and for quality control.

III. Student Services

Preparatory to an examination of standards pertaining to student support services, the following questions were addressed: Are comprehensive learner support services provided? Is advising or tutoring offered? Are motivational strategies employed? What safeguards are in place to ensure quality instructional feedback for students? Best practices should reflect the institution's response to what Wilhelms called "the freest, most fluid medium of teaching and learning that has yet been devised" (quoted in Moore, 2000, p. 22).

i. Independent Study staff demonstrates a strong student-friendly commitment. Students receive prompt, courteous attention in person, by email, phone, postal service, and fax.

ii. Advising or reference services are provided. The Independent Study staff maintains good relationships with the Admissions Office, advising councils, Financial Aid Office, Veterans' Office, and all college departments.

iii. Students receive regular reminders to maintain progress. They are provided with online strategies to maintain progress.

Students have online access to check progress.

iv. Students receive quality, timely feedback from instructors. Individualized feedback is provided, where appropriate.

v. Students take proctored mid-term and final exams in a quiet, secure, convenient testing environment. Proper identification is required. Security measures, such as cameras, are in place whenever possible or appropriate. The Independent Study program maintains a list of approved proctors whose credentials are verified. For added security, multiple forms of exams are administered.

vi. Library services, tutorial services, and technical assistance are provided as needed.

vii. Online registration is provided via a secure Web site. For maximum convenience, 'one-stop shopping' is offered. The Web site contains current course listings, descriptions, and faculty biographical data. Appropriate links are provided to sample lessons. Information on downloads, plug-ins, and learning tools are offered, along with clear instructions for implementation.

IV. Assessment and Evaluation

The ACE defines assessment as follows:

The measurement or evaluation of a student's learning whether gained in the classroom, from prior experience, or through Independent Study. It involves generating evidence and making judgments of an individual's competence by comparing his/her performance against established criteria. Assessment may also refer to the examination of other attributes of the learner and the learning experience, e.g. how people learn, what helps them retain and use their learning, or how we evaluate affective as well as cognitive learning. The term assessment is increasingly being used to reflect the measurement of student learning and related processes. (1990, p. 26)

While still used to refer to student learning, evaluation is being used increasingly to refer to "the process involved in examining and judging educational programs and institutions" (ACE, 1990, p.24). As institutions improve accountability measures, it is becoming increasingly important that the assessment activities of Independent Study programs form an integral part of the institution's overall assessment initiative.

The central concern of all initiatives that approach the subject of assessment and evaluation in education is that of student learning. Accrediting agencies disseminate guidelines on assessment techniques and offer workshops on training for higher education officials. Much has been written about the need for accountability and transparency in public education; however, despite our efforts, how do we know students are learning? What assessment tools should be employed? How do we judge their effectiveness? Should proctored examinations be a feature of assessment? What are the specific challenges involved in attempting the assessment of distant learners?

The subject of interactivity merits special attention in the online learning environment. While traditional educators emphasize student-student interactivity, when it comes to learning acquisition, there are, in fact, endless possibilities. Some educators believe that only one form of interactivity is required for learning to take place (Anderson, 2002). Independent Study offers a number of forms of interactivity; student-instructor, instructor-student, and student-materials. In this regard, it is again worth noting Childs' finding that when it comes to learning, "the brain does not give a damn" (1973).

The following recommendations are appropriate when initiating accountability measures with any educational program:

i. The Program works with the academic units and uses measures consistent with departmental assessment plans

ii. Courses use a variety of objective and subjective assignments: written work, capstone projects, e-portfolios, etc.
iii. Assignments use a range of measures and rubrics from acceptable to unacceptable levels
iv. The Program promotes quality lesson feedback that reinforces learning from course design to delivery
v. Lesson feedback is direct, clear, concise, specific, and constructive, with suggestions for individual improvement
vi. Lesson feedback validates student effort and acknowledges exceptional performance
vii. The Program documents assessment procedures and publishes them on its Web site
viii. The Program provides accountability by posting details of activities on its Web site
ix. The Program invites input from students, parents, and employers
x. The Program submits courses for peer evaluation to professional associations' awards committees

There is renewed pressure on Independent Study programs to provide online exams due to the proliferation of software designed to facilitate that option. Since security is of major concern, in order to preserve the integrity of the process, assurances must be given that software is foolproof. Given that there exist, as yet, shortcomings associated with any commercially available program, proctored examinations remain the recommended approach to the assessment of student learning.

The subject of motivation is of major importance in distance learning. As Rekkedal noted, rapid feedback is important to students for both understanding and motivation to complete courses (1983). At every level of course design, development, and delivery, programs need to incorporate a variety of strategies to help motivate learners. Immediate or prompt feedback is a key factor in keeping students engaged in learning, because delayed lesson feedback can interfere with the process to the extent that learning may not actually take place. In the 1970s, the University of Missouri developed the Computer Assisted Lesson Service (CALS) to address the issue of motivation. Using the telephone hooked to a 'Silent 700' machine, this successful system provided immediate feedback to students submitting multiple-choice test questions, constituting a very effective learning device (Prensky, 2000). The system was successfully replicated by numerous institutions. More recently, using extensive lesson feedback, the University of Missouri revamped and refined the system, adapted for Web-based, online delivery.

Since AACIS has initiated a formal process to set down standards, the independent learner will be the beneficiary as he/she gains confidence that the old 'image problem' is a thing of the past. As is the case with any field, establishing best practices in Independent and Distance Learning remains a work in progress. It is the type of activity that will require constant revision. What we know is that Independent Study practitioners have a wealth of expertise in extending learning opportunities for those Wedemeyer described as "back door learners" (1981). This study represents a basis on which to build a set of quality standards for all branches of distance and e-learning.

CURRENT CHALLENGES FACING THE ORGANIZATION

The AACIS membership is comprised of independent study practitioners from at least 50 United States institutions of higher learning. With the inevitable fusion of the identifiers—correspondence study, independent learning, distance education, online learning, and e-learning—there remains a challenge for the field, and for all those involved in 'non-traditional' education. It is time to adopt a unified approach to open learning. As Higher Education budgets tighten and technology drives educators to expand online access for revenue generation, there will be renewed demand for greater flexibility and a more open learning system. From the number of contracts that exist between universities and software vendors for course delivery, it is evident that online programs are viewed as a creative approach to solving some of the problems—not the least of them financial—in American education. Many institutions are outsourcing what should be considered essential services, crucial to their missions, including the provision of student support services. It is somewhat paradoxical that some of the very institutions that pride themselves on their 'brand' are the first to outsource key functions to the same commercial operators and call centers as their competitors, with little concern for the loss of control over quality standards.

Spearheaded by the Massachusetts Institute of Technology (MIT), the open source initiative is perhaps the most exciting educational development to emerge in recent times. Scores of dedicated professionals are devoting their time to advancing learning opportunities for all. It remains to be seen whether administrators will embrace the benefits of a low-cost movement that has evolved from the ground up, or cling to expensive systems such as Blackboard, Desire2Learn, and others.

What hampered the development of Independent Study in the past—its low status, small staffs, and meager resources—ensured that correspondence study remained suspect by academics. Despite some notable steps forward, there remains a need to develop strategies to achieve parity for Independent Study courses with traditional resident instruction. Hopeful signs include greater acceptance of online courses due to a greater comfort level with technology on the part of new generations of faculty. Furthermore, improved academic profiles of practitioners in the field of Independent Study are leading to greater credibility. The enthusiasm with which the Best Practices Committee accepted the challenge is reflective of this higher level of professionalism among practitioners. This initiative projects a certain level of maturity where AACIS is now poised to engage in self-critical analysis, a significant step for the organization.

Given the phenomenal success of the United Kingdom Open University and institutions that emulated it around the world, progress along those lines seems to be the best way forward. Developments in Europe as a result of the Bologna Accords, designed to ensure greater cooperation among institutions of higher learning on such matters as transfer credit and performance standards, promise sweeping changes that will have a direct effect on American education in the near future. This initiative takes a radical approach to the assessment of student learning and the entire area of accountability. In a global context, United States graduates will find themselves at a distinct disadvantage if programs of study do not take note of the requirements set down in the Bologna Accords.

Despite the excitement generated by emerging technologies, self-pacing remains one of the top priorities for learners (Duff & Quinn, 2004; Greenwald, 2006). What sets Independent Study apart has always been its recognition of the centrality of the learner, regardless of the constraints of time or physical location. What about the lessons of history? Anna Eliot Ticknor, William Rainey Harper, William H. Lighty, Charles A. Wedemeyer, and Gayle Childs demonstrated the fact that, regardless of the technology employed, learners will continue to press for flexibility to accommodate their needs. With the recent decline in global economic activity, it seems certain that learners will demand more asynchronous options as they endeavor to enhance their employability. They should not face unnecessary restrictions as they strive to pursue their educational objectives. Learners will need to be nimble, possibly change careers, and avail themselves of a variety of options at their disposal as they become empowered to take charge of their academic and professional careers.

With an eye to the future, AACIS is considering a name change for the organization that will reflect its noble legacy—one of creative adaptation. As the lines become even more blurred between resident, online, off-campus, independent, and distance students, rigor and quality are of as much concern as ever. At this juncture in its evolution, when opportunities exist for growth in any number of directions, enriched by innovations driven by information technology, the Independent Study organization's new title should encapsulate the organization's mission to preserve the primacy of the independent learner in online distance education. As a matter of urgency, the new organization will need to increase its membership to attract practitioners committed to preserving a space for, and validating, the independent learner who, for whatever reason, cannot, or chooses not to, occupy a seat in a classroom setting. Responding to the expectations of these individuals, as Wedemeyer so wisely urged, educators must "boldly and fearlessly assert" (quoted in Pittman, 1992, p. 6) their high standards and review, revise, and adapt them to embrace a new generation of independent, distance, and online learners.

REFERENCES

Almeda, M. B. (1988). Speaking personally with Gayle B. Childs. *American Journal of Distance Education*, *2*(2), 68–74. doi:10.1080/08923648809526626

American Association for Collegiate Independent Study. (2008). *Awards*. Retrieved in November 2008, from http://www.aacis.org/main/main.htm?page=awards

American Council on Education Task Force. (2000). *Principles of good practice for alternative and external degree programs for adults*. Washington, D.C.: American Council on Education and The Alliance.

Anderson, T. (2002). *An updated and theoretical rationale for interaction* (Athabasca University's IT Forum Paper No. 63). Athabasca University, Albuquerque, Canada. Retrieved on November 5, 2008, from http://it.coe.uga.edu/itforum/paper63/paper63.htm

Berg, G. A. (2001). Distance learning best practices debate. *WebNet Journal,* (5-6), 17. Retrieved on November 26, 2008, from http://www.aace.org/pubs/webnet/v3no2/3=2DistanceLearning.pdf

Bergmann, H. F. (2001). The silent university: The society to encourage studies at home, 1873-1897. *The New England Quarterly, 74*(3), 447–477. doi:10.2307/3185427

Burcaw, S. S. (1981). *Standards for independent study: The American experience in higher education.* Paper presented at the Conference of the International Council for Distance Education, June, Vancouver, British Columbia, Canada.

Burton, G. (2008, November). *Establishing best practices in distance and independent study*. Paper presented at the Sixteenth Annual American Association for Collegiate Independent Study Conference, St. Louis, MO. Retrieved from http://www.aacis.org/pages/conferences/2008conf.htm

Chickering, A., & Ehrmann, S. (1996). Implementing the seven principles: Technology as lever. *American Association for Higher Education Bulletin,* 3-6.

Chickering, A., & Gamson, Z. (1987). Seven principles of good practice in undergraduate education. *AAHE Bulletin, March*, 3-7.

Childs, G. (1973). "Correspondence Study: Concepts and Comments." Speech given at the NUEA Conference: Omaha, NE. 17 April, 1973.

Curran, C. (2008). Online learning and the university. In W. Bramble & S. Panda (Eds.), *Economics of distance and online learning: Theory, practice, and research* (pp. 26-51). New York: Rutledge.

Duff, A., & Quinn, D. (2004). *Benefits of online learning.* Online Learning at UniSA (University of South Australia). Retrieved in November 2008, from http://www.unisanet.unisa.edu.au/learn/learnOnline/?PATH=/Resources/Workshop-ready4online/Learning+Online/&default=Welcome.htm

Edelson, P. J. (1991). Codification and exclusion: An analysis of the early years of the national university extension association (UCEA) 1915-1923. *Continuing Higher Education Review, 55*(3), 176–189.

Gerrity, T. (1976). *College sponsored correspondence instruction in the United States: A comparative history of its origins (1873-1915) and its recent developments (1960-1975).* Unpublished doctoral dissertation, Teachers College, Columbia University, NY.

Greenwald, S. (2008, September 3). *University of Missouri—Columbia campus visit, slide #32.* PowerPoint presented in 203 Whitten Hall on the University of Missouri, Columbia Campus.

Hawkins, B., & Barone, C. (2003). Assessing information technology: Changing the conceptual framework. In P. A. McClure (Ed.), *Organizing and managing information resources on your campus* (pp. 129-145). San Francisco: Jossey-Bass.

Hawkins, B., & Rudy, J. (2007). Educause core data service 2007 summary report. *Educause Connect.* Retrieved on October 21, 2008, from http://connect.educause.edu/Library/Abstract/EDUCAUSECoreDataService20/47414

Mackenzie, O., & Christensen, E. (1971). *The changing world of correspondence study: International readings.* University Park, PA: Penn State.

Mallory, H. (1915). Discussion. *Proceedings of the First National University Extension Conference.* Madison: University of Wisconsin Press.

Moore, M. (2000). *Speaking Personally about Distance Education.* University Park, Pennsylvania: The American Center for the Study of Distance Education.

Murphy, D., Casey, D., & Fraser, J. (2007). Talk about walkabout: Evaluation of a flexible learning initiative. In P. Tsang, R. Kwan, & R. Fox (Eds.), *Enhancing learning through technology* (pp. 11-26). Hackensack, NJ: World Scientific Publishing Company.

National Survey of Student Engagement. (2008). *Promoting engagement for all students: The imperative to look within.* Bloomington, IN: Indiana University Center for Postsecondary Research.

North Central Association Higher Learning Commission. (2000). *Best practices for electronically offered degree and certificate programs.* The Higher Learning Commission. Retrieved on November 5, 2008, from http://www.ncahlc.org/index.php?option=com_docman&task=cat_view&Itemid=229&gid=80

Pittman, V. (1992). Evaluation and independent study. *Adult Assessment Forum, 11,* 6–13.

Pittman, V. (1998). Low-key leadership: Collegiate correspondence study and "campus equivalence." . *American Journal of Distance Education, 12*(2), 36–44. doi:10.1080/08923649809526993

Prensky, M. (2000). *Digital game-based learning.* NY: McGraw-Hill.

Rekkedal, T. (1983). The written assignments in correspondence education. Effects of reducing turn-around time. *Distance Education, 4,* 231–250. doi:10.1080/0158791830040211

Rohfeld, R. (1996). Access and the search for academic respectability in continuing higher education. *Continuing Higher Education Review, 6,* 54–66.

Russell, T. L. (2001). *The no significant difference phenomenon, 5th ed.* Montgomery, AL: International Distance Education Certification Center.

Watkins, B. (1991). A quite radical idea: The invention and elaboration of collegiate correspondence study. In B. Watkins & S. Wright (Eds.), *Foundations of American distance education: A century of collegiate correspondence study*. Dubuque, IA: Kendall/Hunt.

Wedemeyer, C. (1963). *Brandenburg memorial essays on correspondence study, vol. I.* Madison, WI: University of Wisconsin-Extension.

Wedemeyer, C. (1966). *Brandenburg memorial essays on correspondence study, vol. II.* Madison, WI: University of Wisconsin-Extension.

Wedemeyer, C. (1977). Independent study. In A. S. Knowles (Ed.), *The international encyclopedia of higher education* (Vol. 5, pp. 2114-2132). San Francisco: Jossey-Bass Publication.

Wedemeyer, C. (1981). *Learning at the back door: Reflections on non-traditional learning in the lifespan*. Madison, WI: University of Wisconsin Press.

Wedemeyer, C., & Childs, G. (1961). *New perspectives in university correspondence study*. Chicago, IL: The Center for the Study of Liberal Education for Adults.

Wedemeyer, C., & Najem, R. (1969). *AIM: From concept to reality: The articulated instructional media program at Wisconsin*. NY: Syracuse University Press.

Chapter 2
Dual-Design Strategies for Modularizing E-Learning for Academic and Commercial Uses

Shalin Hai-Jew
Kansas State University, USA

EXECUTIVE SUMMARY

This chapter addresses applied strategies for modularizing e-learning along both an academic university track and a commercial one. Academic qualifications and professional certifications have been seen as complementary in some ways, antithetical in others. Another way to visualize both is as one-in-the-same in terms of learning contents, albeit with versioning for the various differences. The chunking of a curriculum for both a formalist college setting and for a business one involves creative applications of the module format, particularly given the disparate needs and learning outcomes of the two (often) different learner audiences. This case examines the differences between the learning needs of both demographics. This case then sets the dual-design scene from an instructional designer point-of-view.

BACKGROUND

This case centers on a Midwestern university that offers undergraduate, graduate and post-graduate studies. This university was established in 1863 as a land-grant, public university. A leader in research, this institution of higher education features 60 academic departments in nine colleges: agriculture; architecture, planning, and design; arts and sciences; business administration; education; engineering; human ecology; technology and aviation, and veterinary medicine. Its graduate programs include nearly 100 master's degree programs and some 50 doctoral programs; it attracts students from around the world. The distance learning program was started in 1974. With a popular distance learning program, the university's Division of Continuing Education (DCE) is finding growing demand for creating trainings for companies, non-profit organizations, and other learners who are non-traditional learners. The interests lie mostly in commercial ventures, which support employee training, continuing certificate, and skills maintenance.

DOI: 10.4018/978-1-60566-870-3.ch002

SETTNG THE STAGE

Traditionally, universities have packaged learning into academic years, semesters, or quarter terms, for over a hundred years. More recent learning terms have involved short courses and even extreme accelerated learning between formal terms through "intersession" courses. The new scheduling of shorter learning chunks and the high investments in multimedia development of electronic learning has led to the popularization of modules. Modular delivery of learning offers more flexibility in accommodating learners (Crynes, 1996). Modules offer ways to disaggregate a curriculum into restructured pieces built around different learning outcomes. These smaller components may enhance the targeted use of assessments with unique learners, and provide "benefits in ease and quality of instruction" (Burgess, 2003, p. 128). This author suggests that modules also make it easier to incorporate new texts to a course. The updating and revision of course materials may also be easier (pp. 132 – 133). "Module delivery is flexible in that it is variably credit rated and can be studied in full-time, part-time and distance modes," observes McClelland (2002, n.p.).

Prior to the initial exploration into modularizing an academic curriculum for commercial trainings, the university had not developed the in-house capacity to use its learning / course management system (L/CMS) to deliver automated trainings. It was working on various technological methods to support self-enrollments into the learning system and to allow external corporate access to their employees' learning records—both non-trivial technological aims. It also had not developed the capacity to strategically extract contents for modularized trainings in a for-profit setting.

Modularity

Modularity involves strategies for organizing and delivering contents for human-facilitated learning or automated learning, through the creation and use of digital learning objects (LOs). Modules are to be interchangeable, with the ability to add context, fresh assessments, customized sequencing, and new cultural contexts to the learning. Modules may serve as stand-alone short-courses; they may complement face-to-face, hybrid, or fully online learning.

Modules in the Research Literature

Modules in adaptive learning systems automate customizing education to specific learners based on their profiles and learning behaviors (Agarwal, Edwards, & Pérez-Quiñones, 2006). In learning business processes, learners benefit from more personalized vs. standardized learning in modules (Hawryszkiewycz, 2005, n.p.). Modularizing a curriculum may offer different ways to version a curriculum for the "auditory, kinesthetic and visual learner" (McNutt & Brennan, 2005, pp. F1H-27 to F1H-31). Indeed, modules may offer powerful analysis of user actions to see what parts of the learning experience may be improved (Poe, McGowan, Hansen, & Singh, 2008).

Modules involve both conceptual knowledge and transfer knowledge (Barr, Pandy, Petrosino, & Svihla, 2005). Theoretical knowledge refers to the conceptual modeling of a particular learning domain. Transfer learning refers more to applied and hands-on knowledge, to a variety of professional settings and situations. In the research literature, there are different ways for the creation of learning modules. There are on-the-fly captures of learning materials in software engineering to lower the human labor intensiveness and to promote more rapid development of the module; there are resulting modules as a byproduct of software engineers' daily work (Grützner, Angkasaputra, & Pfahl, 2002, pp. 533 – 539). A virtual work-based learning environment offers a series of critical thinking exercises in another modular build (Dansberry & Cates, 2004). Theme-based modules are created for deployment over an asynchronous learning network (with distance-enabled collaboration tools) in an informatics course (Nulden, 1999).

A human-computer interaction (HCI) course employs learner dialogue to conduct problem-solving (Iqbal & Every, 2005). Case studies are employed in modules for undergraduate and graduate students in manufacturing engineering, based on the research that cases may focus on relevant details, offer student motivation, consolidate and integrate important concepts and techniques, and offer strong learning transfer (Richards, et al., 1995, as cited in Standridge, 2000).

Complex dynamic mental models that show interrelationships between phenomena in the world may be captured and conveyed in modules (Verleger, Diefes-Dux, Rickus, & Schaffer, 2005). Mental models refer to learners' conceptualizations of phenomena (as compared to subject matter experts' more sophisticated conceptual models in their particular expert domains).

Barbosa and Maldonado (2006) propose a standard process for the development of educational modules and suggest that a formal structure may "ease the distributed and cooperative work to create, reuse and evolve educational modules" (p. T2J-5). Their model offers three categories of processes: primary, supporting, and organizational processes. This draws from the ADDIE model (analysis, design, development, implementation, and evaluation) (p. T2J-7).

Others are built experientially, based on Kolb's experiential learning cycle, which begins with concrete experiences, some reflection, an abstract conceptualization, and active experimentation, to ground the learning (Chattratichart, 2007). Experiential practice may enhance the experiences of reinforcement learning; it may enhance virtual apprenticeship learning. Some modules in learning are embedded in a virtual learning environment (VLE) and enhance judgment, discretion and decision-making in a simulated clinical context (Lee & Lee, 2004, n.p.).

Some take advantage of the human brain's hard-wiring to "surprise" learning into those going through their modules. There is unexpected hissing in the "surprise zone" between the learning modules to reinforce the learning based on the brain hard-wiring of responding to threats (Kock, Chatelain-Jardón, & Carmona, 2008, pp. 183 – 197).

Some design multi-level online modules for mixed audiences with selected contents for different levels (Hermann, et al., Feb. 2003, pp. 196 – 200). In some organizations, modules are developed independently with different versions of course materials for easier inheritance and adaptation by different faculty and scaffolding for different learners (Zhu, Wong, Sturz, Shi, & Tang, 2008, pp. 148 – 156). In fast-evolving course areas of study, modules may be created to effectively address this cutting-edge learning (Helps & Renshaw, 2004, pp. 56 – 59). Modules enable nimble course redesigns for currency and relevance (Helps, 2007). The concept of module strength or "high cohesion" from computer science does apply to the building of modules, with the various components unified with "strong internal relationships" (Churchier, Irwin & Kriz, 2003).

An Environmental Scan

After conducting some initial research, the university now commences with an environmental scan and finds that there are a number of actual and potential stakeholders for modular e-learning. Some suggest that bringing "participant stakeholders" in to the module design process results in a more usable architecture, contents, and effectiveness, particularly in work contexts (Giordano & Bell, 2000, pp. 53 – 60). Figure 1: "Stakeholders for Modular E-Learning" suggests that numerous individuals and entities have input in the development process and outcomes.

One unexpected discovery is that modular e-learning may not only benefit academic and commercial stakeholders but also the general public in terms of making some information and learning more accessible to non-expert and lifelong learners.

Different Approaches to Learning: Academic and Business-Based

The disparities between university learning and business-based trainings for employees rapidly become clearer. These differences involve a range of factors.

Oversight

First, the academic track is overseen by various accreditation agencies, at the university levels as well as at various degree and programmatic levels. The business track often is guided by domain-specific regulatory agencies.

Research and Development

Research and adding new learning is a critical piece of most academic programs at the university level, albeit less so for community or junior colleges. Higher education focuses on innovation and new approaches as a core value. In work places, research and development (R&D) are a specialized part of the overall organization, but the majority of employees will be going through trainings for acquiring knowledge and skills and maintaining various competencies.

Scheduling

College courses are offered in academic years and terms within those years (whether semesters or quarters). These academic long-courses are organized around credit hours. For many businesses, their training is non-credit-based; they are measured in the length of time that goes into the trainings. If "credits" are extended, these are extended as "continuing education units" or CEUs. Many business-based courses are short courses or refresher trainings, with some of them wholly self-paced.

Learning Sequencing

The sequencing of college learning tends to be by degrees, whether major or minor. How a person progresses through a degree program depends on academic criteria, and testing in or out of educational tracks and opportunities. Business-based sequencing is based around outcomes and learning objectives necessary to function in particular work roles. How an employee progresses through a training program may be quite flexible, with open-entry and open-exit, testing in and out, and sometimes just "seat time" (experiencing a training experience, without any related assessment). While the first group may have required courseware, the latter may have more discretion to opt-in to some trainings.

Far- or Near-Transfer

University learning aims for far-transfer learning, so the learning applies to different situations and may be expressed in innovative new ways. The learning is longitudinal (although the half-life of a university degree is said to be about two years, at which point new learning has to be applied to keep a person competent in his or her field). Training in workplaces often have short-term application and require near-transfer learning.

Levels of Difficulty

The curriculums in universities and colleges range from developmental to complex and highly difficult. The trainings offered in business organizations vary in difficulty but rarely cross over into complex learning, for which many turn to the institutions of higher learning. University learning includes principles and knowledge along with skills-based learning, whereas commercial entities focus on knowledge transfer and applied, practical skills.

Context

Academic courses tend to be highly contextualized within a knowledge domain or cross-domain (but related) fields. Short-courses in the business context tend to be decontextualized, for widest applicability.

Facilitation

Many college courses are human-facilitated by professors and graduate teaching assistants; many business-related trainings involve professional trainers from third-party training businesses, or they tend to be automated and delivered through tangibles (like CDs, DVDs, pre-loaded laptops) or learning / course management systems (L/CMSes).

Costs

Comparatively, university and college courses tend to involve fairly expensive tuition, but work-based trainings tend to have "invisible" costs to the trainees. These expenses are folded into the cost of doing business and supporting a nimble and adaptable workforce.

Figure 1. Stakeholders for modular e-learning

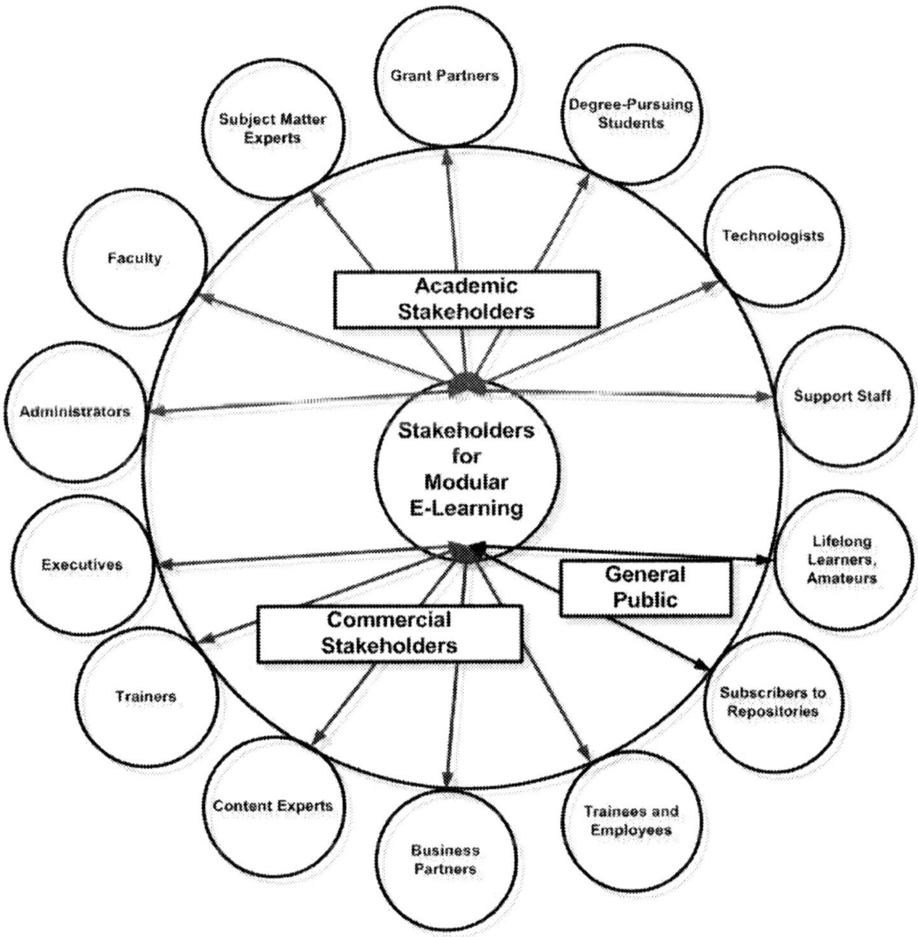

Assessments

Most assessments in higher education are high-value and have an impact on the learners' career choices. In work places, many assessments may be pass / fail, and many have a lower risk to learners. Given the high-risk of academic assessments, many academic dishonesty measures are deployed to ensure the correctness of test results. In work-based trainings, there are varying dishonesty measures—many of which are rudimentary.

Course assessments in universities are open-ended—such as papers, reports, designs, presentations, and digital learning objects. There are plenty of standard exams used, too. In work places, performance and learning are often assessed by close-ended assessments such as multiple-choice, short-answer, T/F, and mix-and-match questions.

Customization vs. Generic Learning

College courses tend to involve customized learning, with faculty and academic advisors aware of the respective student's goals and unique needs and doing their best to accommodate those. In work places, the trainings tend to be generic and not tailorable to individual learners.

College students often have contact with their instructors and supportive instructors and graduate teaching assistants, as well as tutors and related student clubs. By contrast, those in the workplace may have little to no direct contact with the subject matter experts (SMEs) or course creators. They may have some help desk support in the company but no tailored direct access to the expertise addressed in the trainings.

Learning Resources

Those in colleges and universities may have rich access to a variety of learning resources: human expertise, high-end technologies and labs, library-based informational resources, and virtually global loans of materials. In work places, depending on their capitalization, employees may have comparatively more limited resources. Some larger corporations may have in-house digital repositories of contents that are proprietary, and some will subscribe to third-party content providers' resources, but often, companies have fewer resources.

Social Cognition

Based on social cognition, college and university courses may involve study groups, in-class partnering around assignments, and virtual learning communities. There is a value seen in human interactivity and in cohort-building of fellow learners in certain graduate programs. Professors often offer back-channel and informal communications to enhance the learning. In businesses, there may be a much thinner pool of colleagues with whom to study and collaborate. The interactivity that is deployed often is automated, with mostly "store-and-forward" e-learning development. Employees of businesses often have to work hard to create their own sense of a professional community.

Types of Learners

Academic-track learners draw from a fairly young demographic and those who are in mid-career changes. Those on the commercial track tend to be older and to have spent years already in the workforce.

E-Learning Technological Differences

In higher education, e-learning may be delivered by learning / course management systems (L/CMSes), which offer integrated tech systems and features like digital content delivery, learner tracking, digital archival, and live-virtual-interactivity. Digital contents are also delivered via websites and data repositories. For the commercial track, much e-learning is delivered through third-party live interactivity. Materials may be delivered by websites and tangibles like CDs or DVDs.

Legal Differences

While both higher education and business-based learning must comply with various laws (copyright / intellectual property; trademark protections; patent laws; trade secret laws, and libel and defamation laws), each area also is affected by unique laws.

Academia has some copyright exemptions of "fair use" and the TEACH (Technology, Education, and Copyright Harmonization Act) for the performances of particular contents. Creative Commons™ copyright releases offer exemptions to accredited academic institutions for uses of some copyrighted materials. Federally-funded e-learning needs to be accessible. The contents that faculty create also must honor the federal Family Education Rights and Privacy Act (FERPA), which protects student privacy. Business-based trainings do not have "fair use," TEACH Act or non-commercial Creative Commons exemptions. This difference will be seen in the building of commercial modules. This means that there must be copyright releases for all copyrighted contents used in the commercial domain. There must be precautions about information prematurely being used in trainings with wide distributions and a multitude of possible uses.

Also, media releases of people who participate in video and audio captures, photo shoots, and other recorded events will need to sign media releases that include potential commercial use of the final products (and even the raw files). Their permissions have to be attained before any multimedia capture is done. Without clear planning and execution from the beginning, projects may be limited to academic use and may not have any application in a commercial context.

Building for-profit modules from an academic course then requires some awareness of these differences in order not to contravene laws or cause liabilities for the "deep pockets" of the university.

CASE DESCRIPTION

This chapter would be built around the "case" of multi-disciplinary instructional design work at a Midwestern university. The unique challenge of the case is designing modules for both higher education and non-academic–credit delivery. The approach will include clear applied design on live cases involving extant university curriculum and trainings; unique and disparate learning needs for the same domain contents; the technological affordances, and pedagogical strategies. It is understood that the modular learning to be delivered must be high-end and fairly complex given the areas of expertise in a university and the more sophisticated learning needs of many workplaces. This case will explore how to define effective modularization of contents in these dual purposes.

The university wants to put into place a standard process to creating modularized contents from a university-level semester-length course. It has a range of authoring tools, a homegrown learning / course management system, a well visited website, and sophisticated technology staff to support the endeavors.

Of special concern are the challenges of intellectual property and other legal considerations. Also, the university wants to ensure that it meets the needs of professionals in their respective fields by creating learning modules that are accurate, flexible, and well designed.

CURRENT CHALLENGES AND PROGRESS

The university instructional design group has a general sense of the differences between the needs of the institution of higher education and those of the various business organizations. They progress to learning about modular e-learning.

Modular E-Learning

The team approaches the definition of a module in a broad sense, as a chunk of learning.

Underlying Pedagogical Theories

Some pedagogical theories related to modular learning seem to come from the research on digital learning objects (DLOs). The team stumbles right into the debate on the optimal size of learning objects and the basic components that may be necessary.

The "Context and Learning Object Design" model created by Thorpe, Kubiak & Thorpe (2003) shows how granular objects offer experienced educators increased flexibility and usability. Learning objects that are aggregated and non-granular offer more instructional and cultural context; these objects are more easily usable by novice educators because the designed educational application has been integrated (as cited in Haughey & Muirhead, 2003). This model suggests that highly granular learning objects have been stripped of their instructional and cultural context, whereas the most aggregated ones suggest or explicitly define an instructional and cultural context.

McGreal offers a granularity model in which information objects are combined to form lessons; the lessons are combined to form modules. The modules become parts of reusable learning objects (RLOs). These reusable learning objects combine to form topical units. The topical units are combined to form simulations. The simulations may then form the underpinnings of e-learning courses, which then are parts of overall programs (McGreal, 2002).

Metadata models about labeling digital learning objects (such as the Shareable Content Object Reference Model—or SCORM) prove helpful in setting some guidelines for the digital builds.

The concept of the SCORM "ilities" may provide some guidance to the room for goals: interoperability, affordability, durability, reusability, accessibility, manageability, and scalability. These design guidelines suggest that the digital learning objects must be able to work on different Web platforms (repositories, L/CMSes, and sites). It suggests that the design must be streamlined for cost and efficiencies. Durability suggests that the learning must be future-proofed and designed in a way that it may be versioned into the future for long-term use. Reusability means that the learning must be applicable in multiple purposes and approaches. Accessibility means that people should be able to find the contents on a repository or L/CMS. Manageability means that the learning object must be convenient to handle, and scalability suggests that an exponential number of new users should be able to use the materials without causing undue pressures to the system.

A number of pedagogical theories may come into play depending on the particular learning situation. For example, situated cognition informs the building of immersive simulation modules. Social cognition informs the interactivity in a live human-to-human case analysis. Mental modeling and discovery learning theories enrich the creation of digital ontologies. Clark and Mayer's multimedia and cognition theories inform the team's work (2008).

The team then comes up with some core questions:

- What do modules look like? What are necessary elements of a commercial module?
- What is the difference between a nested culturally-situated module vs. a neutral stand-alone one?
- What sorts of transitions may be designed for connecting related modules?
- What technologies will be needed to execute on creating such modules?
- What are the requirements to connect to digital libraries, data repositories and L/CMses?

- How will modules be assessed for learning efficacy?
- What sort of workflow for modular creation would be the most efficient for the university?

BUILDING MODULES

The team decides to figure out those questions by actually walking through the building of a learning module from a specific curriculum. They analyze hundreds of syllabi from a variety of fields to understand course structures. Course schedules may be organized in some of the following ways: in developmental phasing with the easier foundational materials introduced first; by book chapters (a default to author organization); thematic topics, and also as a lead-up to a cumulative project.

The group breaks up the different types of learning they come across. They are focusing mostly on the academic side at first and will focus on the commercial side later. They're looking at the source, not the output yet. They also decide that such commercial modules should not only have contents from the academic course but should draw from other digital repositories and resources. Table 1: "A Brainstorm of the Types of Learning in Academia (and How to Organize the Related Learning Modules" provides an overview of some types of learning in academia and ways those contents may be sequenced.

They decide to apply their findings into a spatial model of how an academic semester curriculum may be evolved into smaller modules. They decide to leave the module as a "black box" currently but focus only on what value may be drawn from the existing course. They categorize the types of learning into three main categories—knowledge acquisition, mixed methods, and applied skills. Figure 2: "The Extraction of Commercial Modular Contents from a Semester Term" provides a visualization of this process of transitioning from academic learning to commercial learning.

It's clear to the group that the particular traditions of the domain field will affect the design of the learning that comes out of it.

Componentizing a Module

The design group decides that it may be a good idea to design a module abstractly. There should be an introduction and a conclusion, just to give the modules a set of beginning and end. There should be a space to prime learners for the learning to follow in the intro, and the post-module learning will help learners segue to the next phase.

Learning objectives will help learners be more cognizant of what they'll be studying. The contents will show high variability—whether these are narrated lectures, audio, video, slideshows, text, or a mix of various types of multimedia.

If there is a social learning aspect, there may be human-mediated interactivity. Or there may be machine-created automation. The assessments need to relate closely to the learning and add value to the learning. There may be optional downloadables to enhance in-world learning; the real-world learning aspect is a critical aspect of professional certifications (Hitchcock, 2007). Work samples and instructor notes may add more richness to the learning.

The team has decided to leave the decisions of what to include in a module to the subject matter experts and those working on the module design.

Techno for Modules

There are plenty of technologies that will help the team to create the modules for Web delivery. It seems that with single sourcing, creating for the Web will allow the modules to be deployed through data repositories or libraries, L/CMSes, websites, and tangibles. Some modules are de-

Table 1. A brainstorm of the types of learning in academia (and how to organize the related learning modules)

Types of Learning in Academia	How to Organize the Related Learning Module(s)
Factual Learning (a piece of a knowledge domain)	Topic or subject-based Example based Classifications and divisions
Knowledge Structures / ontologies / taxonomies / dynamic mental models	Complex systems Entities and relationships Interactions between entities
Process Learning	Sequence or steps in a process (after task decomposition)
Learning Outcomes	Topically organized, fact-based learning Practices Skills Exercises
Analytical Thinking	Case analysis Problem solving Decision-making scenarios Project-based learning (with real-world projects)
Experiential Learning / Situated Cognition	Immersive environments Simulations Interactive live human-facilitated interactions
Emergent Learning / Innovative Problem-Solving / Inventiveness	Project-based learning Problem-based learning
Social Learning	Collaborations, group projects Interactivity
Discrete Learning Methodology	A virtual game show A simulation

livered via inter-operable platforms (Balci & Inceoglu, 2007).

The widely available authoring tools available may output any of the following: slideshows, audio-annotated slideshows, screen captures, screencasts, pedagogical agents, mind-mapping visualizations, text, audio files, video files, automated games, simple simulations, and designed immersive spaces. Plenty of asynchronous and synchronous dialogues may be held using various technologies. Most authoring tools enable the inclusion of metadata and outputting an object that will have a wrapper that will enable that object to "play" on various repositories and learning / course management systems.

Enhancements to the Learning

Also, the building of scaffolding for a particular module may enhance its flexible use by a mixed audience of users; this may include antecedent learning and post-module learning, to help learners segue.

Understanding the respective learners in the field, their learning preferences, and past learning, will be important for the module design. Knowing the developmental phase of the learners will be helpful. Also, designers need to know what the primary languages of the learners are and if there may be cultural issues that may affect the learning. They also will need to consider whether people will go through the course modules as individuals or as groups, as that will affect the assignment design.

Feedback also seems critical to the learning. This may involve "a readable recording of their

Figure 2. The extraction of commercial modular contents from a semester term

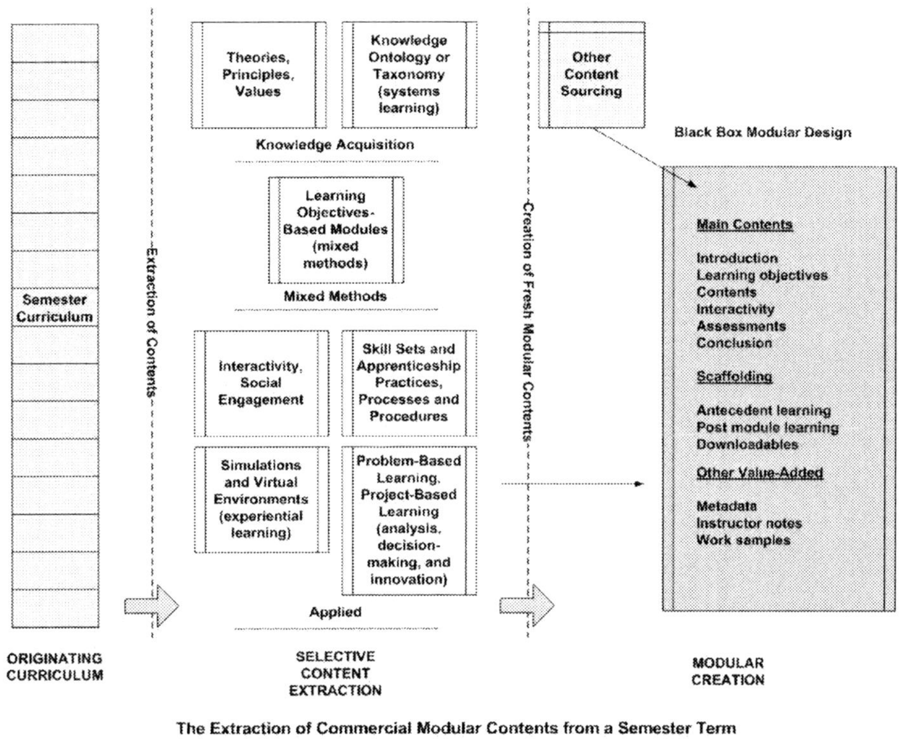

actions" (Plaisant, Rose, Rubloff, Salter, & Schneiderman, n.d.).

Some builds will have to build for "uncertainty," or the interpretive suppleness needed in some fields. Clearly, there must be a fit between the pedagogy, the technologies, and the learning objectives and outcomes.

Commercial considerations may involve the need for "simple English" for global delivery, particularly if language translators may be used for second-language global employees. There may have to be the need to avoid politically insensitive topics. They may have to be some vetting of learning so as not to "over-share" information that may involve competitive advantages in particular R&D applications.

A Module Development Timeline

When they first start working with faculty, they realize that some go to familiar teaching and learning structures, such as building modules around existing lectures (Burgess, 2003, p. 128). This tendency will have to be combated with a light effective sample modules and strategies to build in a fresh way.

As the group progresses, they decide that there are seven discrete steps to the e-learning module development timeline:

1. Learning Needs Assessment
2. Local Resource Analysis
3. Module Design
4. Development of the LOs and Module
5. Alpha and Beta Testing

6. Module Release
7. Planning for Updates

They visualize the steps in a timeline. They note that the academic and commercial tracks diverge from the beginning, with a special focus on intellectual property laws early on and then learning design later in the process. Figure 3: "Module Development Timeline" provides an applied overview of this workflow.

THE INTEGRATION OF MODULES INTO LEARNING

The effectiveness of a learning module is determined in part by how well it is integrated into a learning context. In the granularity arguments about optimal sizes of digital learning objects, the more lithe and context-independent objects are most flexible but will require sophisticated experienced instructor work. The more culturally-contexted modules may be more stand-alone but would not apply as readily to different learning situations. These may be less adaptive to a new learning context. Assessing learner responses and the efficacy of learning should inform how a module is revised.

The group surfaces some questions about increasing the learning value:

- How may modules be designed for human facilitation?
- How may modules be designed for automated learning / self-discovery learning?

Assessing Modules for Learning Efficacy

The team will most certainly be asked about how to assess modules for learning efficacy. From the business side, basic performance is often sufficient based on an applicable standard exam. This question of modular efficacy will depend on who the learning audience is, the types of learning involved, and the desired outcomes.

The User Experience

They decide that user testing is important, and they think some questions such as the following would be important:

- What is the learner experience in the module?
- Is the sequence effective for learning?
- Is the language clear? Respectful of learners?
- If learners seek help in the module, do they find it?
- Do learners get sufficient feedback?
- Would the learners be amenable to learning from other similar modules in the future?
- How well did learners perform on the assessment?
- Was the assessment clearly connected to the learning?
- Were there technology challenges?

General Learning Quality Questions

The group brainstorms some questions about learning quality in a module:

- What are the main takeaway lessons of the module? The unintended messages?
- How rich is the feedback loop in the learning?
- How well does the module "travel" between cultures? Nationalities? Genders? Languages? Educational domains? Identity groups?
- How accurately does the module address the materials?
- How well does the module segue and connect with other learning?
- How well does the module capture the full real-world complexity of the topic?

Figure 3. Module development timeline

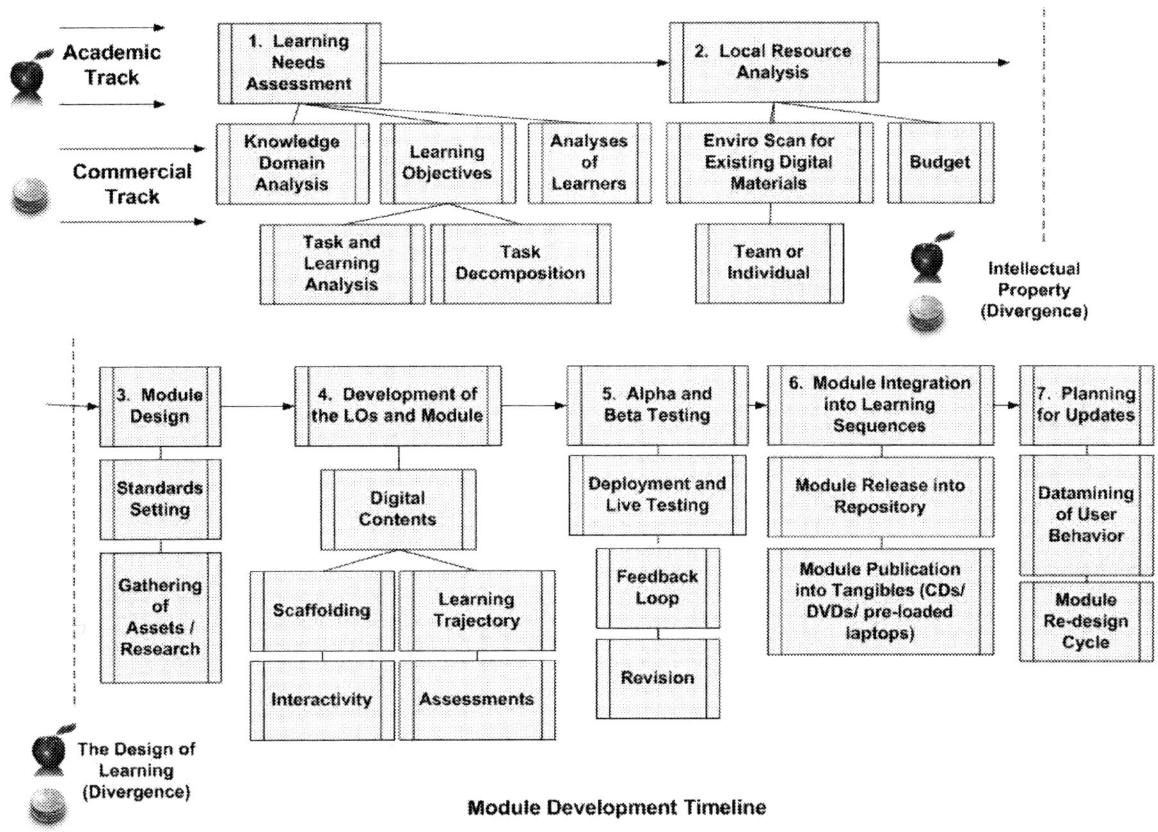

- How applicable is the learning in the module to a real-world work situation?
- How many opportunities does it provide for hands-on learning?
- How original and non-derivative are the learning contents?
- How well does the module promote far-transfer learning? Near-transfer learning?

The next steps will involve creating an instrument to test the efficacy of the learning by asking users about their learning experience.

CURRENT CHALLENGES / PROBLEMS FACING THE ORGANIZATION

What this organization needs is to hone its approaches to extracting curricular contents from academic course curriculums and deploying these in modular format with a greater number of commercial clients. This process will become more nuanced and will be strengthened with more direct hand-on development work.

LESSONS LEARNED AND FUTURE

Creating modules from a semester-long university course for a commercial entity requires plenty of forethought and planning. It requires deep knowledge of the various mixed learners. It requires assessments of the learning and proper evolution of the contents to meet learner needs. This university team has raised more questions than it has answers.

ACKNOWLEDGMENT

Thanks to Dr. Bettie Minshall for the first question about modularizing university curriculums for commercial trainings. Lynda Spire has opened many doors to explore interesting technologies and practices for e-learning. Dr. Sue Maes has provided some helpful direction and funding at critical moments. As always, thanks to R. Max.

REFERENCES

Agarwal, R., Edwards, S. H., & Pérez-Quiñones, M. A. (2006). Designing an adaptive learning module to teach software testing. [Houston, TX: ACM.]. *SIGCSE, 06*, 259–263. doi:10.1145/1124706.1121420

Balci, B., & Inceoglu, M. (2007). A Web-based learning content design platform and metadata editing. *37th ASEE/IEEE Frontiers in Education Conference* (pp. F1H-13-F1H-18). Milwaukee, WI.

Barbosa, E. F. (2006). Towards the establishment of a standard process for developing educational modules. *36th ASEE/IEEE Frontiers in Education Conference* (pp. T2J-5-T2J-10). San Diego, CA: IEEE.

Barr, R., Pandy, M., Petrosino, A., & Svihla, V. (2005). Challenge-based instruction: The VaNTH biomechanics learning modules. 35th ASEE/IEEE Frontiers in Education Conference (pp. SIG–8-IG–13). Indianapolis, IN.

Burgess, S. (2003). Course modularization for improved delivery and assessment. *Consortium for Computing Sciences in Colleges, Northwestern Conference* (pp. 128–134).

Chattratichart, J. (2007). A theory-based approach to designing student learning context. [San Jose, CA: ACM.]. *CHI, 2007*, 1721–1728.

Churcher, N., Irwin, W., & Kriz, R. (2003). Visualizing class cohesion with virtual worlds. *Conferences in Research and Practice in Information Technology* (pp. 24). Australian Computer Society.

Clark, R. C., & Mayer, R. E. (2008). *E-learning and the science of instruction*. San Francisco: Pfeiffer.

Crynes, B. L. (1996). Curriculum granularity. *IEEE FIE '96 Proceedings* (pp. 740–744).

Dansberry, B. E., & Cates, C. L. (2004). A social responsibility learning module for use in cooperative education. *34th ASEE/IEEE Frontiers in Education Conference* (pp. F3F-3-F3F-6). Savannah, GA: IEEE.

Giordano, R., & Bell, D. (2000). Participant stakeholder evaluation as a design process. [Arlington, VA: ACM.]. *CUU, 00*, 53–60. doi:10.1145/355460.355472

Grützner, I., Angkasaputra, N., & Pfahl, D. (2002). A systemic approach to produce small courseware modules for combined learning and knowledge management environments. [Ischia, Italy: ACM.]. *SEKE, 02*, 533–539. doi:10.1145/568760.568855

Haughey, M., & Muirhead, B. (2003). *Evaluating learning objects for schools*. Retrieved on November 1, 2008, from http://www.usq.edu.au/electpub/e-jist/docs/vol8_no1/fullpapers/Haughey_Muirhead.pdf

Hawryszkiewycz, I. T. (2005). *A framework for integrating learning into business processes*. Australian Computer Society, Inc.

Helps, C. R. G., & Renshaw, S. A. (2004). Design of a flexible case-study instructional module for operating systems for information technology. [Salt Lake City, UT: ACM.]. *SIGITE, 04*, 56–59.

Helps, R. (2007). Dancing on quicksand gracefully: Instructional design for rapidly evolving technology courses. [Destin, FL: ACM.]. *SIGITE, 07*, 1–7. doi:10.1145/1324302.1324304

Hermann, N., Popyack, J. L., Char, B., Zoski, P., Cera, C. D., Lass, R. N., & Nanjappa, A. (2003). Redesigning introductory computer programming using multi-level online modules for a mixed audience. [Reno, NV.]. *SIGCSE, 03*, 196–200. doi:10.1145/792548.611967

Hitchcock, L. (2007). Industry certification and academic degrees: Complementary or poles apart? *SIGMIS–CPR '07*. St. Louis, MO: ACM.

Iqbal, R., & Every, P. (2005). Scenario based method for teaching, learning, and assessment. [Newark, NJ: ACM.]. *SIGITE, 05*, 261–266. doi:10.1145/1095714.1095775

Kock, N., Chatelain-Jardon, R., & Carmona, J. (2008). An experimental study of simulated Web-based threats and their impact on knowledge communication effectiveness. *IEEE Transactions on Professional Communication, 51*(2), 183–197. doi:10.1109/TPC.2008.2000345

Lee, K., & Lee, J. (2004). Benefits of a virtual learning environment in enabling collaborative and constructivist learning in a community of specialist nursing practice. *Proceedings of the IEEE International Conference on Advanced Learning Technologies (ICALT '04)*. IEEE.

McClelland, R. J. (2002). Evolving Web-based delivery: Managing the transition from VLE's designed by academics to commercial VLE's, the post-graduate experience. *Proceedings of the 28th Euromicro Conference*. IEEE.

McGreal, R. (2004). *Learning object granularity. Learning objects: A practical definition*. Retrieved on November 1, 2008, from http://itdl.org/Journal/Sep_04/article02.htm

McNutt, L., & Brennan, M. (2005). Work in progress—learning styles and e-learning, what is the connection? *35th ASEE/IEEE Frontiers in Education Conference* (pp. F1H-27-F1H-31). IEEE.

Nulden, U. (1999). Thematic modules in asynchronous learning network: Designing introductory courses. Proceedings of the 32nd Hawaii International Conference on System Sciences (pp. 1-10). IEEE.

Plaisant, C., Rose, A., Rubloff, G., Salter, R., & Schneiderman, B. (1999). The design of history mechanisms and their use in collaborative educational simulations. [Palo Alto, CA.]. *Proceedings of the Computer Support for Collaborative Learning, CSCL, 99*, 348–359.

Poe, D., McGowan, K., Hansen, C., & Singh, G. (2008). Refining educational content through a closed-loop FLOW approach. *Inroads— . SIGCSE Bulletin, 10*(2), 102–106. doi:10.1145/1383602.1383645

Standridge, C. R. (2000). Teaching simulation using case studies. *Proceedings of the 2000 Winter Simulation Conference* (pp. 1630–1634).

Verleger, M., Diefes-Dux, H., Rickus, J., & Schaffer, S. (2005). An online bioengineering learning module: Gene regulation. *35th ASEE/IEEE Frontiers in Education Conference* (pp. SIG–20-SIG-25). Indianapolis, IN.

Zhu, L., Wong, J., Sturz, B., Shi, Y., & Tang, C. (2008). An integrated evaluation method for module-based undergraduate information retrieval education. Consortium for computing sciences in colleges. *JCSC*, *23*(6), 148–156.

Chapter 3
Moving Toward a Blended Course Delivery System

Leah Blakey
Drury University, USA

EXECUTIVE SUMMARY

Today's university students demand a high degree of flexibility in scheduling, as well as a quality educational experience. Many students enjoy the flexibility of online courses, but some have learning styles that make a traditional classroom experience preferable. This chapter draws on the experiences from numerous institutions; however, the particular experiences of Drury University's transition to blended classes will be at the center of this case study. The issues the schools have faced, as well as the resolutions that were designed and implemented, will be discussed in detail. The goal of the chapter is to provide a detailed, practical guide to the best practices in blended education.

INTRODUCTION

Today's university students demand a high degree of flexibility in scheduling as well as a quality educational experience. Many students enjoy the flexibility of online courses, but some have learning styles that make a traditional classroom experience preferable. Soules takes this idea further by pointing out that "those students at a distance from the face-to-face classroom often express the suspicion that they are missing something. Not only are they missing what they might learn from any material presented in class that is not included in the online resource materials, they are also missing the learning that comes from participation in discussion and interaction. Additionally, some people feel that it is not as easy to ask questions online. There is validity to these concerns in my experience: students do seem to benefit from classroom discussions, from the clarification of difficult material, and from interaction with an instructor about assignments, and other matters related to the course." (Soules) All of these needs and desires can be fulfilled through a blended class format. Drury University defines a blended class as a class that has both online and face-to-face classroom components. It is the one

DOI: 10.4018/978-1-60566-870-3.ch003

of the best designs for instructional delivery in today's fast paced, quality driven, highly competitive academic world.

Many schools have turned to blended classes as an optimum response to student demands for high quality and flexibility. (Hybrid Classes, 2002) This chapter draws on the experiences from numerous institutions; however, the particular experiences of Drury University's transition to blended classes will be at the center of this case study. The issues the schools have faced as well as the resolutions that were designed and implemented will be discussed in detail. The goal of the chapter is to provide a detailed, practical guide to the best practices in blended education. As very little has been written on the best ways to manage blended courses, interviews and personal experience are the basis for this chapter.

Drury University officially began the blended class program in the spring of 2008. Leaders of the program immediately turned to published material to find direction. Unfortunately, there have been very few relevant journal articles written on the subject of blended classes and their administration. Therefore, the leaders turned to colleagues with experience in using a blended format for advice and best practices. From their successes and failures, Drury University built a policy for the program and a training model for its instructors.

There are some common issues facing all schools attempting to use blended classes. These two main issues involve: a) communicating a blended course's requirements to potential students and b) training instructors to maximize the best of both online and traditional classroom teaching techniques in a blended class.

The resolution of these issues is directly tied to the culture of the individual schools and their respective student bodies.

There are also issues that instructors in blended courses encounter that differ widely, depending on the schools and disciplines involved. To address these issues and evaluate various responses, the experiences of more than twenty people with backgrounds in teaching and learning through blended classes will be considered. The results of these interviews will provide detailed, hands-on approaches and advice for universities and individuals considering a blended class delivery method.

Competition for students increases annually. In order to compete, schools have to devise new ways to meet their students' demands for flexible scheduling and dynamic classes. Blended classes can combine the best of both the online and traditional classroom to meet these demands. (Hybrid Classes, 2002) By drawing on the experiences of so many instructors who have successfully mastered the blended delivery format, this chapter will help you design and deliver the best possible blended program and class.

BACKGROUND

History

Online education became a reality at Drury University in 1999. Since then, the program has grown at a rapid pace. This chart (Figure 1) shows the growth in the number of classes over the last nine years.

However, the remaining segments of the College of Graduate and Continuing Studies (CGCS), where online education is housed, have not seen similar growth. As Figure 2 demonstrates, enrollment in face-to-face classes has fallen during the same time frame.

Students claim this is because they prefer the online delivery of classes because it fits better into their busy lives. In addition, transportation costs and parking issues are reduced for the students.

While students preferred and thus enrolled increasingly in online classes, the directors of the face-to-face programs struggled to maintain enrollments at a break-even level. Increasingly, face-to-face classes were cancelled due to low enrollment. In an attempt to bring more students

Figure 1.

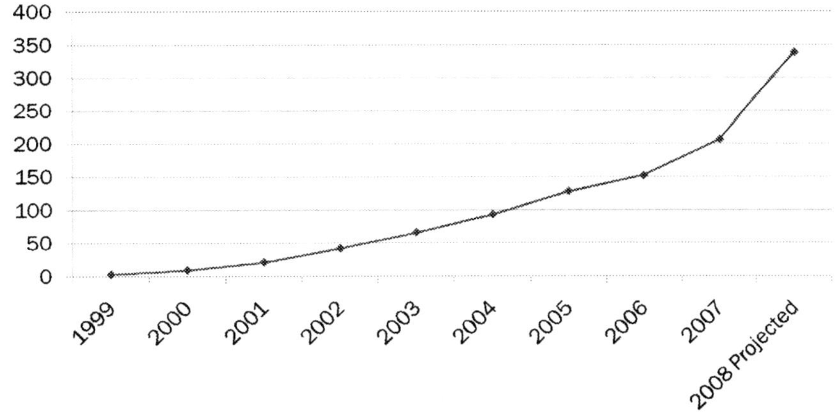

into the brick and mortar classrooms, Drury University moved slowly to a model involving increasing numbers of blended classes.

Movement Towards Blended Classes

Students increasingly stated on their evaluations that they preferred the convenience and flexibility of online classes. Some also stated that, while they would like to be in a face-to-face class, their busy lives really would not accommodate this. In reaction to these statements and decreasing enrollments in the face-to-face classes, Drury University's evening college (CGCS) began brainstorming ways to keep the students enrolling in the face-to-face classes.

Altering the start times of the classes was one of the first things the administration attempted in order to accommodate student schedules. The College of Graduate and Continuing Studies (CGCS)

Figure 2.

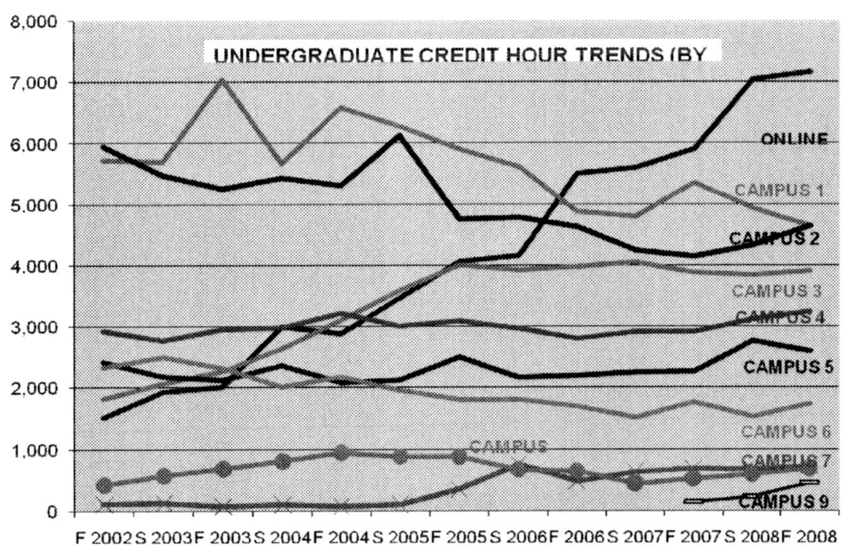

has eight campuses and those campuses that could began offering classes during the daytime hours. However, keep in mind that most of the students at the satellite campuses were non-traditional degree seekers and would need to schedule classes around work and family. The administrators scheduled classes that would begin after parents dropped their own children off at school and would end before the children needed to be picked up. While classes had historically been in the evening only, this manipulation of the normal times has enjoyed some success. Unfortunately, the main campus was unable to offer classes in this timeframe because it is also home to the traditional academy or day-student base of the University.

In a further effort to draw students into their classes in the evening, the main campus component of CGCS started manipulating start times the best they could during the after 5:00 p.m. window in which the classrooms on campus became available. Unfortunately, the change of time measured in terms of minutes was not significant enough to lure students back into the brick and mortar classrooms.

The next level of brainstorming stumbled upon a phenomena that was already happening on campus unofficially to various degrees. This was the idea of taking normal face-to-face class components and posting certain selected components on an online platform.

THE PRESENT

Characteristics of Blended Education at Drury University

Blended education began unofficially at Drury University because instructors wanted online components added to the face-to-face classes for various reasons. The online education office supported this gradual move towards online because it was also a step towards demonstrating the value of online classes to both faculty and students. Keep in mind that this was a time period where some concern and distrust of online education still existed in the minds of some tenured faculty in the day school. These various degrees of online incorporation into face-to-face classes became the basis for the official start to a program of blended learning that occurred in the 2007-2008 school year.

Various Degrees of Blended (enhanced, blended, online plus)

As stated in the ground breaking study done by the University of Wisconsin, "Hybrid (blended) courses show enormous variety in how the face-to-face ratio to online time is distributed." (Aycock, 2002) Blended education at Drury University started with what we now refer to as enhanced classes. These are classes that are primarily face-to-face classes with small amounts of the instruction, assignments, or communication being conducted online. Table 1 shows the growth in enhanced courses from fall 2005 thru spring 2008.

Many traditional instructors begin rather slowly when they enter the realm of an enhanced class for the first time. The most common techniques used by instructors to enhance a face-to-face class are placing their syllabus online, maintaining an electronic grade book, and keeping all e-mail for the class within the electronic platform, rather than using the University wide e-mail. This is still a very popular usage among faculty members for both the traditional day school as well as the CGCS courses. (Note: There are several electronic platforms available commercially including Angel, eCollege, Blackboard, and WebCT. Drury University has utilized the commercial platform called WebCT for several years. For convenience, future references to an electronic platform will simply be as WebCT.)

Enhanced courses are also used by instructors who travel a great deal. The idea of canceling a daily class completely has become outdated because, with a WebCT component, the class

Table 1.

Semester	Fall 2005	Spring 2006	Fall 2006	Spring 2007	Fall 2007	Spring 2008
Approximate Number of Enhanced Courses	35	38	40	44	60	52

can meet asynchronously while the instructor is maintaining his or her professional commitments at conferences around the world. Many times these classes only use the WebCT portion of the class when the instructor is out of town.

However, as the last two winters have dealt the Midwest more than its fair share of winter precipitation, instructors and students both are increasingly realizing the value of having an enhanced class that can continue through ice and snow – particularly when the physical campus is closed.

The next level of combined class is the type of course that Drury University actually refers to as a blended class in their course schedule. At Drury University, a blended class is defined as a class whose meetings occur half online and half face-to-face. Instructors receive specific training regarding how to design classes that use two different delivery models. Particular emphasis is given to classes of varying lengths. This is necessary because Drury University offers blended classes of both sixteen and eight weeks. Classes that run for sixteen weeks are expected to have eight face-to-face class meetings and eight weeks totally online. Classes that run for eight weeks are supposed to have a face-to-face meeting every week coupled with a full week of online instruction every week.

Believing that any class could be offered in a blended format, the fall 2008 schedule boasted 36 blended courses. This constituted 23% of the fall schedule for Drury's evening college on the main campus. It also spanned fifteen disciplines out of a total of twenty-two. The administration was pleased enough with student reaction to these classes to propose a spring 2009 schedule containing 49 classes.

Larger universities have created blended course arrangements that allow them to better address the needs of students in classes with very large enrollment. For example, the University of New Mexico utilizes the blended to model in classes that have "250 students where half attend the class one day per week while the other half attends online. The second scheduled meeting the groups switch." (Hybrid Classes, Bloom, 2006) This type of course division is particularly useful in science courses such as Professor Jim Swan's biology II class. "I anticipate it will be helpful in reducing the tedium as we review anatomy part by part. They will be able to read more on their own and create charts and graphs." (Hybrid Classes, Bloom, 2006) The University of New Mexico currently faces an enrollment issue because more students want to enroll in courses such as biology II than can fit in the classroom. "With 250 students currently enrolled in the course and only half physically attending each class, enrollment can be boosted by 50. By increasing the enrollment to 300, we increase by 25 percent the number of students who can take the class." (Hybrid Classes, Bloom, 2006) Thus, schools of various sizes and cultures are turning to the blended model to help them address the challenges they face.

A third type of blended class is also starting to make its way into Drury University's vocabulary. The associated term is "online plus", which refers to a class that is all online except for one meeting. There are three primary ways this type of course is used:

1. a course that is all online except for a required real life field trip;

2. a course that is all online except for one proctored exam; and
3. a course that is all online except for one formal presentation that is delivered face-to-face.

Online plus courses that are totally online except for a required field trip offer students multiple ways of learning that often cannot be duplicated in a totally online class.

For example, a class on Lewis and Clark at Drury University is very well suited to an online plus format. The class meets online for the first seven weeks of the semester during which time the basic elements of the Lewis and Clark expedition are taught. These elements provide the foundation for understanding the facts of the voyage. The class then takes an all-day field trip to visit various Lewis and Clark museums in the St. Louis, Missouri area. This fieldtrip lets the students touch the replica keel boat, taste smoked deer meat, experience the difference in winter camp life for the captains versus the enlisted men and the Native Americans, and (if they so choose) experience the prickles of a prickly pear. The class then resumes its online dimension for the remainder of the semester with a much heightened awareness of the physical challenges of the voyage.

Online plus classes that meet totally online except for one proctored exam are also very common. In fact, many universities which refer to their classes as "online" actually adhere to this type of arrangement. The exam is a proctored exam that can be taken in numerous locations such as public libraries, schools, branch campuses, etc. The student taking the exam must provide identification, which is considered by many schools to be their strategy to ensuring that the person doing the work in the course is the person actually enrolled in the course.

The final popular version of online plus classes involves a class that is totally online except for when the class physically comes together one time for oral presentations. This is very popular in Communication courses and senior research courses where the major project of the semester must be completed as both a formal paper and formal presentation. These models of online plus classes are most often employed when there is a specific concern about a particular aspect of the course (final or presentation, for example) being conducted online or when an instructor feels that a field trip or field experiment is absolutely vital to the class.

The Role of the Online Education Office in Blended Courses

The manner in which Drury University e-learning is currently configured means that all e-learning at the University relies on the Office of Online Education to support their technical needs and provide technical training to faculty. Considering that this office is responsible for the most prolific growth on the campus, this added responsibility of technical support for all versions of blended classes often strains the staff. The office currently is responsible for 136 totally online classes and 95 additional classes that have some online component. This workload has forced the administration to add an additional staff person to manage the blended courses for the office.

The online office also has the responsibility for technical training for all faculty assigned to teach a blended course and those who have never taught online for Drury University. This is a major undertaking that coincides with fantastic growth and exciting new opportunities for the program as the office is transitioning to a new method for training its faculty, adding and shifting staff around, and considering a new learning management system for the next academic year.

Oversight of Blended Courses

The management of blended classes is divided between two individuals. One is responsible for the faculty training and oversight of the courses, while the other is responsible for the scheduling and staffing. Oversight of the courses involves multiple levels of evaluation conducted by administration, peers, and students.

The first level of evaluation examines the instructor's plan for the course and syllabus. There is an art to dividing a class up where half of it is conducted face-to-face and half of it is conducted online. Appendix A contains sample syllabi of blended courses that demonstrate some successful ways to divide up a blended course between the face-to-face and online meetings. Determining which assignments to put online and which to conduct in the face-to-face meetings takes an understanding of learning styles, teaching techniques, and a firm understanding of the objectives of the course. This must be coupled with an eye for keeping the momentum of the class moving between such varied meetings. All issues must be included in the initial plan for the class and laid out clearly in the syllabus. Drury University has a syllabus template that requires its faculty to designate what parts of the class will be conducted online and what parts face-to-face before the course ever begins.

Peter Sands has outlined five principles to help instructors plan the online and face-to-face aspects of a blended class. These include: "1. Start small and work backward from your final goals. 2. Imagine interactivity rather than delivery. 3. Prepare yourself for loss of power and a distribution of demands on your time more evenly throughout the week. 4. Be explicit about time-management issues and be prepared to teach new skills. 5. Plan for effective uses of classroom time that connect with the online work." (Sands, 2002)

Aycock's findings from the University of Wisconsin study stated, "there is only one effective way to use online technologies in hybrid courses: it is essential to redesign the course to integrate the face-to-face and online learning. The online learning modules are central to a hybrid course's success, and the students' work online must be relevant to the in-class activities." (Aycock, 2002) Thus, these principles must be tailored to fit the culture of the university and the individual needs of particular classes, but taken as general guidelines, they can be very helpful in course design.

The next level of evaluation takes place between the second and fourth weeks of a blended class. It involves an administrator accessing the online part of the class to check on how things are going. The administrator completes a checklist for the class and sends it electronically to the instructor with specific feedback regarding areas that could be strengthened as well as suggestions for how the instructor might do this. A copy of the evaluation sheet is provided in Appendix B. As you can see from this sample evaluation form, the primary items the evaluator is looking for are the clear delineation of what is expected online versus what is expected face-to-face and whether the instructor is actively involved in the class and gives timely feedback.

The next evaluation goes to both faculty and students. It is administered half way through the semester, which means eight week long classes are almost over and sixteen-week classes are at midterms. Because the blended program has only officially been up and running for one academic year, administrators ask a lot of questions to try to discern a number of different aspects of the experiences of both the faculty and the students. Feedback from these mid-semester evaluations is used to plan for the next semester. Samples of the evaluation form for faculty and students can be found in Appendix C and D, respectively.

The final evaluation is the IDEA (Individual Development and Evaluation Assessment). This program is administered through Kansas State University and provides both statistical data and constructive feedback from the students. The results are sent to the faculty member, administrative officials, the dean of the college, and department chairs approximately three months later. While this feedback is often valuable, it is not timely and too often the next semester has started before we realize a change needed to be made based on student or faculty comments. A sample IDEA feedback form can be found in Appendix E.

Scheduling and Staffing of Blended Courses

There is an art to scheduling and staffing blended classes, an art that Drury University is finally mastering. However, the university trashed many masterpieces along the way. The challenge lay in trying to schedule the majority of the blended classes in eight-week blocks, while trapped in a calendar schedule that made it nearly impossible to get two full eight week blocks for every class out of a semester.

Drury University has been offering eight-week block classes in a totally online format quite successfully for years. These intense, short classes are very popular with students and research has shown that the learning assessment outcome is similar to when students take the same class in a 16-week format. (Blakey, 2007) It was a natural progression to go from totally online classes being eight weeks long to blended classes of eight weeks.

In order to cover the same amount of material normally covered in a sixteen week class, instructors of eight-week long courses plan to have both a face-to-face meeting and an entire week online every week. Mathematically, this works; when looking at the calendar, it does not. University holidays such as Labor Day, Fall Break, and Thanksgiving took away one, or in the case of Thursday classes, two of the eight face-to-face meetings and forced instructors into six and seven weeks of face-to-face meetings. Faculty members were not reticent in communicating their feelings about this infringement on their syllabi.

The mistake realized, the administration set out to fix it before the next semester began. The first step towards solving the problem was being aware of the issue months before the semester began. Secondly, the administration took steps to move the blended courses out of the academic calendar of the university. This was necessary because—in fall semesters—the Monday and Thursday classes had two fewer meeting times than all other days and, in the spring semester, Monday classes had three fewer meeting times than all other days. By extending the academic calendar two days—from a Wednesday start to a Monday start—and redefining non-holiday breaks, administrators were able to construct an academic calendar that allowed for eight weeks, a one-week break, and another eight weeks.

For the eight-week blocks and also for the sixteen-week blocks, there is still a certain skill to scheduling the classes. Drury University offers blended classes that have face-to-face meeting times in the evening in two different time formats – basically a 5:00 p.m. class and an 8:00 p.m. class. Building the blended schedule meant taking into consideration the classes a given student would want to take and positioning them in a way where the classes did not overlap or force the non-resident student to potentially have to drive to campus every night of the week. Long-time Drury advisor Teresa Montgomery helps craft the schedule each semester. In discussing the process, she stated "When we schedule for the eight week blended I think that we try to support that eight week class with something that the student would want the second eight weeks at the same time. So if we offered HIST 101, Monday @ 5:20 for the first 8 weeks then it would be nice to offer HIST 102 perhaps Monday night @ 5:20 the second 8 weeks. . . When we look at the first block blended classes I think we need to have supporting classes in the second block since many adults schedule their classes around family or work activities." (Teresa Montgomery, interview, October 10, 2008) Students need a predictable schedule they can rely upon.

Staffing blended classes has not been as challenging as scheduling them. The one rule that we have abided by is that all blended instructors must have taught a fully online class for Drury University before teaching a blended class. This ensures that the individual has been through Drury University's online training and understands the expectations and culture of the university prior to entering a blended class. It also ensures a certain level of comfort with the technology and online resources the university has to offer.

Interest in Blended Courses by Administrators, Faculty, and Students

There are mixed feelings towards blended courses at Drury University. The administrators who oversee the blended courses are very interested in their success and growth. Increasingly, students in the university's evening college have been choosing to enroll in online classes rather than face-to-face courses. This has led to such small class sizes in the face-to-face classes that many of them were canceled or instructors struggled to accomplish their goals with so few students.

On many online student evaluations, though, students would say that they preferred face-to-face classes, but just could not fit them into their lives. Administrators are hopeful that the blended option will give the students the best of both worlds – some face-to-face contact with their instructor and the convenience of part of the class being online. Parris Watts, who is Dean of CGCS and has ultimate oversight of the college's blended program, stated that he believes "the blended approach to learning will become a highly desirable option for students and teachers, alike. It provides an opportunity for them to interact, in person, to make certain that the course content being delivered online is understood. It is also offers a way to reduce the amount of time required to be physically present in seated class each week. This combination of enhancing the way new knowledge is being obtained and processed and the efficiency with which this is being done could render the blended form of learning one of the best things to happen in higher education since the advent of online instruction." (Parris Watts, interview, November 18, 2008)

The idea that the future of blended course delivery is a bright and growing one is echoed by John Bourne, the editor of the *Journal of Asynchronous Learning Networks* who stated that "Within five years, you'll see a very significant number of classes that are available in a hybrid fashion. . . I would guess that somewhere in the 80- to 90-percent range of classes could sometime become hybrid." (Young, 2002) Nevertheless, projected growth does not always equal acceptance.

Faculty members have expressed both devotion to and abhorrence of blended classes. Most, though, see the potential positive and negative issues of the blended format and are trying to accentuate the positive. Consider this English faculty member's response on her evaluation of the blended experience. "I have taught online classes for Drury for almost nine years. This is the second time for this class and the second and third time doing it in the blended format on two campuses. I LOVE the blended format as it gives us the best of both worlds: online and seated. It allows me to post pertinent information on the site and get down to business when we are together. By getting down to business, it allows me the luxury of working individually or in groups with the students who can actually do the work while I am monitoring their progress. This provides guided practice and effective motivational strategies." (Kim Mobley, interview, November 16, 2008)

When asked about the challenges to teaching in a blended format, the same instructor wrote, "Finding the balance between providing the right kinds of strategies and prompts to let the students be in charge of their own learning process (and following our parameters) is vital to this type of process. Too many times we revert to the old school ways of being the kings and queens of our classrooms and not allowing the students to show off the discipline, energy, ideas, time and products of their efforts. We are not the show. The spotlight is on them. We have to still be that guide on the side... constantly monitoring, waiting, prompting, guiding and building." (Kim Mobley, interview, November 16, 2008)

Most of our instructors have positive feelings towards the blended course model. They all agree that they had to find a pattern and strategy that fit their class and worked to maintain the momentum of the class both online and face-to-face.

Student reaction to blended classes has been mixed. Here are some of their comments:

- "I thought it (the blended class) was well planned out and fit well into my schedule. The only drawback was fitting so much information into a short period of time . . . I like the convenience of working online, but also being able to meet with the instructor and other students." (History 101 student)
- "I like the opportunity that we have to discuss with others on the (online) board about the assignments and help each other work through the issues and understand them. With busy schedules this helps a lot. At the same time you have the benefit of hearing a lecture." (Criminology 275 student)
- When asked what they liked best about the blended class format, a business communication and writing student said, "I participate more because I am able to say more online after class. I like the fact that the class does not have to go till midnight because of the time that we spend in class. I also got to know the students and professor better." When asked what type of class they preferred, the same student answered, "Blended. I am not an online person. I like the interaction with students and professor both in class and online. I like the fact that you also have a form of communication going on." (Communication 220 student)
- "I like not having to meet all the time, I can complete my work when it's convenient to me." (Management 321 student)
- When asked what she liked about her blended business communication and writing class, a student replied, "Absolutely nothing." She went onto explain that she very much prefers totally face-to-face classes because they have "more discussion and interaction with the class and the professors. More material covered. A greater sense of learning as opposed to biding time or having busy work." (Communication 220 student)
- "This is the first semester I have ever taken an online and a blended class. Both of them have their advantages; however my preference is the blended class. You still have the option to work at your own pace, yet you get the face to face time with your professor to address any troubles you may be experiencing, and pushes you to complete your assignments in a timely manner." (Communication 220 student)
- When asked what he liked about his blended leadership theory and models class, a student replied, "Flexibility. I didn't feel like I had to wait another week to ask a question. There was the opportunity to ask it online." When asked which type of class he preferred, he replied, "Online. I like the flexibility of not meeting at a set time every week. I live in a small town with a small Drury campus that doesn't offer many classes so I usually end up driving about an hour to Springfield (to class). (Leadership 300 student)
- When asked what she liked about her blended leadership theory and models class, another student replied, "That you could organize thoughts and share them with others online. Then come to class and get reinforcement from lecture." When asked which type of class they preferred, the student responded with "Blended. You can do most of your work at home. The lectures are short and sweet; and you have

someone online to ask questions 24 hours a day." (Leadership 300 student)

- An earth science student voiced the sentiments of many students by stating, "I can do the majority of my work from home but have face-to-face lectures and access to the instructor and classmates. I can complete 6 hours in one semester by attending campus classes only once per week. (Physics 100 student)
- "I have taken a lot of online only and I really like connecting with a teacher a few times." (Communication 220)

That the student evaluations were mostly positive is a great thing for Drury University's young blended program. However, students truly vote with their enrollments, and online classes are still filling up much more quickly than blended classes. Perhaps it will take a few more semesters for the word to spread about the positive experiences students are having in blended classes, or perhaps the total flexibility of online will always prove more important when students choose their courses.

Training Procedures and Concerns

Drury University trains its blended faculty by following a blended format. As stated earlier, the online education office is responsible for all of the technical training of blended faculty. The rule that all faculty members must have taught online prior to teaching blended means that normally very little technical training is necessary. However, refresher courses and introduction to new technologies is still very important.

All blended pedagogy is covered in a face-to-face meeting of blended faculty. Ideally, faculty would attend this training workshop during the semester before they teach a blended class. In the workshop we cover basic pedagogy, blended syllabus construction, guidelines for planning a good blended course, determining what parts of the class to do online and what parts to complete in the face-to-face meetings, testing, assessment, time management, and how to maintain momentum. The workshop is designed to be interactive with best practices shared from veterans to new blended instructors. The second half of the workshop is spent in a computer lab with the instructors working on their syllabi and online class sites while having colleagues at their elbows for suggestions and advice.

Training grows and changes from semester to semester based on the feedback Drury University receives from both faculty and students. It is the University's goal to make blended classes the best option for CGCS students so the administration is flexible in it's guidelines and structure while being deeply committed to maintaining the rigor and excellence for which the university is known.

How Blended Courses Relate to University Culture

Drury University is committed to offering its students a "personalized education filled with new ideas and established truths." (Drury University webpage, http://www.drury.edu/futurestudents/) This means unique classes taught by highly skilled instructors. Drury University's classes are typically not the lecture and note-taking classes of old school academia. The classes often exit the classroom to engage students in their community, environment, and world at large. The blended format fits perfectly into this culture.

Drury University trains its blended instructors to use Drury's blended format to highlight the best teaching practices of both the online and face-to-face models. Many of the instructors take advantage of the flexibility of the blended model to include multiple field trips and projects in the class that a normal face-to-face class would not have time for. They then use the online platform to fill in gaps in the class lectures and discussion to make up for the time spent on these kinetic learning experiences. A survey American History blended class is a good example of this.

The survey of American history class has a multitude of out of the classroom experiences built into it. There are multiple field trips to St. Louis area museums, Civil War battlefields, cemeteries, and other historic sites in the area. There is also a day of learning Native American games and culture where the students actually play the games, eat the food, and try on the clothing of long ago as well as a day of learning primitive skills that were used by American pioneers as well as a night of map making using the stars. All of these experiences combine to make the class very popular with students and a highly engaged learning experience. However, if all of these activities were combined into a traditional face-to-face class, then there would be little time left for lecturing, testing, and other traditional teaching tools. With the blended model, however, the class can have it all. The instructor spends the face-to-face time engaging the students in the hands-on activities and field trips. They then move everything else, except for the final, online. This combination gives the students the best possible educational experience and fits perfectly with Drury University's culture of a personalized education.

Other schools have realized similar benefits when instructors placed course material online and planned the face-to-face time for accomplishing specific course objectives. "During 1999-2001, the University of Wisconsin System Curricular Redesign Grant Program funded a collaborative project involving UW-Milwaukee and four UW-College campuses (Rock County, Sheboygan, Washington, and Waukesha). Coordinated by UWM's Learning Technology Center, the project developed a Web resource of hybrid courses, created a model faculty development program for teaching hybrid courses, and supported 17 faculty in their efforts to design, develop, and teach their first hybrid courses... Instructors reported that the hybrid course model allows them to accomplish course learning objectives more successfully than traditional courses do. Our faculty participants almost universally believe their students learned more in the hybrid format than they did in the traditional class sections. Instructors reported that students wrote better papers, performed better on exams, produced higher quality projects, and were capable of more meaningful discussions on course material." (Garnham, 2002) Studies such as this one support the experiences of Drury University and the movement towards a bright future for blended courses.

THE FUTURE

Challenges to the Future of Blended Classes

Drury University administrators have great expectations for the future of their blended class program. However, there are obvious challenges in the near future that will have to be successfully dealt with in order for the program to grow into its full potential. These challenges include increasing student enrollment in the blended courses, training all blended faculty, and maintaining technical support in the face of a growing online education program. These challenges are in the near future and answers need to be found soon.

Increasing Student Enrollment

Blended student evaluations are predominantly favorable and the number of blended classes offered is 23% of Drury University's Springfield CGCS schedule; however, students are not flocking to these classes. The average enrollment in blended courses was up from 7 students in the spring of 2008 to 12 students for the fall 2008 semester. Comparatively, the average enrollment in online courses for the same semester was 18.1. Administrators are happy with this increase in average class size but would like it to increase even more to a number around 20. Why aren't more students choosing to enroll in blended courses? The answer may be that, even though blended courses offer some flexibility, they still cannot match the flexibility of online classes.

The answer may also lie in the fact that blended classes are relatively new to the schedule, and students may not fully understand them yet and therefore are hesitant to enroll in them. The answers to both of these possibilities will require time and continued improvement of the blended program.

Over the next several semesters more attempts need to be made to move more classes to the online plus format. This normally means the class is online except for one meeting. That meeting may be for a final, a field trip, or field experiment that is critical to the class. It will be vitally important that the day and time of the one face-to-face meeting be clearly communicated to students prior to registration so students are not surprised on the first day of class to find out that they *have* to be available for a class activity on a certain date only a few weeks away.

Time must also be given for students to familiarize themselves with blended classes. Drury University's CGCS program is small, and most advertisement is through word of mouth. Students talk about what courses are good or bad, which instructors are difficult or easy, which classrooms are hot or cold. Drury University needs to give them time to discuss the benefits and drawbacks of blended courses as well. Positive experiences promoted by current students will bring more students into blended classes than any amount of radio advertising could.

When pressure to increase enrollments is on, it is often difficult to wait long enough to give a new delivery model time to grow and mature. However, that time is required in order to allow students to hear about and attempt blended courses. That time is also necessary to allow administrators time to try different versions of classes and see which equation works best for most students.

Training Faculty

Most faculty evaluations said that the training they received prior to teaching their first blended course adequately prepared them for the class *if they attended a faculty training workshop or met one-on-one with the trainer*. However, there were a number of faculty members who did neither. They had had experience teaching both online and face-to-face classes at Drury University before and set out to combine both into a blended class on their own. This did not work out well.

Due to the fact that CGCS staffs its classes with adjunct instructors almost exclusively, a more flexible form of training must be developed. Administrators must develop an alternative to the face-to-face workshops, and the most logical answer is an online training module. There are many options available for this online component including the use of the school's current learning management system, WebCT, as well as virtual world options available on the internet such as Second Life. Regardless of which avenue is followed, it is clear that a goal of the blended program for the next year must be the development of some sort of online training program for its faculty in order to be able to require full participation by all blended faculty.

Maintaining Tech Support in the Face of Growing Online Program

Drury University's College of Graduate and Continuing Studies' brightest shining star is its online education program. The growth of the online program greatly outpaces all other areas of growth in the rest of the university. With this growth and their plan for increased future growth, it is obvious that the four staff people are often overwhelmed with the workload. This could potentially have a negative impact on blended courses because the blended program relies totally on the online education office for all technical support and training.

The online portion of all blended courses is maintained on the online education program's learning management system. While the university has another option, it is not as functional or easily monitored as the WebCT system provided by the online education office. Furthermore, by requiring all blended course instructors to have taught online for Drury University prior to teaching their first blended course, administrators ensure that instructors know how to use this piece of technology as well as the pedagogy that accompanies it. Any blended instructor who has not taught online for Drury University in the last three years must go through the online education office's four-week new instructor training program in order to brush up on their online teaching skills.

The support the online education office provides the blended course program is vital to the success of blended classes. However, the growth of the online education program has greatly outpaced the growth (or even the anticipated growth) of the blended education program. The online education office's first priority must be the totally online courses. Therefore, the blended education program must find a way to work with or around this situation.

CONCLUSION

A mixture of online and face-to-face class time blended into one course offers the best answer to students demanding extremely flexible and rigorous classes coupled with the opportunity for a bit of contact with their instructor and classmates. While the concept of blended classes is a young one that has been slow to grow, the potential is great for instructors and students to realize the best of both the online and face-to-face classroom through this model.

Drury University's young blended education program has experienced many of the same issues as other blended programs. The two main issues have involved communicating a blended course's requirements to potential students and training instructors to maximize the best of both online and traditional classroom teaching techniques in a blended class. Communicating the requirements and benefits of blended classes to potential students is critical to growing the enrollment in the blended courses, which constitutes the final challenge many schools face when they begin a blended education program.

Drury University is not alone in facing this challenge. The University of Wisconsin found a similar problem in their blended courses, "Our instructors learned that students required repeated explanations about the model, explaining clearly what it is and why the instructor chose it. To quote from one student's observation, "There was only one real problem; it was difficult at first to understand how the course was being taught and to get the technology to work properly. After the course got rolling, it did get easier and easier to get a grip on it." (Aycock, 2002) Thus, effective communication regarding the reasoning behind, goals of, and requirements of blended classes is imperative for all schools utilizing a blended model of class delivery.

Overcoming these obstacles is imperative because the potential for blended courses is so high. They can provide nearly the same convenience and flexibility as online courses. However, they can also provide face-to-face time that some students find necessary to learn difficult concepts; ask more in-depth questions; to take part in kinetic learning activities as well as field trips and experiments; or just to have the comfort of an instructor's presence. Today's student body demands flexibility along with the best educational experience possible, and blended courses can provide that experience for them.

REFERENCES

Aycock, A. Garnham, C., & Kaleta, R. (2002). Lessons learned from the hybrid course project. *Teaching with Technology Today, 8*(6). Retrieved on October 22, 2008, from http://www.uwsa.edu/ttt/articles/garnham2.htm

Bonk, C. J., & Graham, C. R. (2006). *The handbook of blended learning: Global perspectives, local designs*. San Francisco: Pfeiffer.

Garnham, C., & Kaleta, R. (2002). Introduction to hybrid courses. *Teaching with technology today, 8*(6). Retrieved on October 22, 2008, from http://www.uwsa.edu/ttt/articles/garnham.htm

Hybrid classes bloom in arts and science. (2006). Retrieved on October 23, 2008, from http://www.unm.edu/~market/cgi-bin/archives/001242.html

Hybrid courses: Maximizing resources and student learning. (2002). Retrieved on October 22, 2008, from http://courses.durhamtech.edu/tlc/www/html/Special_Feature/hybridclasses.htm

Sands, P. (2002). Inside outside, upside downside: Strategies for connecting online and face-to-face instruction in hybrid courses. *Teaching with Technology Today, 8*(6). Retrieved on October 22, 2008, from http://www.uwsa.edu/ttt/articles/sands2.htm

Soules, M. (n.d.). *Hybrid online courses & strategies for collaboration*. Retrieved on October 22, 2008, from http://records.viu.ca/~soules/hybrid.htm

Young, J. R. (2002). Hybrid teaching seeks to end the divide between traditional and online instruction. *The Chronicle of Higher Education.* Retrieved on October 22, 2008, from http://chronicle.com/free/v48/i28/28a03301.htm

APPENDIX A: SAMPLE SYLLABI FROM BLENDED CLASSES

The two biggest challenges about blended classes revolve around communicating to the students the peculiarities of a class that is partially online and partially face-to-face and designing a class that flows well in this format. The following four syllabi (Figures 3, 4, 5, 6, 7, and 8) show how four very successful blended instructors do both.

APPENDIX B

Table 2. Course Visitation Report Sheet

Monitor Name:			
Instructor Name:			
Course Number:			
Course Title:			
Report Date:			
Text Block Usage:	Yes	No	
Calendar Indicating Seated vs. Online Weeks:	Yes	No	
Course Navigation/Structure:	Easy	Challenging	
Course Materials Current:	Yes	No	
Syllabus Indicating Seated vs. Online Expectations	Yes	No	
Grading Scale:	Yes	No	
Discussion Board Read Within:	24 hours	48 hours	72 hours
Grading Rubric:	Clear	Neutral	Unclear
E-mails Read/Responded to Within:	24 hours	48 hours	72 hours
Grades Posted:	Weekly	Bi-Weekly	Once a month
Comments:			

APPENDIX C

Table 3. Questionnaire for Faculty Teaching a Blended Course

Questionnaire for Faculty Teaching a Blended Course
Have you taught **online classes** for Drury University? ______
If so, have you taught the class you are currently teaching blended as a totally online class for Drury University? ______
If so, which format do you prefer – online or blended? ______ Why? ______
Did your blended class meet for the entire semester, 12 weeks or 8 weeks? ______
What were the benefits and drawbacks to the length of the class you taught? ______
Did you meet your goals and objectives for the class? ______ If not explain. ______
What changes would you make if you taught this class again in a blended format? ______
What did you like best about your blended course? ______
What do you see as the greatest challenges to teaching in a blended format? ______

Figure 3.

Communications 215: Foundations of Communication Syllabus
8 Week Blended Course
Myra Miller, M.Ed., ABD, Instructor

Attendance:
This is only an 8 week course! Attendance and participation is extremely IMPORTANT. Your initiative, self-motivation, and commitment are all important aspects of a successful class experience. If you are going to miss due to situations beyond your control, please notify me ahead of time. For the seated classes, you will lose attendance and participation points if you are absent and those points cannot be made up. For the online classes, you will receive an automatic five-point deduction for late work. However, if you communicate with me your situation prior to class, I will work with you to make up in-class work or allow you to submit homework.

COURSE SCHEDULE

	COMM 215 - CLASS OUTLINE (Instructor: Miller) Schedule is tentative and subject to change.		
Week	**Date**	**Chapters/Topics**	**Activity/Assignment**
Week 1	Jan 27-Feb 2	**Introductions, review syllabus, group activity**	**Seated Class** on Tuesday, January 27
	Online	**Discussion Board** / Assignment #1	Assignment #1, Due online February 2
Week 2	Feb 3-9	***Chapter 1-3 - Introduction and Overview***	**Seated Class** on Tuesday, February 3
		Chapters 4-7 - Interpersonal Messages	
	Online	**Discussion Board** / Assignment #2	Assignment #2, Due online February 9
Week 3	Feb 10-16	***Chapters 8-10 - Relationship Development***	**Seated Class** on Tuesday, February 10
		Chapters 11-13 - Relationship Maintenance	
	Online	**Discussion Board** / Assignment #3	Assignment #3, Due online February 16
Week 4	Feb 17-23	***Chapters 14-16 - Influence***	**Seated Class** on Tuesday, February 17
		Chapters 17-18 - Group Decision Making	
	Online	**Discussion Board** / Assignment #4	Assignment #4, Due online February 23
Week 5	Feb 24-Mar 2	***Chapters 19-21 - Organizational Communication***	**Seated Class** on Tuesday, February 24
		Chapters 22-24 - Rhetoric	
	Online	**Discussion Board** / Assignment #5	Assignment #5, Due online March 2
Week 6	Mar 3-9	***Chapters 25-29 - Media & Culture / Media Effects***	**Seated Class** on Tuesday, March 3
		Chapters 30-32 - Intercultural Communication	
	Online	**Discussion Board** / Assignment #6	Assignment #6, Due online March 9
Week 7	Mar 10-16	***Chapters 33-35 - Gender & Communication***	**Seated Class** on Tuesday, March 10
	Online	**Discussion Board** / Assignment #7	Assignment #7, Due online March 16
Week 8	17-Mar	**Final Project - Team Report**	**Seated Class** on Tuesday, March 17
		Final Grades	Final Team Project Report, Due online March 17

Questionnaire for Faculty Teaching a Blended Course
How well did the students understand that they were enrolling in a class that was partly online and partly seated? ______
Please list any suggestions you have for how this communication should be done in the future. ______
Were your students open and receptive to the blended format? ______
What changes could be made to make it better? ______
Did your students understand the technological requirements of the class? ______
If not, were you able to help them? ______ Please explain. ______
How well did the students participate in both the online and seated components? ______

Figure 4.

English 203: American Literature I (1620- 1865) Syllabus
8 Week Blended Course
Kristin Howard, Instructor

Participation/Attendance
Students are expected to attend class each Tuesday evening as well as log in frequently during the week to participate in online discussions/assignments per the schedule below. Much of the learning that occurs during this course happens in the cyber classroom. If you do not log into your course frequently, for whatever reason, you lose the benefit of sharing of knowledge by the faculty and other students. Additionally, the class loses the benefit of your knowledge and experience; thus, the learning experience is diminished and will be reflected in your grade.

Schedule
January 27 – (**In CLASS** in Pearsons, Rm 22) Introduction/Cover Syllabus/Timeline of American Literature/Assign Classical Novel/Background
Read (for Feb 3): Stories of the Beginning of the World:
"The Iroquois Creation Story" (version by David Cusick) – p. 17
Pima Stories of the Beginning of the World (version by Thin Leather & J.W. Lloyd) p. 21
-"The Story of the Creation" p. 22
-"The Story of the Flood" p.2 4
Choose any two of the five Native American Trickster Tales (pgs. 72-98)
Online Be sure to have Introduction posting completed online by Sunday (2/1) and reply to 2 classmates online by Tuesday (2/3)

February 3 – **In class**, discuss previous week's readings. (Each student will be assigned something to discuss.)
Read: Alvar Nunez Cabeza de Vaca-"The Relation of Alvar Nunez Cabeza de Vaca" p.41
John Smith – "The General History of Virginia, New England, and the Summer Isles" p. 57
William Bradford – "Of Plymouth Plantation" p. 105
John Winthrop – "A Model of Christian Charity" p. 147
Anne Bradstreet – choose any four of her poems
Cotton Mather – "The Wonders of the Invisible World" p. 308
"The People of God in the Devil's Territory" p. 308
"The Trial of Martha Carrier" p. 310
Online, be sure to have Week 2 posting completed online by Sunday (2/8) and reply to 2 classmates online by Tuesday (2/10)

February 10 – **In class**, discuss previous week's readings. (Each student will be assigned something to discuss.)
Read the play *The Crucible* by Arthur Miller
Online, be sure to have Week 3 posting completed by Sunday (2/15) and reply to 2 classmates by Tuesday (2/17)

February 17 – **In class**, discuss previous week's readings. (Each student will be assigned something to discuss.)
Read: Jonathan Edwards "Sinners in the Hands of an Angry God" p. 425
John Adams & Abigail Adams: choose 3 letters to read
Thomas Paine – "Common Sense" p. 630
Olaudah Equiano – *From "The Interesting Life of Olaudah Equiano"*
Phyllis Wheatley – "On Being Brought from Africa to America" p. 752
Online, be sure to have Week 4 posting competed online by Sunday (2/22) and reply to 2 classmates online by Tuesday (2/24) ***

February 24 – **In class**, discuss previous week's readings. (Each student will be assigned something to discuss.)
Read: Washington Irving: "The Legend of Sleepy Hollow" p. 965
Ralph Waldo Emerson: "Nature" p. 1110
"Self-Reliance" p. 1163
Nathaniel Hawthorne: "Young Goodman Brown" p. 1289
"The Birth-Mark" p.1320
Edgar Allan Poe: choose 2 poems & 1 short story
Margaret Fuller: *"From The Great Lawsuit: Man versus Men. Woman versus Women"*
Sojourner Truth: "Speech in the Women's Rights Convention in Akron, Ohio, 1851" p. 1695

Online, be sure to have Week 5 posting completed online by Sunday (3/1) and reply to 2 classmates online by Tuesday (3/3)

March 3 – **In class**, discuss previous week's readings. (Each student will be assigned something to discuss.)
Read: Harriet Jacobs: "Incidents in the Life of a Slave Girl" p. 1809
Walt Whitman: "Song of Myself" p. 2210
Herman Melville: "Bartleby, the Scrivner" p. 2363
Emily Dickinson: choose four of her poems
Online, be sure to have Week 6 posting completed online by Sunday (3/8) and reply to 2 classmates online by Tuesday (3/10)

Figure 5.

History 104: Western Civilization II-1550 to the present Syllabus
16 Week Blended Class
Dr. Louisa Rice, Instructor

Format and Expectations—The Blended Class
As you should know, this is a blended class. This means we meet approximately part of the time in the seated classroom, and part of the time online, at our WebCT site. Neither the online nor the seated part of this class is "optional." The Mondays we meet and the dates we're online are clearly marked on the course outline and on the WebCT course calendar. I will also remind you as much as possible. Beyond that, it's your responsibility to make sure you know when we meet, and to come to class/post online as required.

Because we only meet 9 times this semester, you cannot afford to miss any classes. If you miss one class, you will lose 5 points from your final grade; two classes will lose you 15 points from your final grade. Three or more classes missed will result in you failing the class.

Class time will be divided between lecture, discussion and film/image viewing. Make sure you have read and are prepared to discuss the assigned reading. You should bring the relevant books to class so that you have them for reference—**this is especially true of the *Sources, Images and Interpretations* book. PLEASE bring this to every class.**

35% of your final grade is exclusively based on what you do online (10% on quizzes, 25% on discussion postings). The quizzes will be multiple-choice and based on our reading. Discussion postings will work as follows:

On the Course Outline below there are five postings marked. Each week on the Tuesday prior to an online discussion, I will post a short lecture on our topic and create a discussion assignment. During the week between Tuesday and Monday you will post your response to this assignment, and respond to two of your classmates' postings. **More, important information on precisely how to do this can be found in the documents "Discussion board directives" and "Grading rubric for online posts."**

Course Schedule

Note: For each week we meet online I have indicated the dates that you'll be working on a particular topic. Note that it looks as though we "miss" weeks; these weeks are dedicated to you reading in advance of lecture. When we meet on a Monday this is essentially the "end" of a week—you have done a week's worth of reading ready for my lecture on a topic and our discussion of it. Similarly, during online weeks, your final posting date for responses is a Monday, before we move onto new material.
In addition, for most weeks your online post or our in class discussion will also be structured around documents from the sourcebook: *Western Civilization: Sources and Interpretations*. In the outline, the documents to be read are labeled according to what kind of material they are: P= Primary, S=Secondary, V=Visual.

All assigned reading must be completed by the time we meet.

Reading Schedule and Course Outline:
Week One
Monday Jan 26th
First meeting, syllabus, Europe in 1550: Absolutism and Constitutionalism

Week Two: Reading, Textbook Ch. 14
Monday Feb 2nd
Meet: Science, the Enlightenment

Week Three: Reading, Textbook Ch. 15
Monday Feb 9th
Online: Slave Trade, Enlightened Absolutism [Post]
Read
P: Frederick the Great's *Political Testament* (24)
P: *Propaganda and the Enlightened Monarch* (45)
S: *The Problem of Enlightened Absolutism* (49)

QUIZ 1

Week Four: Reading, Textbook Ch. 16
Monday Feb 16th
Meet: French Revolution
Read
P: *What is the Third Estate?* (55)
P: *The Declaration of the Rights of Man and the Citizen* (57)
P: *Declaration of the Rights of Women and the Female Citizen* (58)
S: *France under Napoleon: Napoleon as Enlightened Despot* (75)
S: *Napoleon Bonaparte and the Legacy of the French Revolution* (77)
S: *Women and the Napoleonic Code* (78)

Figure 6.

Week Five: Reading, Textbook Ch. 17
Monday Feb 23rd
Online: Industrial Revolution [Post]
Read **P:** *Testimony for the Factory Act of 1833: Working Conditions in England*(82)
P: *The Conditions of the Working Class in England* (84)
P: *Woman in her Social and Domestic Character* (87)
P: *Women and the Working Class* (88)
S: *Early Industrial Society: Progress or Decline?* (93)
S: *The Family and Industrialization in Western Europe* (94)

QUIZ 2

Week Six: Reading, Textbook Ch.18 and *The Communist Manifesto*
Monday March 2nd
Meet: Ideologies and debating the Communist Manifesto

Week Seven: Reading, Textbook Ch.19 pp.596-603
Monday March 9th
Meet: Mid-Term Exam and Nationalism

Week Eight: Reading, Textbook Ch.20
Monday March 16th
Online: Empire [Post]
Read: **P:** *Does Germany need Colonies?* (116)
P: *The White Man's Burden* (117)
P: *Controlling Africa: The Standard Treaty* (119)
V: All sources (119)
S: *The Age of Empire* (125)
S: *Imperialism as Nationalistic Phenomenon* (125)

QUIZ 3

Spring Break: Sat 21st-Sun March 29th YAY! [Start *All Quiet on the Western Front]*

Week Nine: Reading, Textbook Ch.21
Monday March 30th
Online: Modernity [Post]
Read: **P**: *The Origin of the Species and the Descent of Man* (130)
P: *Why we are Militant* (136)
P: *Foundations of the Nineteenth Century: Racism* (137)
P: *Judaism in Music* (138)
V: All sources (138)
S: *European Women* (142)

Week Ten: Reading, Textbook Ch.22, *All Quiet on the Western Front*
Monday April 6th
Meet: WWI: Reading Ch.22

QUIZ 4

Week Eleven: Reading, Textbook Ch.23, *All Quiet on the Western Front*
Monday April 13th
Meet: *All Quiet* Discussion, Interwar
Read: **P**: *The Doctrine of Fascism* (174)
P: *Mein Kampf* (175)
P: *Nazi Propaganda Pamphlet* (177)
V: All Sources (182)

Figure 7.

Week Twelve: Reading, Textbook Ch.24
Monday April 20th
Meet: WWII and Holocaust
Paper Due: All Quiet on the Western Front
Read: **P**: *The Informed Heart: Nazi Concentration Camps* (179)
P: *Witness to the Holocaust* (179)
S: *Hitler and Nazism* (186)
S: *Hitler's Willing Executioners* (187)

Week Thirteen: Reading, Textbook Ch.25
Monday April 27th
Online: Cold War [Post]
Read: **P**: *The Truman Doctrine and the Marshall Plan* (193)
P: *The Cold War: A Soviet Perspective* (194)
P: *The Berlin Wall* (195)
V: All Sources (202)

QUIZ 5

Week Fourteen: Reading, Textbook Ch.25
Monday May 4th
Meet: The Present/Review
Read: **S:** *Terrorism and The Clash of Civilizations* (218)
S: *The Future After 9-11-01* (219)
S: *Globalization* (223)

FINAL EXAM: Monday May 11th, 5.20 pm.

Questionnaire for Faculty Teaching a Blended Course
Did they favor one over the other? ______ How did you deal with that? ______
Did you have attendance problems with the seated component? ______ If so, how did you deal with them and was this successful? ______
How well were you prepared to teach this class in a blended format? ______
What do you wish would have been covered in the training? ______
What level classes do you feel the blended format works best? 100, 200, 300, 400.
Why? ______
What is your preferred type of class? Seated, blended, or online
Why? ______
Was the technology (WebCT and in the classroom) sufficient for your class? ______
If no, what would you like to see added? ______
What do we need to consider changing regarding the scheduling of blended classes? ______
What do we need to consider changing regarding the training of blended instructors? ______
What do we need to consider changing/adding regarding the support you receive while you are teaching a blended class? ______
What kind of feedback did you receive from students regarding the blended nature of the class? ______
Comments? ______
Name:_________________________Course___________________

Figure 8.

Management 422: Corporate Policy and Ethics Syllabus
8 Week Blended Course
Bill Prince, J.D., Instructor

Participation: Absence from class may result in the loss of participation points. Much of the learning that occurs during the course happens in the classroom. If you are absent, for whatever reason, you lose the benefit of sharing of knowledge by the faculty and other students. Additionally, the class loses the benefit of the sharing of your knowledge and experience; thus the learning experience will be diminished and will be reflected in your grade.

Schedule:

Week 1 (1-21 to 1-25):
In Class: Introduction to course, review syllabus and course expectations.
Watch "The Tragic Case of Dunder Mifflin"
Online: Read Chapters 1, 2, and 3
Discussion responses due on Jan. 25, 2009 by 10:00 p.m.

Week 2 (1-26 to 2-1):
In Class: "The Corporation"
Online: Read Chapters 4, 5, and 6
Discussion Responses due 1-29 by 10:00 p.m.
Discussion Replies due 2-1 by 10:00 p.m.
Response to the Corporation due 2-1 by 10:00 p.m.

Week 3 (2-2 to 2-8):
In Class: "Roger and Me (watch and discuss).
On Line: Read Chapters 7, 8, and 9.
Discussion Responses due 2-5 by 10:00 p.m.
Discussion Replies due 2-8 by 10:00 p.m.
Paper topic due 2-8 by 10:00 p.m.

Week 4 (2-9 to 2-15):
In Class: "A Civil Action" (view and discuss).
Sign-up for presentation date.
Online: Read Chapters 10, 11, and 12.
Discussions Responses due 2-12 by 10:00 p.m.
Discussion Replies due 2-15 by 10:00 p.m..

Week 5 (2-16 to 2-22):
In Class: Midterm (covers Chapters 1-12)
Online: Read Chapters 13 and 14
Discussion Response due 2-19 by 10:00 p.m.
Discussion Replies due 2-22 at 10:00 p.m.

Week 6 (2-23 to 3-1):
In Class: Case Presentations and discussions
Online: Read Chapters 15, 16, and 17
Discussion Responses due 2-26 by 10:00 p.m.
Discussion replies due 3-1 by 10:00 p.m.

Week 7: (3-2 to 3-8):
In Class: Case presentations and discussions.
Take home final distributed.
Online: Read Chapters 18, 19, 20
Discussion Response due 3-5 by 10:00 p.m.
Discussion replies due 3-7

Week 8: (3-9 to 3-11):
Final Examination and paper to be submitted in electronic form 3-11 by 10:00 p.m.

APPENDIX D

Table 4. Questionnaire for Students Taking Blended Courses

Questionnaire for students taking Blended courses Fall 2008
How many classes are you taking this semester? ________
What day and time did your blended class meet seated? ________
Did your blended class meet for the entire semester 16 weeks,12 weeks or 8 weeks? ____
Have you ever taken a totally online class at Drury University? ________________
Please indicate the title of your blended class.______________________________
Rate the following on a scale of 1 to 3. 1=very good, 2= satisfactory, 3= unsatisfactory
This course fit into my overall schedule. ______
The course was stimulating and challenging.______
The seated portion of the class was well attended._____
The online portion of the class was helpful to the overall understanding of the material.__
The course was well organized._______
I had access to and understood the required technology for the course. ______
Please rate the following on a scale of 1 to 3. 1= yes, 2= not sure, 3= no
Would you like to see more blended classes offered? ________
Would you like to see blended classes offered in 8 week blocks during the semester?____
Would you like to see blended classes start at 6:00 pm or later? _________
Would you be likely to take a blended class offered from 8:10-9:50? ______
Please answer the following questions: (Use the back if needed)
How well was the format (approximately ½ online and ½ seated) of a blended class communicated to you when you enrolled? Please explain.
How well did your instructor identify what you were to do online versus what you were to do for the seated classes? Please explain.
What have you liked best about your blended class(s)?
What improvements to the blended format would you suggest?
What was the most difficult part of this class? Did the difficulty have anything to do with the class being blended instead of totally online or seated?
Did you learn the material you think you should have learned in this class?
What level classes (100, 200, 300, 400), do you feel works best with the blended format and why?
Is this your first blended class ever? If not what was your previous experience?
Will you enroll in blended classes in the future?
What is your preferred type of class (seated, blended or online) ? Why?
Circle your classification: Visiting Student, Freshman, Sophomore, Junior, Senior.

APPENDIX E: SAMPLE IDEA EVALUATION FEEDBACK FORM

Figure 9.

Faculty Name: Blakey, Leah — Term: Spring 2006–2007

Course: Liberal Arts & Sciences, General Studies & Humanities GSTU 493 — Page 3

Section II. Student Ratings of Progress on Relevant Objectives

This graph shows student progress ratings on the objectives you chose as ***Essential*** (Part A) and those you chose as ***Important*** (Part B). To the degree that students make progress on the objectives you emphasize, your teaching has been effective.

Part A. Essential Objectives	T Score Unadj. / Adj.	2% of all classes	28% of all classes	40% of all classes (Avg. range)	28% of all classes	2% of all classes	Your Average * (5-Point Scale) Raw	Your Average * (5-Point Scale) Adjusted	IDEA Average
1. Factual knowledge	55 / **52**						4.3	4.1	4.0
8. Oral and written communication skills	57 / **54**						4.3	4.1	3.8
Part B. Important Objectives									
6. Creative capacities	51 / **47**						3.9	3.6	3.9
7. Intellectual/cultural understanding	58 / **54**						4.3	4.0	3.7

20 30 40 45 50 55 60 70 80

T Score—Comparison with the IDEA Database where the objective was selected as "Essential" or "Important" **

├✕┤ Unadjusted T Score ± one standard error of measurement.

├◆┤ Adjusted T Score ± one standard error of measurement: adjusted for student work habits (item #13); student desire to take the course regardless of who taught it (item #15); and instructor reported class size.

Similar to Section I, you may wish to assign ratings to categories. Simply assign T Scores to categories as follows: **Low** (lowest 10%) = T Score below 37; **Low Average** (next 20%) = T Score 37–44; **Average** (middle 40%) = T Score 45–55; **High Average** (next 20%) = T Score 56–63; and **High** (highest 10%) = T Score above 63.

These graphs are intended to help you identify a focus for improving your instructional effectiveness. If student progress ratings on ***Important*** or ***Essential*** objectives are disappointing, you are encouraged to discuss improvement strategies with your department head, the campus faculty development specialist, or a colleague. Such strategies could focus on matters such as teaching methods/styles, class activities and assignments, the text and other readings, assessment/feedback, and the need for course pre-requisites. You might also consider using the IDEA Diagnostic Form the next time you solicit student ratings, since it is designed to help identify specific teaching methods to use in improvement efforts.

Note: Students in your class also rated their progress on the objectives that you classified as being of ***Minor or No Importance.*** These ratings are considered irrelevant in judging your teaching effectiveness. However, a review of student ratings on these objectives, found in **Section IV** (Statistical Detail), may provide you with insights about some "unintended" or "additional" effects of your instruction.

* Statistically, adjustments can exceed 5.0 on the 5-point scale. If this occurs, "Your Average," reported in the table above, will be rounded to 5.0. However, the T Score reported will reflect the actual adjusted score, which may exceed 5.0. Therefore, identical adjusted scores of 5.0 may have different adjusted T Scores.

** Normative information (T Scores) and the process for adjusting scores were recently updated. See page 1 for "A Few Words of Caution."

Figure 10.

IDEA Student Ratings of Instruction

Blakey, Leah
GSTU 493
Senior Seminar
online
DRURY UNIVERSITY
04-23-2007 -05-04-2007
Local Code: 4279

Comments: Use the space provided in the text area below for your comments.

- Dr. Blakey is an extremely passionate teacher and she really gets involved. You can tell she puts alot of effort into it and she makes us excited to learn.
- I enjoyed taking this class.
- I thought this class was great, I was concerned when it first began, but I was able to do everything required. I ran into a problem with my internet connection, since I had dial-up, it was difficult for me to view some of the Power Point presentations unless I went to another computer. I live in a very rural area and had to drive 50 miles to find one with the program I needed. People should be warned. I also think putting a face to the names would be nice. I met some online classmates at a recent banquet, maybe a place to post pictures during our introduction process would make the class more tied together. Just some thoughts, thanks!
- Enjoyed this class alot!
- This class could have been better if the instructor had picked out books that are intersting to modern day students. Reading about Africa and meat packing industries does not leave an excitment that I can think of. Most part this class was very boring and bland. The writing for answers could have been lowered a little bit.

Chapter 4
Objective-Oriented Assessment in Desire2Learn for Quality Matters

Haomin Wang
Dakota State University, USA

Mingming Shao
Dakota State University, USA

EXECUTIVE SUMMARY

This chapter is a report on using a course management system Desire2Learn® to facilitate the implementation of the quality assurance standards recommended by Quality Matters. After a brief introduction to the Quality Matters standards, the chapter describes how the Desire2Learn learning environment can be structured to build an interconnected framework to promote objective-oriented, criterion-referenced, dynamic assessment. A pilot study was conducted to explore how learning activities can be assessed in alignment with measurable learning objectives, with reference to criterion-based rubrics. The chapter discusses issues found through the study and lessons learned in using the competency structure of Desire2Learn to enforce objective-oriented assessment.

LITERATURE REVIEW

The value of learning objectives and measurable learning outcomes has long-standing support in the literature (Bloom, Engelhart, Furst, Hill, & Krathwohl, 1956; Mager, 1962, 1975). Assessment criteria and assessment constraints are usually key determinants of the teaching and learning strategies chosen (Australian Flexible Learning Framework, 2004). A considerable amount of research findings indicate that learning objectives should be clearly stated and assessment criteria clearly specified in order to help students see what to achieve and how to achieve it (Koszalke & Ganesan, 2004; Murphy, Mahoney, & Harvell, 2000; Pawan, Paulus, Yalcin, & Chang, 2003; Shea, Swan, Fredericksen, & Pickett, 2002; Stewart, Hong, & Strudler, 2004; Youngblood, Trede, & DeCorpo, 2003).

Online assessment has been found most effective when learning objectives are clearly linked to assessment activities (Hyde, Booth, & Wilson, 2003). It is generally suggested that learning objectives should be aligned with learning activities (Garrison & Anderson, 2003; Garrison & Cleveland-Innes,

DOI: 10.4018/978-1-60566-870-3.ch004

2005; Jensen, Self, & Rhymer, 2002; Kidney & Puckett, 2003; McLoughlin, 2001; McVay Lynch, 2002; Lewis, 2002). And assessment strategies and tasks should be tied to the learning objectives and learning process (Koszalke & Ganesan, 2004; Koszalka & Bianco, 2001). With the connection between assessment and learning process made explicit, students are expected to better see the relevance of the assessment to the learning tasks (Leask, 2001; Hyde, Booth, & Wilson, 2003). To help students relate learning objectives to specific learning activities and assignments, learning objectives should be specified at unit or module levels (Koszalke & Ganesan, 2004; Sims, Dobbs, & Hand, 2002; Stewart, Hong, & Strudler, 2004; Trigano & Pacurar-Giacomino, 2004).

Emphasis on using behaviorally observable terms to describe learning objectives has been criticized for paying excessive attention to behavioral performances and inadequate attention to affective and cognitive processes. In contrast, constructivism calls attention to learners' mental activities and processes. Jonassen (1994) proposed a set of eight rationalized criteria for evaluation mental models. The criteria include the following characteristics: coherence, personal relevance, fidelity with real world, imagery, complexity, applicability/transferability, and inferential ability. In order for the online learning environment to accommodate constructive learning, some changes in instructional design are believed necessary. Among other things, instructional objectives should be negotiated, not imposed; and evaluation of learning should become less criterion-referenced (Jonassen, 1991). However, objectivists argue against the constructivist view and believe that non-objectivist approach is inoperable. Many instructors teaching foundational subject matters tend to identify with the objectivist view, particularly math and computer science instructors (McVay Lynch, 1998). Jonassen (1994) later modified his constructivist position stating that constructivist approach is more appropriate for advanced learners.

Online assessment is still very much in an embryonic state. Assessment in Web-based instruction (WBI) is different from traditional classroom techniques. While the emphasis of traditional assessment tends to be on the mastery of the body of knowledge underpinning a discipline area, online assessment focuses more on collecting evidence of competence across skills and underpinning knowledge (Hyde, Booth, & Wilson, 2003). With computer-assisted administering, online assessment has some notable advantages such as instantaneous feedback to learners, enhanced by well-designed learning loops, opportunities for multiple attempts as well as reduction in grading time by the instructor.

Online assessment can also provider greater flexibility as to when, where, and how assessment is undertaken. However, online assessment usually needs to be planned from the inception of the teaching program or courseware design to determine how learning and assessment can be best integrated. Effective online assessment is also an integrated process in which learning experience and assessment activity build upon each other and are mostly interdependent (Hyde, Booth, & Wilson, 2003). Currently, online is used more for summative assessment than for formative assessment (Anderson, 2001). More efforts are needed to promote formative assessment in online learning. Formative feedback and progressive monitoring are beneficial to learners (Macdonald & Twinning, 2002; Shea et al., 2002; Thurmond, Wambach, Connors, & Frey, 2002).

BACKGROUND

This study took place in a mid-western public university established in 1881. In the later 1980s, the institution went through a mission change to become a pioneer in technology applications to advance teaching and learning, and integration of technology into curriculum. The University started offering distance courses in the early 1990s, and delivering online courses a few years later, using a combination of password-protected Web content, an in-house developed quiz engine, a third-party discussion board and e-mail services. In 2004, the University started using WebCT as the institutional course management system. A year later, the University adopted the quality assurance standards recommended by Quality Matters (2008) to help instructors use an objective-oriented approach to the design, development, and improvement of their online courses. In the spring of 2008, the University switched to Desire2Learn as the state-wide course management system.

SETTING THE STAGE

Originally developed by MarylandOnline, Quality Matters (QM) consists of a standards-based rubric for reviewing the design of online courses through a peer review process in order to assure course quality. The framework of QM includes eight broad standards with 40 elements. A prominent feature of QM is the emphasis on learning objectives to be specified at unit levels and associated with learning activities. Specific terms are to be used to describe what learning activities are to be performed and how they are to be assessed. Grading policy should be open and easy to understand for the learners. In addition, there should be clear indications as to how learning activities, assignments, assessments, and course tools should work together to achieve the desired learning objectives. Because much of the focus on learning objectives grows out of behavioral learning theory, the Quality Matters standards include annotations that direct course reviewers to check for indications that learning objectives are meaningfully assessed for affective and cognitive outcomes (Quality Matters, 2008).

While the value of specifying learning objectives with measurable outcomes has long-standing support in the literature and instructional practice (Bloom, et al, 1956; Mager, 1962), it remains a major challenge for course designers and instructors to have learning activities aligned with specified objectives and progressively assessed. This is where Desire2Learn can help.

Desire2Learn (D2L) Enterprise eLearning Suite 8.3 is a course management system developed by Desire2Learn Incorporated. (2008). D2L has a built-in competency structure for specifying learning objectives and assessing student accomplishments of the learning objectives through associated learning activities. Hierarchically, the structure starts with competency at the top. A course can have multiple competencies. A competency consists of one or more learning objectives, and each of these leaning objectives is associated with one or more activities. Both competencies and objectives can be embedded. A competency can have one or more sub competencies. Likewise, a learning objective can have one or more sub learning objectives (see Figure 1). Learning objectives can be attached to competencies at both course level and unit level. While course objectives are typically general, learning objectives at the unit level are usually associated with learning activities and should be specified in measurable terms. D2L course designers can enforce evaluation methods for learning objectives. Such evaluation methods include discussion, dropbox, quiz, and survey. Enforcing evaluation methods is optional. Once they are selected for a learning objective and the

Figure 1. Hierarchical structure of competency

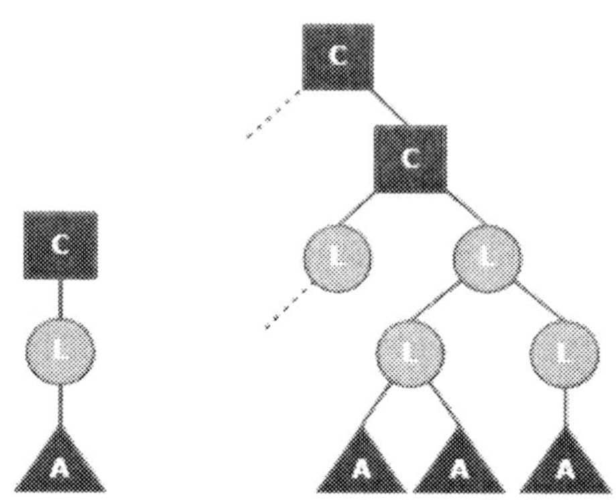

learning objective is saved, the selected evaluation methods cannot be changed.

Competencies do not have grades or levels of achievement associated with them; they are either complete or incomplete. Learners do not complete competencies directly; a learner achieves a competency when the learner completes all the associated learning objectives. In turn, a learner meets a learning objective when the learner accomplishes all the associated learning activities, whether the learner achieves the highest score or just meets the thresholds on the associated activities. Activities include visitation to content topics, discussion participations, dropbox submissions, quiz attempts, checklist completions, survey submissions, and grade items. While most activities are performed through course tools, activities such as lab sessions and field trips can also exist independently without being tied to any tool in the D2L Learning Environment.

Most activities, except for quiz and grade item, need to be assessed through rubrics. A new activity cannot be saved unless it is associated with a rubric (see Figure 2). So, rubrics need to be created before activities. Using rubrics, instructors can set up specific, criterion-referenced performance benchmarks and provide guides so that assessment and grading are open and transparent to learners. When a learning activity is associated with a rubric, a threshold or minimal level of achievement must be set as the passing grade for the learning activity.

Learners' progress in meeting learning objectives can be monitored by both the instructor and learners themselves. For learners, the results can be seen in a complete list (see Figure 3). For the instructor, there is a View Results link for each competency, learning objective, and activity in the Competency List. Clicking a link will bring up a list of all learners with their progress in meeting the competency, learning objective, or activity (see Figure 4). To meet a learning objective, a learner has to pass all of the activities associated with the objective. Likewise, to complete a competency, the learner has to meet all of the objectives associated with the competency.

Advanced planning is needed in building alignments between learning activities and assessment rubrics. A rubric cannot be modified when it is tied to a learning activity; it must be unassociated from the associated learning activities before it can be modified. If the thresholds of some rubrics are

Figure 2. Desire2Learn learning environment 8.3: Advanced search functionality

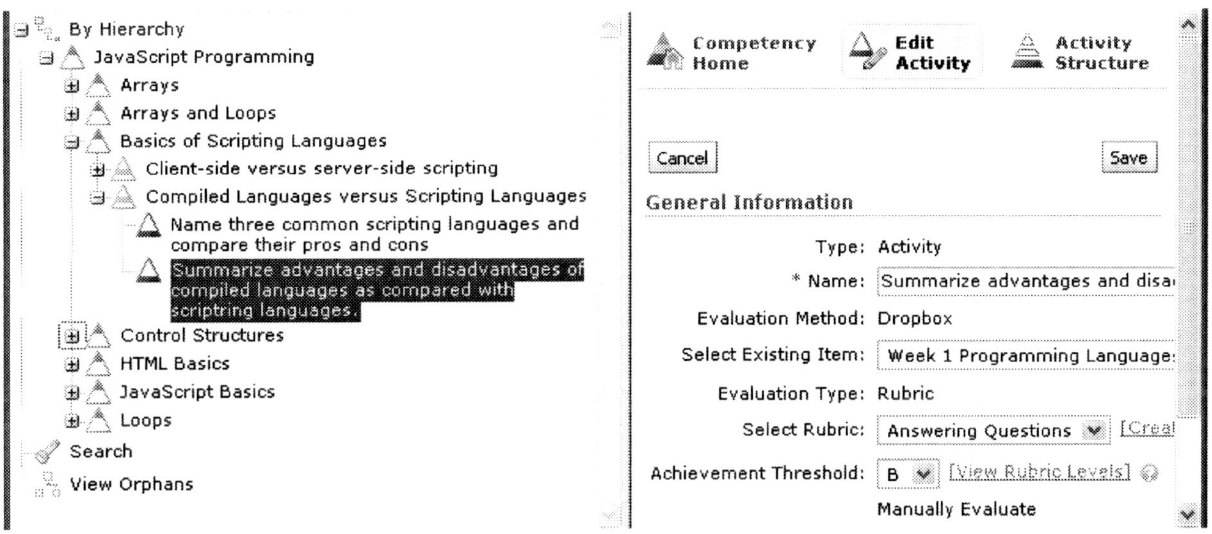

Figure 3. Learner's view of competency results

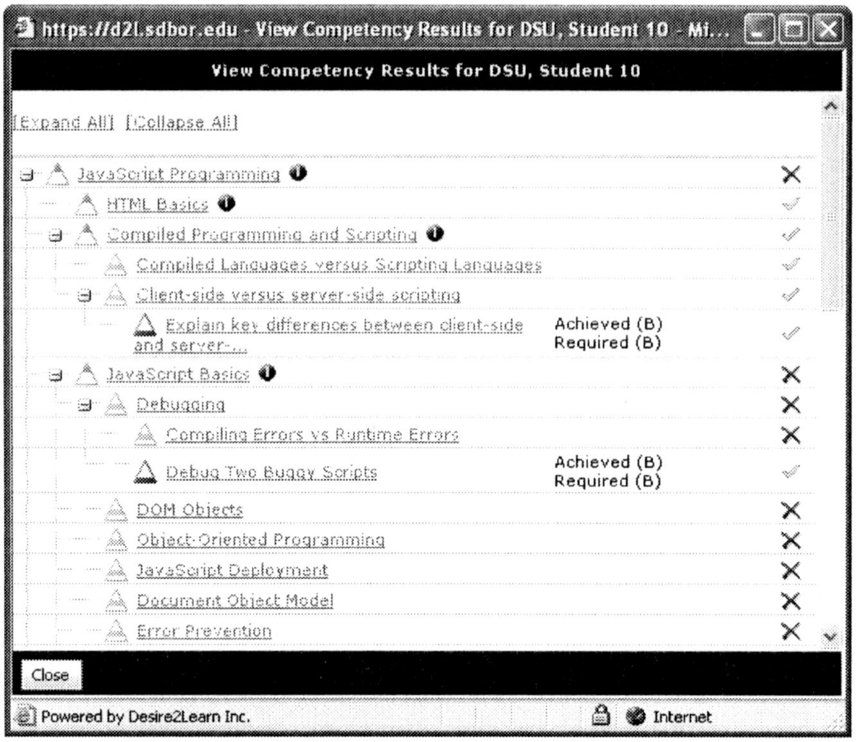

changed or activities re-associated with different learning objectives, the related competencies will need to be re-evaluated.

CASE DESCRIPTION

A pilot study involving a graduate level online class has been undertaken to experiment the use of D2L to enforce the alignment between learning activities and assessment measures. A survey and an interview were conducted near the end of the semester to collect students' and the instructor's perceptions of the assessment tools and their functions in the D2L learning environment. The following is what the pilot study has found recommendable in using the Competency Structure of D2L to help implement Quality Matters standards: (a) building the Competency Structure in advance, (b) defining learning objectives in specific terms, (b) putting granularity into rubrics, (c) monitoring learner progress and providing feedback, and (d) using release conditions to manage learning paths. A pilot course that teaches JavaScript for instructional programming is used for illustration in discussing each area below.

Building the Competency Structure in Advance

Competencies and learning objectives can be interconnected to become a highly interwoven structure. Changing one component in such a structure can be constrained by its association with all associated components. At the higher end of the hierarchical structure is competency. A learner achieves a competency when the learner completes all the sub competencies and associated learning objectives.

A competency in D2L has a status which can be *in draft* or *approved*. An approved competency can be evaluated, but cannot be edited; no additional

Figure 4. Desire2Learn competencies: Instructor's view of learning objective results

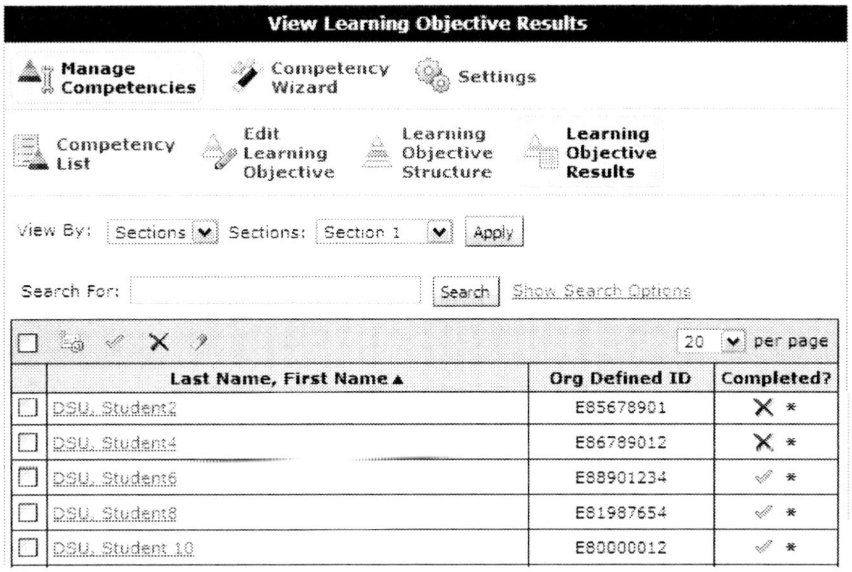

learning objective can be associated with it and objectives associated with it cannot be removed. To add a learning objective to a competency or remove an objective from it, the status of the competency must be in draft. Once a course has started and assessment is in progress, changing the status of a competency and adding or removing associated learning objectives can be disrupting and troublesome. It is therefore a good practice to plan in advance and build the competency structure before the course starts. However, pre-course building of learning objectives could be a challenge if a constructivist approach is adopted to allow negotiation of instructional objectives and dynamic, less criterion-referenced assessment.

Although embedding competencies is allowed in D2L, it can be pragmatically cumbersome. With embedded competencies, the completion of the top competency cannot be evaluated until all sub competencies are evaluated, which is not possible if any of the sub competencies is in draft. The instructor of the pilot course started with two layers of embedded competencies, and soon became aware of the difficulty and inconvenience in evaluating students' achievements, updating the objectives, and managing the embedded competency structure. The instructor then removed the top layer of competency and put the second layer into separate competencies.

Defining Learning Objectives in Measurable Terms

Learning objectives can have varied levels of specificity. Krathwohl and Payne (1971) identified three levels of learning objectives: global, educational, and instructional. Global objectives are complex, multifaceted learning outcomes that require substantial time and instruction to accomplish. They are broadly stated and encompass a large number of specific objectives. The very generality of global objectives is necessary to inspire, but they are difficult to use in planning learning activities and defining appropriate assessment measure. For effective assessment, global objectives need to be broken down into more focused, delimited form (Anderson & Krathwohl, 2001, Sosniak, 1994). Specific learning objectives are defined in measurable terms and assessed through observable learning activities.

In D2L, learning objectives are subsumed under competencies. Most competencies are to be achieved through multiple learning objectives. However, a learning objective does not have to be confined to a single competency. When needed, a learning objective can be associated with two or more competencies. Because objectives have to be assessed through associated learning activities and all learning activities in this pilot course are specified at the unit level, there is no course level objective in the Competency Structure in the pilot course. Nevertheless, overall course objectives are included in the syllabus of the course, as required by the University.

As the accomplishments of learning objectives are assessed through associated activities, learning objectives need to be defined in specific and measurable terms so that there are clearly identifiable alignments between learning objectives and learning activities. Defining learning objectives can help clarify the intent of instruction and guide the formation of instructional activities. Specific learning objectives can inform the students of the exact expectations and the types of evidence that the students need to produce to demonstrate their mastery of the content or skills. Each learning objective usually targets at a single learning outcome, as it is usually easier to assess the accomplishment of a learning objective if the objective focuses on one learning outcome and the instructor does not have to distinguish between complete and partial accomplishments.

In D2L, the course designer needs to plan associated activities when determining learning objectives. Activities in the D2L learning environment can include visitation to content topics, discussion participations, dropbox submissions, quiz attempts, checklist completions, survey submissions, and grade items. When a learning objective is created in D2L, it is important to decide whether to enforce evaluation methods, which can include discussion, dropbox, grade item, manual assessment, quiz, and survey. Once evaluation methods are selected for a learning objective, the methods cannot be changed. In the pilot course, as the instructor was not sure, he chose to leave the evaluation methods open for most objectives so that the learning objectives can be evaluated with any type of learning activity. The instructor selected some evaluation methods for only a few learning objectives where he was quite sure as to what evaluation methods were feasible, primarily discussion participation.

Most courses in higher education focus on the cognitive domain, thus it is important to examine various levels of cognitive understanding. The guidelines followed by the instructor of the pilot course are Bloom's taxonomy (Bloom, et al. 1956) and the revised model proposed by Anderson and Krathwohl (2001), particularly the cognitive process dimension which includes six categories, and related cognitive processes:

1. Remembering: recognizing and recalling.
2. Understanding: interpreting, exemplifying, classifying, summarizing, inferring, comparing, and explaining.
3. Applying: executing and implementing.
4. Evaluating: checking and critiquing.
5. Creating; generating, planning, and producing.

To ensure that learning objectives are measurable, it is best to avoid using verbs that are vague and difficult to observe evidence of outcomes. Such vague verbs include "appreciate", "comprehend", "know", "learn", and "understand." More measurable objectives would be "categorize", "compare", "express", "describe," "identify", and "recognize." Specific learning objectives do not have to be for lower-order skills only. Complex, high-order learning outcomes can be assessed through observable and measurable actions such as "classify", "critique", "design", "differentiate", "explain", "formulate", "justify", "defend", and "summarize."

Building Granular Rubrics

Most activities in D2L, except for quiz and grade item, need to be assessed through rubrics. A rubric is typically a bulleted list or a tabular form specifying levels of performance and grading criteria (see Figure 5). There are many benefits in using rubrics for assessment. Rubrics with specific criteria can provide performance benchmarks, clearly showing the learners what is expected of them and how their work will be evaluated. In this way, a rubric can provide a working guide for both instructor and learners and help assess both learning success and teaching effectiveness. A well-crafted rubric can help implement criterion-referenced assessment (Popham, 2000, 1971) and add consistency and objectivity to assessment (Chatterji, 2003).

A rubric can serve as an implicit contract between instructor and students on performance assessment. When students use rubrics regularly to judge their own work, they begin to accept more responsibility for the end product. Rubrics can also help reduce the time instructors spend grading student work and makes it easier for instructors to explain to students why they got the grade they received and what they can do to improve. Such clear and specific guides are especially important in online education where written communication is the primary means of class interaction and should be as clear and specific as needed.

A good rubric needs to have enough granularity and specificity to accommodate variances in learner performances. Because a learner must pass all the associated learning activities in order to meet a learning objective, the threshold of a rubric should be set at the minimal level of performance that a learner must pass.

In D2L, a rubric can be associated with as many activities as needed. In the pilot course, four primary types of learning activities are identified: (a) answering questions, (b) debugging, (c) discussions, and (d) programming. Four rubrics are developed accordingly. One of the rubrics is for assessing programming work. Six areas are identified for this rubric: accuracy, organization, efficiency, documentation, format requirements, and time requirements. Five levels are defined to differentiate performances and quality of work. The following is a general summary of the differentiations:

Figure 5. A sample rubric

	A 90 or higher	B 80-89	C 70-79	D 60-69
Accuracy & Clarity	Covered all key points. Well informed & supported. Clearly and effectively presented.	Covered most key points. Generally accurate & clear. Support not always appropriate.	Covered about half of the points. Some minor misconceptions or misinterpretations. Little support	Missed more than half of the points. Some major misconceptions or misinterpretations. No support.
Coherence & Focus	Consistently coherent and focused	Mostly coherent, with occasional digression	Noticeable lack of focus, with irrelevant points.	Much of the content is irrelevant.
Text	Well organized. Text flows smoothly. Free of mechanical errors.	Text is generally organized & easy to read. Lacks some transitions. Some mechanical errors.	Text is readable, but does not flow. Lacks transitions. A number of mechanical errors.	Text lacks organization & is hard to read. Some major mechanical errors.

1. **Level 1:** exceptional work in all six areas, with few negligible errors. Well organized and adequately documented. The program compiles and runs as expected without any error. Coding accuracy and efficiency are satisfactory.
2. **Level 2:** descent work that may miss one or two minor requirements, or have one or two minor shortcomings in any of the six areas. The program compiles and runs properly in most regards. Coding accuracy and efficiency need some improvement.
3. **Level 3:** acceptable work that has three or more minor deficiencies or shortcomings. Not very well organized. Inadequate documentation. The program compiles but has some minor runtime errors. Coding accuracy and efficiency are fair, but not satisfactory.
4. **Level 4:** poor work that has some major deficiencies or shortcomings. Disorderly code, with little or no documentation. The program compiles but has some major runtime errors. There are noticeable code inaccuracies and/or inefficiency.
5. **Level 5:** unacceptable work that fails to meet the key requirements. The program fails to compile or run.

There are some good resources online to facilitate rubric development. Two popular sites are the RubiStar at http://rubistar.4teachers.org and Rubrics for Teachers at http://www.rubrics4teachers.com/. For an inexperienced designer, it is usually easier to start with the best and worst levels of performances, and then fill in the middle levels based on the designer's prior knowledge of common issues and problems that students tend to have with the particular type of learning activity. The instructor may even want to encourage students to contribute to the creation of a rubric to promote student commitment and involvement.

Monitoring Learner Progress and Providing Formative Feedback

Learning activities should be assessed in a timely manner so that students can monitor their learning progress and make adjustment when needed. Instructors can monitor learner progress from multiple places in the Competency Structure, either holistically from the Structure Summary of the Competency List, or individually from the Results page of each learning objective or activity (see Figure 4).

The primary goal of monitoring learner progress is to provide formative assessment through timely feedback. To accomplish the goal, the instructor needs to remain informed of students' progress in course work, including assignment submissions, discussion participations, email communications, and project development. In D2L, updates appear on the course homepage for instructors when there are new assignment submissions, unread discussion postings and new email messages. These updates can help instructors stay current in monitoring and assessing learning activities. To help learners better meet the learning objectives, it is a good practice for the instructor to read students' assignment submissions before due dates and provide formative feedback so that students can revise and improve their work.

An effective way to help students monitor their own learning progress is self assessment. In the pilot course, weekly self quizzes were posted to let students assess their mastery of the concepts and techniques covered each week. The instructor made a point in aligning the self quiz questions to the learning objectives of the related unit. In their answers to the survey near the end of the semester, the students indicated that the self quizzes were very useful in helping them to assess how well they had met the learning objectives and grasped the course content. For each question in the self quizzes, the instructor provided specific feedback either explaining what the correct answer is or directing students to related resources to find the correct answer. The students surveyed also indicated that the self quizzes helped them to determine their strengths and weaknesses, and helped them to get prepared for subsequent formal quizzes of the course.

In addition to the usual function that lets instructors make comments in grading assignment submissions and quiz attempts, D2L has a special feature that lets learners give feedback to instructor by commenting and rating on content topics anonymously. The instructor can view the comments and ratings in a central location. With the feedback from the learners, the instructor can continuously adjust and modify the content to meet the learner needs.

Another method to get learner feedback found useful in the pilot study is to poll student opinions through surveys. In D2L, surveys can be attached to learning objectives and assessed through a rubric. Survey submissions can be anonymous or identifiable. Other tools that can help instructors monitor student progress and provide feedback include course mail, chat room, and self assessment, though they are not classified by D2L as activities that can be assessed through a rubric.

Using Release Conditions to Manage Learning Paths

Learning paths are the routes that learners can take in getting through a course and navigating their way around. Learning paths can be highly flexible and non-linear or strictly linear and controlled by the course settings and content sequence. A permeating function of D2L that can help instructors manage learning paths and enforce assessment procedures is Release Conditions. In D2L, Release Conditions can be set on almost any course components and based on either learner performances or group membership or both. Group-based conditions can be useful in promoting diversified learning activities and facilitate learner collaboration. Performance-based conditions can be used to enforce strictly sequential learning based on prerequisite knowledge and skills. D2L also lets the instructor set special access permissions to support individualized learning pace and flexible learning paths.

A major limitation with the current setup of Release Conditions found in the pilot study is that a component with a Release Condition attached is not visible to a learner until the condition is met by the learner. Consequently, the learner will not see what to come next until the learner completes the prerequisite. It would be more user-friendly if such components are visible to the learners with an indication of the attached Release Condition or prerequisites. To make up for the shortcoming, a course outline is used in the pilot course to show the learners the learning paths and Release Conditions attached.

CURRENT CHALLENGES

Through the interconnections between learning objectives, learning activities, and assessment rubrics, the Competency Structure in D2L can serve as a highly pragmatic framework for aligning learning activities with learning objectives, monitoring learner progress, assessing learner accomplishments with criterion-referenced rubrics, and providing students with formative feedback. The framework can help the instructor meet a key requirement of the Quality Matters standards: providing clear indications as to how learning activities, resources, assessments, and course tools can work together to help learners achieve the desired learning objectives.

There are some major challenges and requirements in implementing the Competency Structure. The course designer needs to have a clear overview of the course objectives and unit objectives and know what learning activities to assign in alignment with the objectives. Rubrics need to be set up and released before the course starts so that learners can see what they are expected to accomplish and how their work will be evaluated.

It will significantly increase the workload for the instructor or course designer if the instructor has to develop and revise the Competency Structure as the course evolves. Most of the design and development work needs to be done before the actual teaching and learning start. The effectiveness and sophistication in the use of the online assessment tool is determined to a large extent by the designer's familiarity with the Course Management System and technical skills in using the related components. With multiple components interconnected and interacting with each other, it can be a huge challenge for the instructor to be functionally clear as to how the different elements can work together, and at the same time, continuously monitoring student progresses, providing guidance and assistance when students need them, and updating and adjusting the learning paths to support flexible and individualized learning.

REFERENCES

Anderson, L. W., & Krathwohl, D. R. (Eds.). (2001). *A taxonomy for learning, teaching, and assessing: A revision of Bloom's taxonomy of educational objectives*. New York: Longman.

Bloom, B., Englehart, M. D., Furst, E. J., Hill, W. H., & Krathwohl, D. R. (1956). *Taxonomy of educational objectives: The classification of educational goals. Handbook I: Cognitive domain*. New York, Longman.

Chatterji, M. (2003). Designing and using tools for educational assessment. Boston, MA: Allyn and Bacon.

Desire2Learn, Inc. (2008). *Competencies and rubrics user guide (8.3)*. Retrieved on November 28, 2008, from http://community.desire2learn.com

Garrison, D. R., & Anderson, T. (2003). *E-learning in the 21st century: A framework for research and practice*. London: Routledge Falmer.

Garrison, D. R., & Cleveland-Innes, M. (2005). Facilitating cognitive presence in online learning: Interaction is not enough. *American Journal of Distance Education, 19*(3), 133–148. doi:10.1207/s15389286ajde1903_2

Hyde, P., Booth, R., & Wilson, P. (2003). The development of quality online assessment in VET. In H. Guthrie (Ed.), *Online learning: Research readings* (pp. 87-105). Leabrook, Australia: National Centre for Vocational Education Research.

Jensen, D., Self, B., & Rhymer, D. (2002). A rocky journey toward effective assessment of visualization modules for learning enhancement in engineering mechanics. *Educational Technology & Society, 5*(3). Retrieved on March 11, 2009, from http://www.ifets.info/journals/5_3/jensen.html

Jonassen, D. (1991). Objectivism vs. constructivism: Do we need a new philosophical paradigm? *Educational Technology Research and Development, 39*(3), 5–14. doi:10.1007/BF02296434

Jonassen, D. (1994). *Operationalizing mental models: Strategies for assessing mental models to support meaningful learning and design-supportive learning environments*. Retrieved on March 20, 2009, from http://education.uncc.edu/jagretes/instructional%20design%20articles%20-%20&%20constructivist%20articles/operationalizing%20mental%20models.pdf

Koszalka, T. A., & Ganesan, R. (2004). Designing online courses: A taxonomy to guide strategic use of features available in course management systems (CMS) in distance education. *Distance Education, 25*(2), 243–256. doi:10.1080/0158791042000262111

Krathwohl, D. R., & Payne, D. A. (1971). Defining and assessing educational objective. In R. L. Thorndike (Ed.), *Educational measurement* (pp. 17-45). Washington, D.C.: American Council on Education.

Lewis, B. A. (2002). Learning effectiveness: Efficacy of quizzes vs. discussions in online learning. *Dissertation International Abstracts, 63*(3), 911A.

Macdonald, J., & Twining, P. (2002). Assessing activity-based learning for a networked course. *British Journal of Educational Technology, 33*(5), 603–618. doi:10.1111/1467-8535.00295

Mager, R. (1975). *Preparing instructional objectives, 2nd ed.* Belmont, CA: Fearon-Pitman Publishers.

Mager, R. F. (1962). *Preparing instructional objectives*. Palo Alto, CA: Fearon Press.

McVay Lynch, M. (1998). *Constructivism in instructional design for distance education*. Retrieved on March 11, 2009, from http://web.pdx.edu/~mmlynch/constructivist.html

Murphy, K., Mahoney, S., & Harvell, T. (2000). Role of contracts in enhancing community building in Web courses. Educational Technology & Society, 3(3).

Pawan, F., Paulus, T. M., Yalcin, S., & Chang, C. (2003). Online learning: Patterns of engagement and interaction among in-service teachers. *Language Learning & Technology, 7*(3), 119–140.

Popham, W. J. (Ed.). (1971). *Criterion-referenced measurement: An introduction*. Englewood Cliffs, NJ: Educational Technology Publications.

Popham, W. J. (2000). *Modern educational measurement: Practical guidelines for educational leaders, 3rd ed.* Boston, MA: Allyn and Bacon.

Quality Matters. (2008). *Research literature and standards set support for quality matters review standards*. Retrieved on November 28, 2008, from http://www.qualitymatters.org

Shea, P., Swan, K., Fredericksen, E., & Pickett, A. (2002). Student satisfaction and reported learning in the SUNY learning network. In J. Bourne & J. Moore (Eds.), *Elements of quality online education* (pp.145-156). Needham, MA: Sloan Center for OnLine Education.

Sims, R., Dobbs, G., & Hand, T. (2002). Enhancing quality in online learning: Scaffolding planning and design through proactive evaluation. *Distance Education, 23*(2), 135–148. doi:10.1080/0158791022000009169

Sosniak, L. A. (1994). The taxonomy, curriculum, and their relations. In L. W. Anderson & L. A. Sosniak (Eds.), *Bloom's taxonomy: A forty-year retrospective, ninety-third year-book of the national society for the study of education* (pp. 103-125). Chicago: University of Chicago Press.

Stewart, I., Hong, E., & Strudler, N. (2004). Development and validation of an instrument for student evaluation of the quality of Web-based instruction. *American Journal of Distance Education, 18*(3), 131–150. doi:10.1207/s15389286ajde1803_2

Thurmond, V., Wambach, K., Connors, H., & Frey, B. (2002). Evaluation of student satisfaction: Determining the impact of a Web-based environment by controlling for student characteristics. *American Journal of Distance Education, 16*(3), 169–189. doi:10.1207/S15389286AJDE1603_4

Trigano, P. C., & Pacurar-Giacomino, E. (2004). Toward a Web-based environment for evaluation and design of pedagogical hypermedia. *Educational Technology & Society, 7*(3), 21–37.

Youngblood, P., Trede, F., & DeCorpo, S. (2001). Facilitating online learning: A descriptive study. *Distance Education, 22*(2), 264–284. doi:10.1080/0158791010220206

Chapter 5
Cost Effectiveness in Course Redesign:
The Transformation toward E-Learning

David Kendrick
University of Northern Colorado, USA

EXECUTIVE SUMMARY

As Web-based technologies march forward, improved access to higher education by traditional and non-traditional student, alike, is a certainty, but such technologies as a mechanism for lowering costs are still subject for further exploration. Course redesign from traditional to electronic delivery serves not only to grant access or improve achievement for the student, but can offer a cost savings for the institution. Educational leaders in higher education may consider the Web-based redesign alternative as not only a learning instrument, but a means to cut instructional costs. An explanation and application of a cost-measuring instrument, as well as reviews of literature and Web-based instructional models or strategies, is at the heart of this examination of course redesign. Educational content has become a commodity. Improved networks provide rapid and flexible dissemination of course content, opening up numerous options for organizing programs. Rather than designing content delivery around the schedule and resources of the provider, the institution, it can be customized around the needs of the recipient (Lovett, 1996). Courses, programs, and even degrees, can be organized around a combination of flexible course modules to accommodate particular student/client needs. Technology-mediated instruction, taken to its anywhere-anytime extreme, makes traditional academic calendars and curricular structures irrelevant or even a barrier to effective education.

THE ISLAND UNIVERSITY AND THE COST OF LEAVING THE ISLAND

Texas A&M University-Corpus Christi (TAMU-CC), located on the Coastal Bend of the Texas coast, is a doctoral-granting institution with a strong emphasis in education, nursing, and coastal studies. It has a tradition of serving primarily as a regional institution for the citizenry of South Texas; however, as the fastest growing public institution in the state, it has attracted interest nationally and

DOI: 10.4018/978-1-60566-870-3.ch005

internationally. How does a once regional-serving institution expand cost-effectively and technologically without compromising the integrity of the classroom environment?

There are a number of considerations for the design and development of the online classroom. First is the issue of economics. Unless there is a minimum, sustainable number of students at a distance to benefit from the virtual classroom implementation, the university administration will view the initiative as an inefficient waste of resources. There must be institutional policies and procedures in place. There should be a realistic long-range plan considering equipment, skilled personnel, resources, and instructional impact (Hsu, 1999). Furthermore, as part of the planning and development strategy, there should be emphasis on cultivating an atmosphere of connectivity between sites and participants (Schrum, 1996).

Course development, whether for traditional classroom environments or high-tech settings, has been viewed as a significant drain on education funds. For web-based strategies, the cheaper and easier route has been to purchase pre-developed courses, much like a textbook or telecourse, and adapt the materials for campus use. The Florida Community College Distance Learning Consortium has successfully consolidated the licensing of instructional content since its creation in 1996 (FCCDLC, 2004). The Consortium has saved approximately 50 per cent over the individual institution costs by combining purchases and leveraging resources to make upfront buyouts of high-use course content. Where the course content changes rapidly, such as the case of information technology training, the Consortium has achieved favorable results in state-level master agreements for IT courseware by working with collaborating institutions, pooling their resources towards larger, otherwise unaffordable, purchases (Opper, 2002). Participating institutions benefit by large volume discount pricing. There is a key distinction between licensing and development activities: Licensing curricular content for use becomes an ongoing expense while development is a one time, up front cost. The reality is that any course, whether developed from scratch or licensed and adapted, requires revision within a year or two of its introduction; therefore, in the long run, development can cost more up front and, yet, still obligate the institution to later maintenance costs. The critical factors to consider are the stability and durability (the test of time) of the content and the number of students served (Opper, 2002).

While shared resources bring about greater choice and supply of curricula, effective instructional technology collaboration at the higher education levels may not necessarily translate to success in cost-effectiveness terms in K-12 school systems. One study that investigated the implications of distance education for cost-effectiveness and equity in the allocation and use of educational resources involved participants from nine New York school districts utilizing interactive television as their distance education modality. The results revealed that distance education did, indeed, allow small and rural schools to expand their curricula; however, interactive television distance education courses cost substantially more to offer than traditional on-site courses (Brent, 1999). Districts varied considerably in their utilization of the network and participating students did not benefit equally. In question is the reliance on interactive television, as opposed to web-based learning, as the choice of distance learning modality. Some forms of distance education can result in lower unit costs than those of classroom-based instruction. There is, however, little evidence about the relative costs of information-technology-based distance education (Rumble, 1999).

The costs and benefits of distance education must be weighed. The costs of providing distance education must be analyzed and stakeholders identified. Solutions may be pursued that include focusing on niche programs for which a market exists, understanding the competition, being wary of public education administrators who offer distance education programs for mainly financial gain, and tracking the real costs of technical program support and infrastructure (Willis, 2003).

The Library Education Experimental Program (LEEP) at the University of Illinois, Champaign-Urbana, defies traditional efforts to perform economic analysis as it was never intended to operate as a profit center. Research into its creation suggests a need to rethink the cost-benefit model as applied to distance learning. Although the program does not reveal a dramatic profit, it has resulted in a number of non-monetary benefits that may be difficult to quantify but are of great value to the university and its students (Lorenzetti, 2002). Choosing an accurate yardstick by which to measure the value of distance learning is no easy task.

Institutions engaged in the investment process in distance education are hard pressed to understand the costs and benefits of distance courses. Among the cost factors involved in such courses are capital and recurrent costs, production and delivery costs, and fixed and variable costs. The benefits of distance education relate to student/instructor satisfaction, learning outcomes, and return on investment. The more effective systems excel in access, flexibility, ease of use, and the potential to reach new markets (Bartolic-Zlomislic & Bates, 2002).

One study attempted to identify criteria on which to assess technology's cost-effectiveness. The participants were involved in an Interactive Distance Learning (IDL) initiative in Ohio titled the Telecommunity Project. Results indicated that the value of IDL cannot be measured in the traditional way and that perception of the value of educational technology is heavily influenced by the focus and mission of the institution (Hawkes, 2000).

NO SIGNIFICANT DIFFERENCE

Technology utilization is born out of necessity and in the state of Texas where population centers are spread across wide expanses of rural territory, this is particularly true. The TAMU-CC College of Education has developed a strong program in higher education administration, among other areas, and has sought to develop a cost-effective, as well as an instructionally effective, means of delivering courses to new audiences. This case study takes the first step in examining such feasibility. Again, it seeks to do so without compromising the learning process; that is, there can be no significant difference, regardless of cost, regardless of technologies, in the learning outcomes, in student achievement.

Throughout the past decade, e-learning or distance education has established itself as a viable means of instructional delivery with achievement levels rivaling and, in some cases or disciplines, surpassing traditional course delivery in terms of student achievement. The majority of evidence has revealed no significant differences in learning outcomes and perceptions, attitudes, or satisfaction with online modes of delivery. A number of courses of varying disciplines have been subjected to comparative studies involving a traditional or lecture course and a form (usually web-based) of e-learning course. The "No Significant Difference Phenomenon" transcends any boundaries of discipline; that is, the course of study or the nature of the course, in most cases, is not necessarily the determining factor of success in the online environment (Carey, 2001).

In order to control college prices and, simultaneously, avoid compromising the educational outcomes and quality, there has been a fair degree of redesigning the higher education system at the state levels. Changes vary from a comprehensive review aimed at eliminating program redundancy to the increased usage of instructional technology – an effort popular among politicians and their constituents, but received with some degree of caution by the traditional higher education community (Blumenstyk, 1995). It is not unusual that instructional technologies to provide higher education to greater and, more importantly, more diverse student populations would be welcomed by government leaders; but to do so at a lower cost, coupled with the fact that there will exist no significant difference in the users' achievement level, makes it that much more attractive. Such programs are perceived to offer less expensive means to provide educational services and to alleviate the pressures of increased enrollment, as well (Institute for Higher Education Policy, 1994).

A CASE STUDY FOR A NEW MODEL

Within university systems, collaboration of online resources and delivery has taken hold for either the sake of learning effectiveness or perceived cost efficiency. At Texas A&M University-Corpus Christi, courses and programs once considered confined to traditional instructional delivery have been remodeled for the e-learning environment. For example, in a radically different approach in teaching/learning strategy, the College of Nursing has developed and implemented an innovative e-learning system of modules, eLine, to further its reach and effectiveness within the community, in collaboration with the Del Mar College, and across the state. Furthermore, the College of Education, expanding its reach in a growing doctoral program in Educational Administration and Research, has redesigned what were once traditionally-delivered courses for asynchronous online delivery.

A pilot study, ex-post-facto in design, was conducted to compute and compare the cost associated with the on-line and traditional delivery of selected doctoral courses in the College of Education at TAMU-CC. The following research questions guided the study:

1. What are costs associated with delivery of on-line and traditional doctoral courses at TAMU-CC?
2. Is there a difference in instructional costs between traditional delivery and delivery of redesigned virtual courses?

This study examined two doctoral courses in the College of Education: EDLD 6331 Educational Innovations and EDLD 6306 Higher Education in a Democratic Society. Each section of traditional and e-learning courses was a product of the same instructor teaching the same course as to control for course type and delivery effects. Due to non-probability nature of the sample, no generalizations to other populations or disciplines were made. However, the successful implementation of this study within the College of Education can be replicated outside of the college for the examination of more radical transformations in course redesign as is taking place with the "eLine" modular course program within the TAMU-CC (e.g., College of Nursing).

Being a pilot study and a pilot design for the outreach to new clients, the study was welcomed by the administration of the department and college. The department has in the past served outlying clients by bringing them to the instruction. The new model is to export the instruction to the clients. Of course, web-based instruction is not new, unusual, or unique; however, institutions of higher education infrequently or never directly measure instructional costs, breaking them down in accordance with the procedure and instrument provided by the National Center for Academic Transformation.

The Course Planning Tool, the instrument designed by the National Center for Academic Transformation at the Rensselaer Polytechnic Institute for the PEW Program in Course Redesign was used to measure cost differences. Permission to use the instrument has been granted and encouraged from Carol Twigg, Ph.D., Executive Director, the National Center for Academic Transformation (NCAT). The instrument had been pilot-tested with one redesigned course in the College of Nursing at TAMU-CC.

The results of the pilot study indicate, on the surface, what initially appears to be a cost increase, rather than a savings. The redesign for EDLD 6306 Higher Education in a Democratic Society revealed an 85.9 per cent increase in instructional costs while EDLD 6315 Multicultural Analysis resulted in an even greater increase in instructional costs at a 187.0 per cent increase over traditional delivery. Why would the redesign of these two courses seemingly result in an increase in instructional costs? It is a matter of scale. Doctoral courses, such as these, fill less than twenty seats – physically and virtually; whereas, those programs examined by the NCAT address undergraduate courses, often multi-sectional, catering to hundreds of students. Does this mean that doctoral courses are not cost effective for redesign? To the contrary, not included in the equation is the matter of state formula funding, raising the value of the doctoral student up to ten fold that of the undergraduate; therefore, a redesign bringing access to potential doctoral students otherwise previously denied becomes profitable and necessary. It should be noted that the study concerns itself with measurable instructional costs and does not address what is often perceived as "values" (student/instructor satisfaction or learning outcomes, for example): nor does it address overall departmental or institutional return on investment.

Course redesign is at the heart of the efforts of The Center for Academic Transformation (NCAT) at Rensselaer Polytechnic Institute (CATa, 2005) which has sought new strategies for cost effective delivery – some through e-learning and some not, but all intentionally diverging from traditional teaching and learning structures - at over thirty institutions. Under the aegis of the PEW Foundation, the Center has examined new ways that a limited number of disciplines in higher education can take advantage of the capabilities of information technology to transform their academic practices. If universities could redesign their instructional approaches using technology to achieve cost savings, could they experience quality enhancements to learning, as well? After all, a redesign resulting in cost savings with diminishing academic achievement or poor retention would be of little value. Redesign projects focused on large-enrollment, mainly introductory courses, in an effort to impact the greatest student numbers and generate significant cost savings (CATb, 2005). This case study at TAMU-CC is a minor example of the transformations taking place at dozens of institutions across the country, the difference being that this particular case study focuses on doctoral courses and the NCAT study examines undergraduate courses significantly larger in scope.

CHALLENGES BEYOND THE CASE STUDY: LOOKING BACK, THINKING FORWARD

Historically, distance education initiatives have targeted the nontraditional student. The implementation of two-way, inter-active television networks has brought courses and entire programs to the un-served or underserved. Constantly improving web-based technologies expand the effort further. The aim is no longer toward the nontraditional learner, exclusively. The technology has allowed instructional and course designers to expand the scope of distance education to include more traditional learners. Indeed, as the technology has improved, there has been tremendous growth in distance education capabilities. Looking back, in 1987, only ten states offered a distance learning program. Only five years later, every state had a program in place (IHEP, 1994). The original hope has been that technology could address a number of higher education challenges, including the rising costs of education. Support continues as 95 per cent of educational leaders believed that their legislatures would support future educational technology endeavors (Ruppert, 1996). In times of tight budgets, technology may be one of the few educational investments that states are willing to support (Blumenstyk, 1994).

The governors of the western states were the first to agree to the creation and acceptance of virtual universities to deliver courses through computer networks, television, and other technologies, and award its own degrees. Unlike traditional universities that offer credentials, the virtual university may simply award credentials on the basis of an assessment of mastery of subjects (Blumenstyk, 1995). The western states are experiencing unprecedented growth, particularly in school-age populations, where funding for new campuses may be unavailable. Alternative delivery systems have enormous appeal to legislators who perceive higher education as labor intensive. Utah, for example, projecting a doubling of growth from the mid-nineties so 2010, is depending on distance education to accommodate the expansion. Its commissioner of higher education committed to the replication of nine existing campuses, but, just the same, pledged to keep building of brick-and-mortar campuses to a minimum (Blumenstyk, 1994).

Some state and educational leaders have been unpleasantly surprised by the initial costs of implementing technology-based solutions that were originally aimed at addressing financially-based instructional problems. They questioned whether technology-based instruction actually reduced costs. Virtual campuses appeared to be less expensive than physical buildings. There were, however, significant start-up costs - millions of dollars for microwave transmission systems, fiber networks at $10,000 per mile, and the installation and accommodation of an otherwise "unwired" classroom can run $40,000 (Institute for Higher Education Policy, 1994). A more cost-effective solution over telecourse (interactive television) is becoming obvious: web-based technologies utilizing broadband Internet connectivity. This modality has its expense, as well. There is the cost of virtual student services, library databases, and the royalties for copyrighted online teaching materials (IHEP, 1994). Improved access to higher education by traditional and non-traditional student, alike, is a certainty, but distance learning technologies as a mechanism for lowering costs are still subject for further exploration.

Programs, course offerings, and delivery methods have evolved over the years, over the decades in the cases of some institutions; however, it will take a significant perceived advantage of a new delivery system to convince a large number of institutions to adopt it. The question remains: Will institutions embrace online education as a permanent system of instructional delivery? The evidence would answer in the affirmative as eighty-one per cent of all institutions of higher education offer at least one fully online or blended course. Complete online degree programs are offered by 34 percent of the institutions (The Sloan Consortium, 2004). The statistics are more compelling at public campuses with 97 per cent offering at least one online or blended course and 49 per cent offering a complete online degree program. When asked about the role of distance learning for the future of their institution, 67 per cent answered that they see it as critical long-term strategy (2004). Evidence does not suggest that this trend will reverse any time soon.

REFERENCES

Bartolic-Zlomislic, & Bates. (2002). The costs and benefits of investing in online learning. *Distance Education Report, 6*(9), 3.

Blumenstyk, G. (1994). Networks to the rescue? *The Chronicle of Higher Education*, A21.

Blumenstyk, G. (1995). Campuses in cyberspace. *The Chronicle of Higher Education*, A18.

Brent, B. (1999). Distance education: Implications for equity and cost-effectiveness in the allocation and use of educational resources. *Journal of Education Finance, 25*(2), 229–254.

Carey, J. (2001). Effective student outcomes: A comparison of online and face-to-face delivery modes. *DEOSNEWS, 11*(9). Retrieved from http://www.ed.psu.edu/ascde/deos/deosnews/deosarchives.asp

CATa. (2005). *Center for Academic Transformation at the Rensselaer Polytechnic Institute*. Retrieved in March 2005, from http://www.center.rpi.edu

CATb. (2005). *Center for Academic Transformation, Program in Course Redesign*. Retrieved in March 2005, from http://www.center.rpi.edu/PewGrant.html

Hawkes, M. (2000). The cost factor: When is interactive distance technology justifiable? *T.H.E. Journal*, *28*(1), 26–32.

Hsu, S. (1999). How to design a virtual classroom: 10 easy steps to follow. *T.H.E. Journal*, *27*(2), 96–98.

Institute for Higher Education Policy. (1994). Assessing distance learning from a public policy perspective. *Policy Steps*, *3*, 7–9.

Lorenzetti, J. (2002). Toward an expanded cost-benefit analysis for distance programs. *Distance Education Report*, *6*(24), 1–6.

Lovett, C. (1996). How to start restructuring our colleges. *Planning for Higher Education*, 19.

Opper, J. (2002). Funding and cost containment of educational technology: Shifting policy and practices. Retrieved in September 2002, from http://www.wcet.info/projects/tcm/whitepaper2.pdf

Rumble, G. (1999). Cost analysis of distance learning. *Performance Improvement Quarterly*, *12*(2), 122–137.

Schrum, L. (1996, March). Teaching at a distance: Strategies for successful planning and development. *Learning and Leading with Technology*, *23*(6), 30–33.

The Florida Community College Distance Learning Consortium. (2004). Retrieved in October 2004, from http://www.distancelearn.org/mainPage.cfm

The Sloan Consortium. (2004). Entering the mainstream: The quality and extent of online education in the United States, 2003 and 2004. Retrieved in November 2004, from http://www.sloan-c.org/resources/survey.asp

Wills, B. (1993). *Distance education: A practical guide*. Educational Technology Publications.

Chapter 6
When Distance Technologies Meet the Student Code

Peg Wherry
Montana State University, USA

Deborah Lundberg Windes
University of Illinois at Urbana-Champaign, USA

EXECUTIVE SUMMARY

This case study outlines problems with student conduct in an online undergraduate program and explains how a student code was applied to resolve the issues and institute procedures to reduce future incidents of academic dishonesty and incivility. The study describes several instances of student misconduct and explains how online program administrators responded by improving communication with both students and faculty and by modifying course design and development processes as well as instructional practices. It also reports on how other administrators assisted in handling resolution and discipline. While technology itself may both complicate the maintenance of conduct standards and provide new ways to protect academic integrity, this study demonstrates that the introduction of technology should not change the rules.

BACKGROUND

Administrators and instructors in distance and online education programs often encounter the assumption that technology changes everything, whether the change is due to the technology itself or due to new or different administrative structures supporting technology. For example, most online learning administrators have received the panicked question from an online instructor: "What do I do if I think a student in my online class is cheating?" The response, as we will discuss, should closely replicate the answer to the question: "What do you do if you suspect cheating in your campus classroom?" The following case study is based on the assumption that the more we hold online students to the same standards required in the classroom, the more students will benefit, and thus the reputation of online education will be enhanced. While technology itself may offer challenges to maintaining academic and conduct standards as well as providing new ways to protect academic integrity, we hope to

DOI: 10.4018/978-1-60566-870-3.ch006

demonstrate that the introduction of technology should not change the rules.

Our purpose in this study by administrators of a distance learning program is to outline problems with academic integrity and student conduct that emerged over a two-year period in an online undergraduate program. This series of incidents made us familiar with "chapter and verse" of our University's *Student Code* of conduct (http://www.admin.uiuc.edu/policy/code/) and helped us see how valuable such a student code can be in addressing conduct issues. We will explain how we applied our *Student Code* both to resolve the issues and to institute procedures to reduce future incidents. Our case study will present instances of two types of student conduct: academic dishonesty and incivility. We then discuss how the student code is applied and the changes we made to our procedures as a result.

Academic integrity in both on-ground and online courses receives a great deal of attention in professional circles; a wide range of sources are available on topics such as preventing cheating, whether and how use of technology encourages or defeats cheating, etc. This chapter will touch on those issues, but we will also discuss "what happens next" when student misconduct occurs, in cases of both civility and integrity. We will discuss the importance of dealing proactively with these issues, both for the sake of our program and to improve the image of academic integrity in online learning. To give some indication of the level of difficulty inspiring us to share our experiences, we will mention one case, not used here, in which we consulted the office of the Dean of Students. She not only recognized the names of the students (on a campus of 42,000) but added that "no matter what you do, it will be a nightmare." This was not exactly comforting, though she did offer excellent advice. Details of all instances have been modified in the interest of anonymity.

Guided Individual Study (GIS) offers continuous enrollment, self-paced undergraduate courses through the Office of Continuing Education at the University of Illinois at Urbana-Champaign. Starting seventy-five years ago as a correspondence program with print-based courses mailed to non-traditional students, it is now a predominantly online program focusing on general education courses in multiple disciplines. Enrollment is open year-round, and a student has six months from the date of enrollment to complete the course. Thus, each student works independently, often—but not always—in isolation from classmates. While students need not be admitted to a degree program at the University of Illinois in order to enroll, 90% of enrollments are from degree-seeking students at one of the three U of I campuses (Urbana-Champaign, Chicago, and Springfield). Roughly 40 courses are currently offered, with enrollments at around 1,200 per year. GIS courses are designed and taught by instructors appointed by the academic department which "owns" the course; all academic credit is given by the department itself. However, GIS staff provide guidance in course development and instruction, and mediate when student disciplinary problems arise. Instructors are paid per capita based on student enrollment, with payment occurring at three points during the student's enrollment (in most cases, at the beginning, mid-term and final). All instances we present in this study occurred within the GIS program.

In most institutions, distance learning faculty and staff have many resources to draw on when dealing with student conduct issues At Illinois, we encourage our instructors to work with the academic department head, or departmental director of undergraduate studies, and to adhere to departmental policies. At the same time, we offer to have a GIS staff member present at any meetings between the student and instructor. This assures that the two offices are acting in concert, reinforcing the institutional commitment to good conduct, and eliminating the opportunity for a student to play offices against one another. As we articulate advice for faculty who encounter academic dishonesty, we also work with the Associate Dean for Student Academic Affairs in the academic college that offers the bulk of our courses. In cases of incivility, we have worked with the Dean of Students Office, especially the Director of the Office for Student Conflict Resolution. As we will discuss in this chapter, a proactive approach to handling student conduct issues may serve to prevent or defuse a problem before there is a need to involve campus administrators.

SETTING THE STAGE: OUR CODE AND ITS CONTEXT

Before we proceed, it may be helpful to define what we mean by academic integrity. According to our UIUC *Student Code* (http://www.admin.uiuc.edu/policy/code/), infractions of academic integrity include the following acts:

- **Cheating:** "Using or attempting to use in any academic exercise materials, information, study aids, or electronic data that the student knows or should know is unauthorized."
- **Fabrication:** "Unauthorized falsification or invention of any information or citation in an academic endeavor."
- **Plagiarism:** "Representing the words or ideas of another as one's own in any academic endeavor." This includes failing to attribute any direct quotes as well as paraphrasing another's material without prompt acknowledgement.
- **Facilitating Infractions of Academic Integrity:** "Helping or attempting to help another to commit an infraction of academic integrity…"

It should be noted that in the case of plagiarism, infractions can occur in a casual fashion, where the original source is cited somewhere in the paper, but it is not clear which passages reflect original thought, or in a blatant fashion where "part or all of the paper is taken from another source without attribution," (Braumoeller & Gaines, 2001, p. 837). It is important to make this distinction not because one is acceptable and the other is not, but because they require different prevention efforts and different disciplinary actions when infractions occur.

Mary McManus Ramsbottom, Associate Dean of Student Academic Affairs in the College of Liberal Arts and Sciences on the Urbana campus, points out that all students, regardless of how their course is delivered, are members of our academic community. We operate within this community with certain standards and expectations, and we have an obligation to educate students about why these standards are important. She explains that we cannot count on students knowing what plagiarism is prior to their enrollment. Whether a student is in the classroom or online, we have a responsibility to explain what academic integrity is and to uphold the standards we have set for the community.

Upholding these standards in online courses requires a proactive approach. It is one thing to respond to incidents; it is another to create an atmosphere that discourages incidents from happening. The decentralized structure of a large university can make this a challenge, heightening the need for continuous communication about the standards of our community. While this may be easier at smaller institutions, we do not recommend taking this sense of community for granted or waiting until a major issue arises before implementing practices to both reduce the number of incidents and facilitate problem solving and conflict resolution when they occur.

Brian Farber, Director of the Office for Student Conflict Resolution at the University of Illinois at Urbana-Champaign, reports a general increase in disturbing or alarming student behaviors, both electronically and in person, explaining that the UIUC *Student Code* was revised last year to clarify the definition of conduct subject to discipline. It now includes "harassment or behavior which is so persistent, pervasive, or severe as to deny a person's ability to participate in the University community" (http://www.admin.uiuc.edu/policy/code/article_1/a1_1-302.html). The previous edition of the code had mentioned only "intimidation, harassment or coercion." This new, somewhat more legalistic, language provides more guidance in defining harassment, so it is no longer just in the eye of the beholder.

Farber also advised us as to which students are covered by the *Student Code*. It is clear that students do not have to be in residence on campus, nor must they be enrolled for credit. The UIUC *Student Code* specifies that: "(c) Individuals subject to student discipline include but is not limited to all persons: (1) taking courses at the University; (2) who cancel, withdraw, or graduate after committing behavior which may violate the code; (3) who are not officially enrolled for a particular term but have a continuing relationship with the University; and (4) who have been notified of and accepted their admission." (http://www.admin.uiuc.edu/policy/code/article_1/a1_1-301.html)

Academic Integrity

There is a general assumption, both within higher education and in the public at large, that cheating is easier or more common in online courses than in traditional courses. In a recent study examining perceptions of cheating in online courses (King, Guyette, & Piotrowski 2009), 73.6% of business students queried believed that it is easier to cheat in an online course than in a traditional course. While the truth of this may be debatable, this perception no doubt led to inclusion in the Higher Education Reauthorization Act (HR 4137) approved by Congress in August, 2008 of a provision requiring accrediting agencies to ensure that students who enroll in distance learning courses are in fact the people completing assignments and exams. Fair or not, this perception means that distance learning administrators and faculty must be vigilant, thorough and fair in pursuing conduct cases; this chapter will explore how to handle them.

Assumptions aside, there does not seem to be extensive or conclusive research comparing cheating behavior in online versus classroom-based courses. As Lori McNabb of the UT (Texas) Telecampus observed during a Sloan-C listserv discussion of HR 4137, "I believe that we (distance educators) have done ourselves a disservice by not looking at and addressing academic integrity in online learning. There has been very little published research into the cheating behaviors of online students, especially as compared to on campus students. There have been a few small studies published, but I am not aware of any published results of large, diverse research projects." One of those published articles is a well-designed study of nearly 800 undergraduates which found that the likelihood of cheating in an online course is not statistically different from the likelihood of cheating in the traditional classroom (Grijalva, Nowell, Kerkvliet, 2006). On the other hand, a survey of 1,262 students at a large state university indicated that cheating was much more prevalent in online classes than in traditional lecture courses (Lanier, 2006). Whether or not it is more common in distance education courses than in the classroom, it is a fact of life that academic dishonesty will occur in distance learning programs, just as it does in on-ground programs.

Incivility

In this age of social networking, when students communicate incessantly and very informally online, boundaries around appropriate behavior in cyberspace have grown blurry. New technology applications allow one individual to play many different roles, all from the same chair. Students who sit at the same computer to update their Facebook profile, play online games behind the mask of an avatar, download music, exchange videos, *and* do actual coursework may not remember the most effective ways to present themselves in an academic setting. As we all know, without being able to view facial expressions and body language, it is easy to misinterpret the meaning behind words in electronic communications. It is not just online students, moreover, who have a new understanding of faculty-student communication: a colleague with twenty years' teaching experience tells of a face-to-face conference with a student in which the instructor's explanation about an assignment grade elicited the response from the student, "that's not good enough, bud." In addition, one GIS instructor reported receiving such serious (anonymous) threats from a student in a classroom section that the campus police were asked to investigate.

In the online environment, however, there is a special complication. One of us is married to a professor of psychology who reminds us that "frustration leads to aggression." (Miller, 1941) We need to remember that distance and online students have more opportunities to become frustrated and thus, in some cases, aggressive. Problems with technology itself can make heavy demands on one's patience, and the distance learning environment can compound problems. For example, students who have questions for a classroom professor know they will see him or her at a specific time and place. An online instructor, on the other hand, is perceived by the student to be somewhere out in the ether, and thus cannot be counted on to respond immediately. Not knowing when the instructor will respond and not knowing how else to reach the instructor can be frustrating. This is significant for both course development and ongoing teaching in the online environment.

CASE DESCRIPTION

In the last two years, when both the Program Director and Assistant Head of the division were new to their roles, our office dealt with conduct issues that have prompted us to go back to the UIUC *Student Code* and establish procedures to ensure that all guidelines are followed. These cases, though few in number, represent a variety of outcomes, and all have been instructive. For those involved, any single case of academic misconduct is difficult, so it did not take a large number of cases to prompt our attention and action. We will first discuss incidents in which academic integrity was questioned, how we managed these incidents, and what was learned. Following this, we will discuss instances of student incivility, how they were handled and the lessons learned.

Academic Integrity Incidents

A student contacted the GIS office saying he had information that a student in one of our courses was cheating. After some debate (including the caveat that we should be alert to what the accuser's motives might be), we decided to meet and hear what he had to stay. The instructor then contacted the student charged with cheating to hear her side of the story. While we had no knowledge of the relationship between the accuser and the accused, we did learn from a student affairs administrator that this was not the first case in which the two names were linked. The cheating charge, as it turned out, pulled us into the middle of a very personal conflict. Due diligence required that we investigate the original charges, however, and the instructor ultimately determined that there was no way to prove the charge of plagiarism in this case.

In a second case, an instructor noticed that two students turned in remarkably similar work on a couple of assignments. One of the students had been previously suspected of cheating, but fact-finding led to no further action in that instance. Upon the second possible occurrence, the instructor was tempted by impatience growing out of the first (unproved) instance to consult his department head and identify a penalty for the new instance. This penalty was immediately communicated to both students, who then noted that they had not received "written notice of charges" and been

Figure 1. UIUC GIS Academic Integrity Flowchart 1

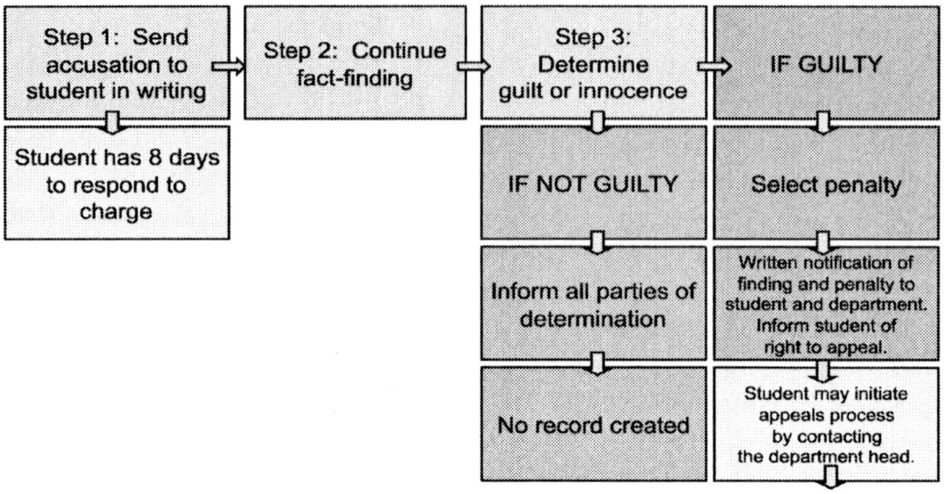

given an opportunity to respond to the charges before the penalty was imposed; this is a violation of the process outlined in the UIUC *Student Code* (Figure 1). The campus administrator to whom this was appealed was forced to conclude that skipping steps in the communication and appeal process meant the charges had to be dropped, despite the fact that the students had clearly colluded (the case did not reach the stage of determining who copied from whom). The students were disenrolled from the course, and the rest of us—GIS staff, instructor, department head—learned an important lesson about following procedures.

In yet another case, while grading lessons in an online course, a faculty member noticed suspicious similarities between two students' papers. While the wording had been changed to a degree, it seemed likely that the lessons were not written independently. In an instance of technology helping to solve these problems, a quick look at those students' profiles on Facebook showed that they both belonged to some of the same campus organizations. As a first step, the instructor reviewed previous lessons to see if this was a one-time incident or a recurring problem. After determining that there was only one set of suspicious lessons, the instructor emailed the two students with a note pointing out the similarities between the papers. He did not pronounce guilt; he just asked for an explanation. Within days, he had spoken with both students and received the full story. One student had "borrowed" the other's paper and admitted he had plagiarized. The plagiarizer fully exonerated the other student, who knew nothing of the incident. The guilty party received a "zero" for the paper and was required to write a paper (with sources) on the *Student Code* and why plagiarism is a serious issue.

What immediate lessons did we learn from these examples? With the first example, we learned the importance of developing and maintaining collegial relationships with administrators who handled conduct issues on campus. They made us aware of key aspects of the story we would otherwise not have known. Given that one situation had involved the law and possible physical danger, we found ourselves in a situation that could have gone very badly had we reacted without all the facts.

In the second and third instances, we learned that following our *Student Code* is crucial if we want the situation to be resolved in a satisfactory manner. In the second case, the code was not followed and two students who most likely were in violation got away without a black mark on their records. However, following the code in the other example allowed for a satisfactory outcome that was, we hope, a learning experience for one student.

Response

As a result of these cases occurring in rapid succession early in our managerial appointments, we decided to be proactive. We had previously handled such issues on a case-by-case basis, relying on "gut" feeling (with results ranging from satisfactory to disappointing), or by searching through the *Student Code* in the midst of a crisis. We determined that we needed to establish procedures, based on the *Student Code*, which would guide us when we faced suspected infractions. We were fortunate that the Student Code in place at our institution was fairly clear and detailed, making our work easier. After developing step-by-step procedures with the cooperation of campus administrators, we proceeded to do more groundwork on two fronts.

First, we determined to look at how we *communicate* the code and academic policies to both faculty and students to see if there were ways to address academic integrity issues before problems arise. Second, we examined *course design and development* as well as testing procedures to make cheating more difficult and less tempting.

Communication

We began by exploring how policies and procedures are communicated to both students and faculty in our program. We found that communication was vague, and made assumptions that students knew what plagiarism was and that faculty knew where to go and what to do (and what not to do) if they suspected an infraction.

To educate students, we developed an online orientation which students are required to complete —and pass—prior to beginning their course. The orientation takes a comprehensive look at what is involved in taking an online course. A substantial portion educates students on what plagiarism is—both blatant and casual—and why it is important to correctly cite a source. The orientation draws attention to the fact that our online students are subject to our *Student Code* of the University, even if they are not in residence on campus. Direct links to the UIUC *Student Code* are provided in the orientation, as well as in the syllabus and course website for every course.

While most students know that *copying* another person's work without attribution constitutes plagiarism, our experience demonstrates that there is confusion about whether it is plagiarism when a student *paraphrases* someone else's work. The GIS orientation not only explains paraphrasing and documentation, but also gives guidelines for students on how to use information from other sources legitimately. This serves to both educate and set the standards students are expected to abide by as they proceed through their academic work. It places instructor and student on the same page so ignorance will not be accepted as an excuse when questions of academic integrity arise.

To educate faculty, we developed and added to our faculty website a step-by-step flow chart drawn from our *Student Code* [Figures 1 and 2] for handling integrity issues, accompanied by a more in-depth discussion of the Code. The chart, which translates text from the code into a visual image, makes it clear to our instructors that they must first inform the student in writing of the suspicion of plagiarism, and then allow the student eight days to respond to the charge. As was clear in one of the cases outlined above, determining guilt and punishment prior to this step of allowing the student(s) to respond to the charge prevented us from disciplining students appropriately because the *Student Code* was not followed. (We also have recommended to campus administrators that language in the code be clarified on this point the next time the code is revised.)

Figure 2. UIUC GIS Academic Integrity Flowchart 2

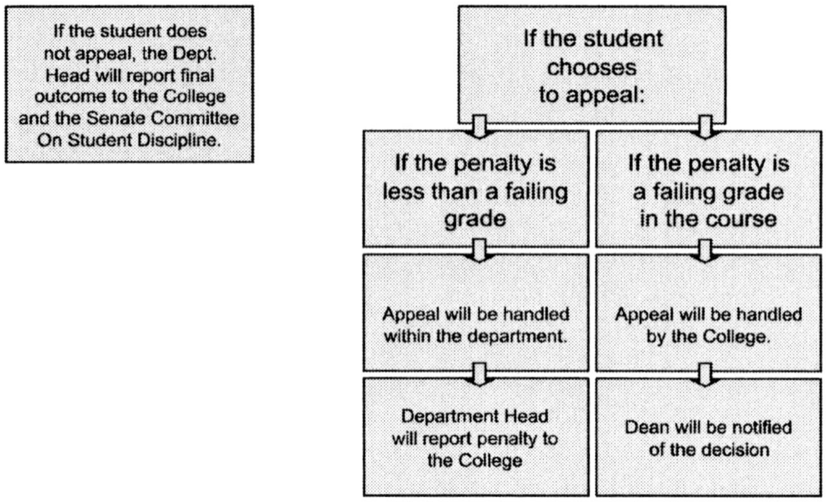

Given the unpleasant task of confronting a student with allegations of inappropriate actions, we found that breaking the process down step by step makes it less daunting. We also repeatedly remind faculty via monthly updates and reminders on the faculty website that if they have any suspicions of academic dishonesty, they should contact our office so we can assist in walking them through the process. We can also assist by providing documentation as necessary and presenting a consistent university position on the issue. Involving a GIS administrator who is detached from the immediate circumstance can also be helpful, particularly when a faculty member might be feeling angry or betrayed, and thus may not be able to judge the situation calmly and objectively.

Course Design and Development

It became evident that the way courses are designed and developed has a bearing on the temptation and ability of students to cheat. As a result, we have taken a long hard look at the way we design and teach our courses and made a number of modifications, such as increasing the number of individualized and context-sensitive assignments, and renewing our focus on active learning strategies.

One strategy is to encourage our faculty to ask questions that require application to the student's life. For example, rather than re-stating definitions of conflict resolution strategies in a Management class, students must discover (through assessments and reflection) and discuss their own conflict resolution styles and give examples of leaders they have known who have used each style. They also discuss a conflict they have experienced that ended badly, identify why it went badly, and then revisit the issue using an appropriate conflict resolution style to fit the circumstance. In their responses it is clear whether they know the basic concept and how it is applied, and they are given the opportunity to challenge the assumptions of the text. As a check against plagiarism, this kind of narrative also makes it easier for the instructor to remember if that example has been used previously.

Another strategy is to create assignments that require the student to look at a recently published journal article or recent event and tie it into course material. For example, a political science course may ask the student to apply a concept to a news event that has happened in the previous month, or a psychology course may ask the student to identify a study that has been published in the previous three months that addresses the topic at hand. This strategy is particularly important for continuous enrollment courses that are difficult to modify each semester. By tying assignments in with recent events or publications, the course stays fresh, and students realize what they are studying has application outside of the course.

Cheating can also be discouraged with the use of such technology applications as discussion boards, blogs and wikis, which allow an instructor to have students post their work online for others to critique and build upon. This introduces an element of peer pressure, as many other students will see the product, not just one hard-working professor. With the openness of the internet, students will be less likely to plagiarize, as they do not know who is reading the paper, and who may report them. Scott Warnock discusses a case in which a student plagiarizing on a message board was caught by two other students: "While digital technologies may contribute to a rise in plagiarism among students, these technologies can help teachers develop constructive, rather than punitive, course environments that discourage plagiarism." (Warnock, 2006, p. 178). The use of blogs, wikis and message boards can foster a sense of community in which students are part of the "check system against plagiarism." Indeed, Warnock attributes the willingness of students to expose the culprit to the sense that the community's mores had been violated. Not surprisingly, according to our LAS Dean of Student Academic Affairs, students who serve on the appeal committee dealing with *Student Code* infractions are often more harsh with their peers than are faculty or staff representatives.

One final strategy to reduce incidents of cheating involves proctored web-based testing. Distance learning programs that have evolved from older formats, like correspondence study, have long relied on proctored or supervised examinations. This practice does not seem to be universal in newer programs that may rely on timed testing conditions, "open book" kinds of questions and an honor code instead of proctors. While most of our GIS courses are online, tests were still given in a pen and paper format but under proctored conditions. We have now moved to web-based testing, while retaining the proctored environment as an important layer of security and authentication. Photo ID materials are presented to the proctor at test time. The proctor makes the exam available to the student on a computer in a testing room. During the exam, test-takers are visually monitored by the proctor, and students' screen activity is also closely tracked using RealVNC, a free, open-source application that provides remote desktop access and prevents students from navigating away from the exam. Web-based testing also contributes to academic integrity by making it easy to randomize questions selected from a larger test banks so no two tests are alike and the likelihood of cheating is further reduced. Other very new technologies, such as Kryterion and Securexam Remote Proctor, provide remote electronic proctoring of exams with a camera and microphone. Because this technology requires the student to purchase hardware, it may be viable for online or distance degree programs; however, in a program such as GIS, where a student may only take one course, this solution may be too cumbersome or expensive. Monitoring and record-keeping for systems like this also can be costly to the program.

Incivility Incidents

In the Fall of 2007, we had our first experience with cases of incivility. Due to the self-paced nature of our program, these cases involved interactions between students and instructors rather than among students. In all instances, student emails to instructors were first impatient, then rude and aggressive, then almost abusive. We provide the following examples to illustrate the point.

- Student emails to instructor used words and phrases like "childish," "your nasty comments," "you are a jerk."

- Students communicated to faculty in all caps: "NO I DID NOT SUBMIT A BLANK ASSIGNMENT ON PURPOSE. IS THIS A JOKE?" and "THIS LEADS ME TO THINK YOU ARE A VERY SIMPLE PERSON WHO HIDES IN YOUR IVORY TOWER."
- One student first reported a number of editorial lapses in course material, then experienced (but did not report) technical difficulties with some links in the course, and finally began SHOUTING (writing in all caps) and accusing the instructor of unfairness and unresponsiveness.
- In another case, in a course that attracted a number of students from our residential campus, we began to worry for the safety of an instructor receiving comments like those above. Fortunately, the "distance" in distance learning reduced that concern.

Context helps us see how the roots of this kind of aggressive, uncivil behavior lie in frustration. In the third of the above instances, a student who was frustrated by the number of "mistakes" she encountered in course materials was compiling them into a file of "evidence" of the instructor's shortcomings. In another case, the instructor repeatedly explained the difference between summarizing the assigned text and offering the student's own analysis and assigned mostly B+ grades to student work that was primarily summary. The student, however, did not understand this issue and began name-calling and patronizing the professor.

Response

As with academic integrity, we found it important to focus on both communication and course development (as well as instruction) to be more proactive in our dealing with uncivil behavior. Our goal was not only to reduce incivility, but also to teach students the importance of conducting themselves professionally in their communication.

Communication and Boundary-Setting

Since these cases were our first experience with uncivil behavior, we found no mention in our program materials of expectations of civil discourse or the standards of academic community. For example, we found one instructor (an emeritus professor with more than 20 years' experience teaching GIS) dealing with an incident above was not aware that our *Student Code* applied to distance students. The professor was delighted to learn that the Code did apply, since a simple reference to the section on intimidation and harassment almost magically de-escalated the issue and the student apologized.

Through these experiences, we worked with our student conflict resolution officer to identify faculty resources for dealing with uncivil behavior. These resources are included on our faculty website. In a recent website posting, we summarized our findings about how to respond to a specific instance of incivility or aggression:

1. Remind the student of the *Student Code*, and that his or her behavior may be in violation of the code.
2. Disengage from any personal comments and address only matters of course content and the student's work; tell the student your communication will respond to only those academic issues. In a case of administrative or technical difficulty, a program staff member would only respond to those issues—not to personal comments.
3. Ask the student to terminate the exchange and not make any contact for some set period of time (like 24 or 48 hours). This can be considered a reasonable request from a university official, and student compliance or non-compliance with this request would

have bearing on subsequent administrative or disciplinary actions.

The same procedure would be followed if the exchange involved distance learning program staff in an administrative or technical support role.

To assist students in this area, the GIS orientation course mentioned earlier addresses civil discourse in online communication. It emphasizes that in a college level course, professional and courteous communication is the standard. First of all, we stress that correct spelling and grammar are always expected, whether in lessons or in email communication with instructors or our office. Abbreviations commonly used in text messaging are not appropriate for professional communication, and messages in ALL CAPS to stress a point can be easily misunderstood as shouting. It is the students' responsibility to write in a manner that gets the point across professionally. In addition, we discuss "ownership" of problems, avoiding accusations, and inappropriate attribution of motives when communicating a problem with instructors or staff. We make clear, as well, that uncivil communication or harassment will not be tolerated. In short, we stress that in addition to the course curriculum, students will learn how to professionally communicate in an online environment. Of course, instructors and administrative staff should model appropriate conduct themselves.

Some of the communication difficulties noted above can be addressed by setting clear boundaries. In an environment where students are often isolated even from other students and never meet the instructor, the idea of boundary-setting may seem unnecessary. But an isolated student can become frustrated; an isolated student writing in the comfort of her home may forget how to behave in class—may not even feel she is *in* class. So the first boundary is establishing the course as a form of professional activity, as we have discussed above. Clearly stating expectations for spelling, punctuation and grammar will help, as will defining the course as a pre-professional experience where the level of discourse should be appropriate to a workplace.

But here we come to another boundary-setting difficulty. Electronic communication, in addition to removing visual and social cues, blurs the boundary between public and private communication styles. An independent learning mode, where each student communicates directly with the instructor and usually not with other students, is inherently more intimate than that of a paced online course where all students move on the same schedule and interact with each other as well as the instructor. Despite efforts to establish the norm of professionalism in communications, as we noted earlier, a student writing in the same physical space and using the same technologies with which he plays games or sends text messages to friends may forget the demeanor appropriate to academic or professional settings.

Similarly, but perhaps less obviously, the instructor writing one-to-one may become more familiar than is appropriate. Many online instructors say they get to know their online students better than those in their physical classrooms, which is usually a positive development. The instructor too, however, may be communicating about course matters with a single student from a home computer surrounded by family matters and kids' photos, and thus their correspondence may become too familiar. As our Student Conflict Resolution Officer points out, it is an awkward and stubborn reality that so long as the instructor is officially creating a permanent credential for the student, there is a power imbalance between the two (B. Farber, personal communication, October 16, 2008). The respect balance can be tilted by either the student or the instructor; as the party with the upper hand, so to speak, the instructor is obligated to clearly articulate *and maintain* the boundaries between appropriate and inappropriate conduct. Instructors shouldn't SHOUT or lose their tempers either.

Finally, and this is a matter of both ongoing communication and course development, it is crucial for instructors to be clear about the time frame in which they will respond to student work and student inquiries. Whether the time period is 24 hours or three business days is less important than being specific about just what the span will be. Instructors should also indicate when they will be away from email for a period of time. There is no need for detailed explanation--references to caring for an ill family member or going on vacation may provide too much information and invite snide comments like "if your kid's health is so important, maybe you shouldn't be teaching this course." Identifying professional absences for conferences or research, on the other hand, helps reinforce the professional setting. In any case, spelling out these expectations for response time can help reduce frustration, and set a conduct standard for the instructor as well.

Many online programs use a paced or calendar-based format that allows—even requires—students to communicate with one another through message boards, discussion groups, and other interactive tools that create opportunities for students to become uncivil or aggressive with one another. In one survey of online instructors on the topic of online bullying, between 30 and 45 percent of respondents reported experience with various forms of uncivil conduct (confrontational, lewd or vulgar, and humiliating postings) among students. Most of the respondents (86%) stated that their institutions do not have policies that define online bullying or outline how to handle it (Reigle, 2006). It may, in fact, be the case that policies addressing appropriate and inappropriate student conduct do exist at those institutions, as they do at Illinois, but that online instructors are not aware of how to apply existing conduct codes in the online environment. Asking about specific policies regarding *online* bullying assumes that online delivery requires separate rules of conduct rather than the application of existing codes of behavior. Yet even without explicit policy or administrative guidance, online instructors responding to this survey had developed ways to handle this form of misconduct (Reigle, 2006, especially unpublished raw data). Overall, these results point to the necessity of informing both students and faculty that standards expected of face-to-face interaction apply to all forms of communication. It may not be necessary to develop new policies to prevent or handle bullying online if instructors are aware that existing conduct codes can be explicitly applied in cyberspace.

Course Design and Development

Whether we are administrators, instructors, or instructional designers, we must act on the maxim that frustration leads to aggression and do everything we can to prevent such frustration. Course content should be professional in appearance and meet the highest standards of editing. If we expect students to spell and punctuate properly, we need to uphold the same standards. For this reason, we have re-vitalized the proofreading and editing steps in our course development process. This includes not only grammar and spelling, but also making sure that web links are live and up to date.

In addition, directions—for everything—should be clear, direct and easy to find. Students should understand exactly how to get technical support, and such help should be reliable. Before and during anticipated down-time and upgrades, we must use email and announcement tools to let students know what is going on (e.g. a system upgrade that will take the course site offline at a specific time), what precautions (if any) they should take, how to get the most current information, etc. And we should remind instructors to be alert to signs of frustration, so that the frustration may be addressed before it turns into aggression. An individual email to a single frustrated student may, in a paced course using discussion tools, need to be supplemented by course-wide email to all students to nip a problem in the bud.

Administrative Responses

In addition to improving communication and refining our course design and development process, we examined our administrative processes to see how they supported or undermined our attempts to ensure academic integrity and civility. In doing so, we realized that our instructor compensation model did not align with the complexities and unpredictable outcomes of working through student conduct cases. If a student withdraws or is dropped from a class due to either integrity or civility issues, our established compensation model meant that the instructor did not receive as much pay (i.e. the final installment) for dealing with a difficult student as he or she would have received had the student completed the course normally. In fact, instructors dealing with conduct issues probably deserve *more* compensation rather than less, and the problem is compounded if, as in one of our instances, two students leave the course as the issue is resolved. Thus, in instances of a student dropping the course for conduct reasons, we have adopted the practice of paying instructors the full amount they would have been paid if the student had completed the course. This is a small token of recognition of the extra time and emotional energy spent dealing with unpleasant conduct issues. We are also piloting a model to share revenue with academic departments, in recognition of the work they do to deal with student conduct and to assure that courses are well-developed and maintained.

Another administrative issue we identified was the need to determine how to handle a student from another institution taking our course. We have not yet threaded through all the complications and interconnections that obtain between sister campuses of the University of Illinois; we may need to consult legal counsel on this. But a couple of points are clear. Every student enrolled in one of our courses is governed by our UIUC *Student Code*, regardless where else the student may be enrolled. Should someone who is not a degree-seeking Urbana student be involved in a conduct incident, the GIS program would fulfill the role of the student's academic college in terms of working through the process and retaining records as outlined in our *Student Code*. As we understand it, we are not obligated to contact another institution where the student may be enrolled, but we are not prevented from doing so. In one case of incivility involving a student at a sister campus, where we had a working relationship with an advising office on that campus, we did consult with that office. We learned that the student's behavior in our course was not typical, and the adviser there helped in calming things.

How Have These Changes Worked?

Communication

First of all, our faculty have been grateful for increased guidance and support in these areas. The step-by-step, clear instructions have reduced anxiety when setting out to conduct a task which is awkward at best. In one case, a clearly unpleasant exchange with a student was quickly stopped when it was pointed out that the student was in violation of the *Student Code*. In a recent case involving suspected cheating on a test, the flow chart allowed us to communicate the required step-by-step process to the faculty member. The immediate panic was alleviated when we explained that initially the instructor only needed to communicate his or her suspicion to the student. When this was done, we moved on to the next step, and so on, until the issue was resolved.

Moreover, it is our belief that most students do not set out to plagiarize or to behave in an uncivil manner. Often, their actions are simply the result of laziness, disinterest or time pressures. Or they become frustrated—even due to non-academic issues like technical difficulties or down time–and allow their emotions to over-ride their common sense. Addressing these issues proactively allows everyone to know what the expectations are, will hopefully prevent some students from acting inappropriately, and will allow for common understandings when problems do occur.

From an administrative point of view, having standards and methods for dealing with these issues in place prior to their occurrence has led to more consistent application of the *Student Code* and reduced time and stress in dealing with these problems. No longer do we "reinvent the wheel" when a cheating or civility issue occurs. We look to our policies, and implement them, and work within positive relationships with key campus administrators when problems do arise.

Course Design and Development

We have found that the re-design of assignments not only reduces the temptation to copy others' lessons, it also increases the degree of active learning and higher order thinking in our courses. Students may have to work a little harder, but the end result is a better educated student. Also, from the perspective of faculty, these papers are far more interesting to read than those that only ask students to repeat information. At least one faculty member, who teaches a management course, reports she learns much from her students as they apply theoretical concepts from the class to different situations she has not experienced. Learning from her students how management principles work in different organizations such as the military, a community center, or a Fortune 500 company has broadened her understanding of organizations, and has enabled her to bring that knowledge back to her own research.

The move to web-based testing has been a success by all accounts. Not only does it decrease time delay for feedback and test results, which students appreciate, but it has reduced the worry that students can walk out of the test, write down the questions, and then pass them on to their peers.

The long-term impact of these changes in civility cases has yet to be determined. In the short-term, we seem not to have had a recurrence of incivility, but we have not, as of this writing, been through the high-stress end-of-year April through June period since these changes have been implemented.

CURRENT CHALLENGES

We are pleased with the improvements we have made in both communication and course design. Our faculty report that they value the information we have provided. Nevertheless, challenges remain.

- Typically, program management staff (including instructional designers) have opportunities to intervene in the course development phase with suggestions that can forestall cheating, frustration, and aggression. In the actual teaching phase of an online course, on the other hand, there are fewer chances for program staff to intervene. While we believe a more engaged approach is essential to good teaching and learning online, and can forestall cheating, we also need to remind our instructors of their role in maintaining a professional and civil atmosphere in the virtual classroom. We can do our best to orient instructors to the complications that can arise, and we must also do our best to prevent frustrations with technology or procedures that can distract both students and instructors from pursuing their academic work in the most professional way.
- A challenge we have faced with web-based testing in our program is the lack of computers with internet access in all proctored locations. Some pen and paper tests are still given in these locations. We anticipate, however, that this problem will diminish with time, especially with the growth of web-based testing for such standardized exams as the GRE and MCAT.

- Communication is always a challenge. No matter how many times you communicate something, busy faculty and students may forget. Thus as administrators, we must continually reinforce our expectations in terms of both academic integrity and civility. For our faculty, we send a monthly update to remind them of faculty website resources available if they suspect plagiarism or have a problem student. With students, we are upfront with expectations through the Orientation and course website. We provide multiple links to the *Student Code*, but again, the message needs to be reinforced continually. It may be that only after the line has been crossed will students understand the importance of academic integrity and civility.
- Many academic units on our campus are making significant inroads into online learning, but the reasons are related more to resources rather than to pedagogical issues. Thus, there might be a temptation among some faculty to treat online course development as a way to stretch resources rather than as an opportunity to address effective teaching practices, whether online or in the classroom. We must be sure to work with all stakeholders from the beginning to ensure that course design encourages academic integrity and that students are aware of their responsibility to work and communicate with respect in the online environment.

Recommendations

We have learned many lessons as we have addressed issues of academic integrity and incivility. While different institutional contexts require different solutions to these problems, we make the following recommendations to those facing similar challenges in their distance education programs:

- We must change the way teachers teach. This applies not just to issues of student conduct, but to the challenges of moving classroom based material online. In a lecture hall setting, much of what is expected of the student is passive absorption of material. The expert speaks, the student listens and learns. Because of sheer numbers of students, assessment is often done in a multiple choice format that can be easily scored electronically. Whether online or in the classroom, this is an invitation to cheat. An emphasis on active learning in which the student must seek out information, apply it and assess it is an important part of our online course development process. This may present a challenge to a faculty member who has grown accustomed to large lecture hall courses, and may involve political battles as well.
- In addition, while assignments that require active learning and higher order thinking may be more interesting to grade, they are more time-consuming than multiple choice/ TF/ short answer tests. This might require increased compensation or a reduced number of students. These recommendations may hit resistance, but we have found faculty on campus who are eager to approach their instruction in a new fashion.
- Distance learning practitioners should become deeply familiar with the conduct codes at their institutions and the processes by which those codes are implemented. There should be consistency between the instructor's policies and practices and those of the institution (and program). If the policy on academic honesty is not explicit and its enforcement not clearly assigned, distance learning administrators should work for clarification.
- Many institutions are grappling with the question of how to evaluate the many technologies now available that seek to verify

student identity. This is particularly true in light of the demands of HR 4137. While we have been quite successful using proctored web-based testing in our program, other institutions may choose to look at technologies that allow a student to take an exam without having to go to a proctor's location. For example, Securexam's remote proctor device uses biometrics (fingerprint and photo ID) to authenticate the identity of a test-taker and audio and video capture that monitors activity in the room to make the instructor aware if there is any potential for cheating. Another product, Acxiom, maintains a database of consumer public records, then asks the student questions based on information in the database that pertains to that student. We encourage further research and discussion on this topic as we grapple with issues such as student expense for technology and privacy issues.

- Another technology that seeks to ensure academic integrity is plagiarism detection software. These programs compare a student's work with both published material and with papers previously submitted by other students, looking for commonalities that might indicate plagiarism. To some, the biggest drawback to using this software, besides the cost, involves copyright issues. If the software programs claim copyright to the student's work once it is submitted, students may rightfully protest giving ownership of their own intellectual property to an organization that may use it for their own profit.

Some faculty find that "Googling" suspicious phrases from student's work is just as effective. One faculty member told of reading a paper that included words that he doubted a particular student knew. He "Googled" a sentence from the paper and quickly found that sentence in a published article. This, he insisted, was much cheaper and much quicker than any software on the market!

Whether we agree with the choice to use plagiarism detection software, one study did find that the very threat of its use was effective in curtailing plagiarism. Braumoeller and Gaines (2001) conducted a study in their Introduction to Political Science course in which they compared the plagiarism rate between classes that were warned orally and in writing not to plagiarize with classes that were told their work would be submitted through a plagiarism-detection application. They found that "warning students not to plagiarize, even in the strongest of terms, appears not to have had any effect whatsoever" on whether or not the students chose to plagiarize. When students knew their work would be examined by the software, however, they found that students were deterred from submitting plagiarized assignments.

- We recommend that if institutions have not yet integrated proctored web-based testing into their programs, they pursue it. This well-established practice can go a long way to meet the demands of HR 4137. To get started, a program that does not use proctors at all should consult with those who do. It is not necessary to pay proctors and, once the program is in place, it is an effective way to undergird the integrity of testing in distance learning situation.
- We also recommend further research into issues of academic integrity and student conduct in online courses, as noted above. The effectiveness of web-based testing, plagiarism detection, even the efficacy of traditional proctoring programs all need to be scrutinized. Our case-based approach to civility issues is at best anecdotal; perhaps there are opportunities to partner with campus student judicial officers to investigate civility issues through more structured research. Furthermore, it is worth finding out whether civility issues are more or less

common in virtual classrooms, and whether it is in fact online interaction itself that leads to less civil behavior in face-to-face settings.

- As on-ground courses incorporate web-based technologies, whether in the creation of blended courses and programs or as supplements to traditional instruction, campus administrators will encounter the kinds of issues we in distance learning are already experiencing. It is now past time to review every student code, faculty handbook, and administrative procedures manual for the way words like "campus" and "classroom" are used—does academic freedom exist only in the physical classroom, to name just one obvious instance? Do references to "documents" include electronic formats? Some institutions have already done this kind of review; those who have not should do so as soon as possible. In addition to bringing the text of the code into the twenty-first century, such a review will identify the exact sections of the code that apply to student conduct in cyberspace as well as those that may be problematic. Administrators of distance learning programs that include non-credit activities should determine whether their student conduct policies and procedures apply to non-credit enrollees.

CONCLUSION

It has long been the norm in distance education, even before the existence of the internet, for the rigor and academic standards of a distance-delivered course to be equivalent to those of traditional classes. We maintain that those academic standards include the expectations for appropriate conduct outlined in every student code, and that those student codes should be applied to all instances of misconduct in online classes.

We strongly encourage all distance learning administrators to work proactively on student conduct issues at their institutions. Learn the policies and procedures before a case occurs; identify and meet relevant campus administrators. It is important to the profession as a whole to improve the image of academic integrity in online learning. On the local level, our work with campus administrators charged with managing conduct issues has improved our relationships with those offices and improved the image of our program. At the academic department level, we have found our working relationships strengthened by not only our support of instructors as they work through conduct issues but also by our willingness to abide by and enforce our campus code of conduct.

Finally, much of the discourse about online technologies in education has shifted recently, away from a commitment to access for non-traditional students that underlies distance learning to a focus on using technology to connect with "net gen" students who may be more tech-savvy than faculty and administrators of an earlier generation. As more and more institutions move toward online content delivery for all students, it will become increasingly important for campuses as a whole to address student conduct issues in the context of online learning. In addition, we must design our courses and programs so students understand they are part of an institutional community, and that even as online participants, there are standards of conduct they are required to adhere to. It is our hope that this case study will begin a discussion on campuses about the best way to adapt or supplement current codes of conduct to improve the learning experience and enhance the effectiveness and reputation of online learning. For more than a century, a long series of new technologies have been used to expand access to education. As distance education professionals, we should focus on technology for the way it extends access and should not expect technology to rewrite the rules of conduct.

ACKNOWLEDGMENT

The authors gratefully recognize the wisdom, patience and generosity of our Illinois colleagues Brian Farber, Associate Dean of Students and Director of Student Conflict Resolution, and Mary MacManus Ramsbottom, Associate Dean of Student Academic Affairs in the College of Liberal Arts and Sciences, as they assisted us through the incidents reported here and contributed to our program improvements and the thinking behind this article. We also would like to recognize the several GIS instructors who allowed us to disguise their experiences for presentation in this case study. We would like to, but we cannot, without compromising the identity of others involved. They know who they are, and we want them to know we are grateful for their efforts.

REFERENCES

Braumoeller, B. F., & Gaines, B. J. (2001). Actions *do* speak louder than words: Deterring plagiarism with the use of plagiarism-detection software. *PS: Political Science and Politics*, *34*(4), 835–839. doi:10.1017/S1049096501000786

Grijalva, T. C., Nowell, C., & Kirkvliet, J. (2006). Academic honesty and online courses. *College Student Journal*, *40*(1), 180–185.

Higher Education Reauthorization Act. (HR 4137, Title IV, Part H, Sec. 495). (n.d.). Retrieved on November 17, 2008, from http://www.govtrack.us/congress/bill.xpd?bill=h110-4137

King, C. G., Guyette, R. W., & Piotrowski, C. (2009). Online exams and cheating: An empirical analysis of business students' views. *The Journal of Educators Online*, *6*(1), 1–11.

Lanier, M. M. (2006). Academic integrity and distance learning. *Journal of Criminal Justice Education*, *17*(2), 244–261. doi:10.1080/10511250600866166

McNabb, L. (2008). Chronicle article focuses on distance-ed academic integrity. Retrieved on October 18, 2008, from Listserv: Sloan-C.

Miller, N. E. (1941). The frustration-aggression hypothesis. *Psychological Review*, *48*, 337–342. doi:10.1037/h0055861

Reigle, R. R. (2007). *The online bully in higher education.* Educational Resources Information Center (ERIC). Retrieved on April 10, 2009, from http://www.eric.ed.gov/ERICWebportal/detail?accno=ED495686

Warnock, S. (2006). Awesome job!–or was it? The 'many eyes' of asynchronous writing environments and the implications of plagiarism. *Plagiary*, *1*, 178–190.

Chapter 7
Alignment of Course Objectives and Assessment Items:
A Case Study

Kenneth Lightfoot
Thomas Edison State College, USA

David Schwager
Thomas Edison State College, USA

ABSTRACT

This case study examines how achieving close alignment between course objectives and course assessments should be an essential goal in the course design/revision process regardless of what mode of delivery is involved. By examining the revision of two courses (Western Civilization I & II) offered at Thomas Edison State College, the authors demonstrate how the application of sound instructional design principles to achieve this alignment resulted in the measurable improvement of student learning outcomes. The major issue examined in this study is how to achieve a close correlation between what a course states that a student should be able to do after successful completion, what practice it offers them to achieve this proficiency, and how they are assessed to determine what level of proficiency they have attained.

BACKGROUND

The advent of the computer has led to a sustained revolution in education in the United States, giving educational institutions the ability to reach a student body literally around the world. This online revolution has enabled large numbers of people who for various reasons could not otherwise pursue their education in a traditional setting to participate in an ever increasing range of educational programs. This steady increase in the number of people now able to receive an education regardless of their location has in turn led to a large-scale effort on the part of educational institutions in the United States to expand their curricula to meet the needs and demands of an increasingly diverse student population. The result has been a tremendous proliferation of online courses at every level of education.

While the computer has opened up a whole new horizon in education for students, the pressure to produce courses and programs placed on those responsible for actual production has often led

DOI: 10.4018/978-1-60566-870-3.ch007

to quality control issues that are unfortunately common throughout the community of education providers. Two of the principle factors affecting quality control in many course production units are limited numbers of staff that can be assigned to a project coupled with tight production schedules driven by demand for both course development and revision. All too often, in addition to content experts, course development teams involve only two or three members of staff who are responsible for all aspects of development and maintenance. Since course authors or content experts are seldom versed in sound instructional design principles and practices it falls upon members of a development team to make sure that these principles and practices are incorporated into the course design and revision process.

Of the myriad issues involved in the production and maintenance of a good-quality course, this case study seeks to examine one in particular that is within the control of course developers: the alignment of the course objectives and assessment items.

Good quality courses don't just happen; they are the result of planning and an involved process of selecting appropriate material for a target audience balanced with devising assignments and assessment items that will give that audience ample scope to master the chosen material and demonstrate that mastery.

One of the key elements in this relationship is the alignment between what a course states that students should be able to do after successfully completing it, the learning exercises it features to prepare them to achieve this level of competency, and the assessment items it uses to test this competency. In most courses this can be expressed as the relationship between course (and if applicable, modular) objectives and assessment items, such as projects, written assignments, term papers, examinations, etc. When there is a clear correlation between what is expected of a student, the exercises they are asked to do in order build towards meeting those expectations, and what they are asked to do to demonstrate proficiency, then a course can be said to be well aligned. If on the other hand, students are asked to demonstrate a proficiency for which they were not sufficiently prepared or at levels different than what they have been prepared for, then a course can be said to be poorly aligned.

The recent complete revision of Western Civilization I and II at Thomas Edison State College provides a case study of how course/modular objectives and assessment items were closely aligned with one another in an effort to enhance students' learning experiences and maximize their learning outcomes. This case study strives to demonstrate the importance of the alignment between course/modular objectives and assessment items as part of the course development and revision process. Student performance in these two courses will be analyzed to gauge how successful this effort has been.

SETTING THE STAGE

Western Civilization I and II have long been a part of the Thomas Edison State College's curriculum. (These courses are available for students to take as either online courses or in the Guided Study program which is print-based. Since the online version is organized into modules and therefore contains modular objectives not featured in the Guided Study version, this case study focuses on the former.) Each course is taken over a twelve-week semester and is organized into six modules. Each course requires students to complete six written assignments, one for each module, and take a midterm and final examination. In addition, students in each course must also take part in three graded discussions as part of their assigned work.

For several years students at Thomas Edison State College used a single textbook containing thirty chapters covering all the material presented in both courses. While the span of the material in the textbook ranged from the earliest civilizations to the end of the Cold War, the organization of the respective courses was dictated by Study Guides that were part of a leased course package that also included video programs.

The parameters set by the Study Guides presented a major problem in course organization; an inequitable distribution of work between the courses. Whereas the Western Civilization I Study Guide covered the advancement of European civilization from the earliest times to the Renaissance, which required students to view thirteen hours of video programs and read ten chapters in the textbook, the Western Civilization II Study Guide assigned the same amount of viewing but a full twenty chapters in the textbook, covering the Reformation to the end of the Cold War. Since most of the chapters in the textbook were of relatively equal length this meant that students progressing on from Western Civilization I were required to do twice as much reading in Western Civilization II. Not surprisingly, retention rates were traditionally low.

Another problem was the quality of assessment in the two courses, particularly the examinations, which generated a number of student complaints over time. Exam questions sometimes focused on relatively trivial items such as specific individuals or events while inadvertently ignoring more major trends and developments of which these people and events were part. This focus on more mundane information was in part facilitated by the nature of the exams themselves. While the exams contained some short and long essay questions that gave students an opportunity to demonstrate their knowledge of broad historical topics, they contained a larger proportion of true-false, matching, and fill-in-the-blank types of questions which generally concentrate on isolated facts rather than more complex issues, such as the examples below, which are taken from an actual exam:

1. In 1517 Martin Luther nailed his 95 theses to the door of the cathedral church in the German town of ________________.
2. ______________was the first president of the Czechoslovakian Republic.

While these types of questions certainly have valid applications, given the expansive nature of the material presented in Western Civilization I and II coupled with the breadth of the time span covered, the consensus of opinion was that they focused too much on small details at the expense of ignoring larger more important historical developments. To make matters worse, none of the six written assignments in either of the courses contained any of these types of questions; rather they consisted of a series of both short and long essay questions. Student complaints that their course work did not prepare them for the exams were, therefore, justified.

Furthermore, a large proportion of examination questions were not aligned with the course objectives. For example, the two examination questions listed above were taken from the older version of Western Civilization II, which had the following course objectives:

After completing **Western Civilization II**, the second semester of a two-semester course, you should be able to understand and describe:

- The economic background of the Reformation and the centralization of European states.
- The similarities and differences between the French *Fronde* and the English Civil War.
- The patterns and effects of economic, social, and political growth in the eighteenth century.
- The role of the American Revolution in international politics.

Table 1. Hierarchy of cognitive skill levels according to Bloom's taxonomy

1. Knowledge	Remembering previously learned material.
2. Comprehension	Understanding the meaning of remembered material.
3. Application	Ability to use learned material in a new way to solve problems, perform new task, etc.
4. Analysis	Ability to break down material into its component parts and explain the relationship between those parts.
5. Synthesis	Ability to put the parts together to form a new whole.
6. Evaluation	Ability to judge the value of material by using a defined set of criteria.

- The most important changes in warfare during the Revolutionary and Napoleonic Wars.
- The outlines and social conclusions that were drawn from Darwin's work.
- The causes of World War I and World War II.
- The changes in daily life stemming from medical and scientific discoveries.

It will be noticed that there is no correlation between the first objective, the only one relating to the Reformation, and the exam question asking where Luther presented his 95 theses, generally taken by historians to mark the beginning of that movement. (The answer to that question, Wittenberg, is a very minor detail in the old textbook's overall treatment of the Reformation.) As far as the second exam question is concerned, there is no course objective that even remotely relates to it.

These examples are representative of the type of misalignment issues that diminish the overall quality of a course. Although attempts were made over time to improve the alignment issues with the Western Civilization courses at Thomas Edison State College it was recognized that a major revision was necessary.

In 2007 a major revision of the two Western Civilization courses was commissioned. The revisions were carried out by a team of two; an instructional designer from the College's Department of Instructional Design and Development and a content expert – in this case a Ph.D. in history.

After reviewing the textbook that had been used for several years it was recommended by the content expert to completely rebuild the courses using a new set of textbooks that were selected because of their superior organization and focus on major historical themes without dwelling on minor details (Palmer 2007). One of the principal aims of this redevelopment program was to create assessment items based on the major historical issues emphasized in the textbooks and ensure a good correlation between them and the course/modular objectives, thereby greatly improving the overall quality of the courses.

CASE DESCRIPTION

The first step in the revisions was to determine what cognitive skill levels should be assessed and then follow on from there to create appropriate objectives and assessment items in conjunction with one another. The lists in Tables 1 and 2 show the hierarchy of cognitive skills as ranked in Bloom's taxonomy and provide useful tools that can be used when designing appropriate measurable activities to be included in a course (Lightfoot, 2006).

Although Western Civilization I and II are 100 level courses which in traditional college settings would normally call for the assessment of cognitive skills only at the lower levels of Bloom's taxonomy, the unique nature of Thomas Edison State College's student body warranted some degree of higher level assessment as well. The

Table 2. Measurable activities associated with the various cognitive skill levels in Bloom's taxonomy

1. Knowledge	Identify, list, name, show, define
2. Comprehension	Explain, compare, summarize,
3. Application	Apply, solve, interpret, modify
4. Analysis	Analyze, organize, plan, devise
5. Synthesis	Justify, design, support, theorize
6. Evaluation	Evaluate, defend, criticize, estimate

college's students, as in most distance education programs, are generally older than students in a traditional college setting and therefore have more life experience to bring to their classes. Accordingly, part of the assessment featured in the two Western Civilization courses needed to provide some scope for students to work at the higher cognitive skills level of Bloom's taxonomy. However, since no prior knowledge of history is required or assumed for prospective students of Western Civilization I and II a careful balance between the levels of assessment items had to be achieved whereby the introductory nature of the course materials would be respected while the student level of maturity could also be accommodated.

In order to arrive at a workable solution to these requirements it was first necessary to determine which methods of assessment would be most appropriate to achieve these aims. Given the nature of the textbooks with their focus on themes and problems of continuing historical importance as well as on specific timeframes while avoiding too much detail, it was felt that the most appropriate method of assessment to give students practice in mastering the course materials at the knowledge and comprehension cognitive skill levels should be in a form where they could draw on the information in their reading assignments while bringing their own judgment to bear. It was decided therefore, to retain the model used in the earlier versions of the courses with a combination of short answer questions (150-200 words each) which ask students to briefly summarize, explain or discuss an issue, etc., and essay questions (300-400 words each) which ask students to discuss and explain topics, events, etc. in greater length and detail.

Certain refinements to this model were introduced during the revisions. Each assignment now features a number of these types of questions which between them aim at covering all the main topics presented in the assigned readings. Additionally, students are given a certain number of each type of question to answer and submit for grading, which means they have a choice, but they are cautioned that they should be able to answer any of them since the examinations may contain questions based on any or all of those featured in their assignments.

One of the strategies employed in designing the new written assignments was to create as much correlation as possible between essay questions that deal with broad issues and short answer questions that focus on more specific themes. Although the short answer questions in many cases relate in some way to the broader issues covered in the essay questions, they focus on distinct topics that can be commented on separately in their own right. This was done deliberately to achieve a more comprehensive coverage of the major topics and themes presented in the courses by offering a range of questions with some isolating specific important topics together with others dealing with a combination of those topics and thus requiring a more complex answer. To avoid redundancy and too narrow a focus however, the questions featured in the written assignments were designed in such a way that it would not be appropriate for

Table 3. Relationship between short answer and essay questions in written assignments dealing with correlated themes

Short Answer Questions:
• What was the idea of 'balance of power' as practiced in Europe in the seventeenth and eighteenth centuries?
• Explain the doctrine of "absolutism", or "absolute monarchy", which affirmed the "divine right of kings."

students simply to incorporate into an essay question answer verbatim or paraphrased answers to short answer questions that deal with correlated topics or issues.

The example below in Table 3, taken from a written assignment in Western Civilization I, illustrates the type of relationship between short answer and essay questions described above. Among the short answer questions in this particular assignment is one that asks students to explain the concept of the balance of power in European politics during the period 1600-1800, and another asking them to explain the concepts of "absolute monarchy" and the "divine right of kings." The textbook discusses all of these topics separately, first in a general sense, and later collectively in the context of the reign of Louis XIV of France, for which it provides in-depth coverage. The three questions can therefore, be answered separately with the essay question dealing with King Louis designed to elicit a comprehensive answer not only requiring students to know the meaning of the balance of power and absolute monarchy-divine right of kings, but also to express this knowledge in a different way than in answering either of the short answer questions while also drawing on a mastery of additional material; i.e. political, economic, and religious, developments in France as well as the reactions of other European states to the rise of Louis XIV. Since students have a choice of which short answer and essay questions to complete and submit, they do not have to choose this particular trilogy of written assignment questions, but by choosing at least one of them they will have some practice in mastering material that relates to the others.

Having based the written assignments in Western Civilization I and II as often as possible around what can be loosely described as a 'hierarchy of questions' devised to test a student's mastery of the course material at the knowledge and comprehension cognitive skill levels of Bloom's taxonomy, the next challenge was to include assessments at the higher levels that enabled them to draw on maturity gained through life experience. The most effective way to do this was to include discussion questions designed to give them ample scope for analysis, synthesis, and evaluation while expressing their own opinions. An example of this type of question is shown below in Table 4. It is taken from the same module in Western Civilization I as the written assignment questions shown above in Table 3. In order to give an adequate answer to this discussion question students not only have to know something about the topics covered in the assignment questions above, they must also be able to analyze the achievements of Louis XIV through his authoritarian style of government in contrast to those of contemporary more representational styles of government; topics that are also discussed in the readings for this module. In that way this question completes the 'hierarchy' of questions featured in this particular module.

The next step in the instructional design process used in the revisions of Western Civilization I and II was to ensure good alignment between the written assignments and discussion questions, the examinations, and the course/modular objectives. This was achieved by closely aligning written assignment questions with the readings from the textbook using keywords and stock phrases in order to avoid ambiguity, and in turn basing

Table 4. Discussion question completing the 'hierarchy of questions

Discussion Question
Although the idea of an "absolute" ruler is not acceptable in modern Western society, what, in your opinion, was Louis XIV able to achieve by having close to absolute power that he could not have achieved through representational government?

Table 5. Objective-assignment alignment; Western Civilization I

Course Objective:
After completing **Western Civilization I,** you should be able to discuss: The reasons for the Reformation and the fundamental religious, social, and political changes brought about by this movement and the resulting 'Wars of Religion.'

the examination questions on written assignment questions. Course and modular objectives were then based on the written assignment/exam question(s) alignment so that there would be a close recognizable correlation between all three elements in each course. An example of this type of alignment, focusing on one major historical topic (the Reformation) isolated from others presented in Western Civilization I is given below in Table 5. (For reasons of security, corresponding examination questions are not listed.)

The example of course/modular objective(s), written assignment/exam question(s) alignment given above is typical of the approach used in redeveloping the two Western Civilization courses and is one that can be used in the development or revision of any course at any level. A useful tool in the form of a chart that can be used to organize this kind of relationship is given below in Table 6, which features the course assessments for Western Civilization I given in Tables 3 and 4 as they relate to the relevant course and modular objectives:

This chart can be repeated to accommodate as many course objectives and associated modular objectives and assessments as necessary when developing or revising a course. Conversely, it can be adapted to have a different focus such as the chart below in Table 7 which emphasizes an assessment item as the main element with the course/modular objective(s), exam/final projcct item(s) alignment relationship subordinated to it.

Table 6. Objective-assessment alignment organization chart

Course Objective
After completing **Western Civilization I,** you should be able to discuss: The struggle between kings who believed their power was absolute and granted by God, and representative bodies, such as parliament.

Table 7. Adapted alignment organization chart

Coursework Assessment Item **Written assignment question(s), quiz questions(s), discussion question(s), problem(s) etc. relating to course objective /modular objective(s) and exam/capstone project item(s).**
1.

Table 8. Breakdown of assessments and the grading weights assigned to each

Assessment Type	Percentage of total grade		Individual percentage contributing to total grade
Written Assignments (6)	40%	50%	6.67% each assignment
Online Discussions (3)	10%		3.33% each discussion
Midterm Examination	30%	50%	30%
Final Examination	20%		20%

How successful were the revisions of Western Civilization I and II at improving the overall quality of these courses? While it may be possible to demonstrate on paper a close alignment between carefully crafted course/modular objectives and assessment items that sound reasonable the only way to answer this question from a student-centric point of view is to determine whether there is any noticeable improvement in overall student performance. In order to gauge how student performance was affected by the revisions described above student grades for both the old and revised versions of the courses were analyzed and compared. The comparisons and results are outlined in the following two sections. In the first of these the nature and type of assessments are discussed in detail, followed by the results of the comparison of student performance.

THE ASSESSMENTS

The revised Western Civilization I and II courses each contain eleven assessments of varying weights that are required throughout the duration of the twelve-week course. These assessments are six written assignments, three discussion topics, a midterm examination and a final examination. A written assignment is completed at the end of each course module, which covers one or more chapters in the textbook. The discussion topics require students to comment about specific questions and then to respond to their peers' comments. The midterm examination covers modules 1 through 3 and the final examination is based on modules 4 through 6. A student's overall course grade is based on their performance on the course assessments, weighted as shown in Table 8.

The written assignments allow students to choose which questions they wish to submit for grading (any 3 short answer questions and any 2 essay questions). However, the course materials specifically state that students should be able to answer any of the assignment questions since there may be similar questions on the examinations. The assignment questions are designed to focus on the most important issues and themes from the recently-completed module and all are tied directly to the course objectives. Grade variation is relatively narrow and generally students perform very well on these assignments (usually scores range between 80% and 100%). There are two main factors that contribute to this grading phenomenon: (1) students can choose which questions to submit for grading and (2) they may utilize all of their course materials to fully answer the questions. Failure to complete a written assignment results in a grade of 0%.

Students must participate in three online discussions during the semester. These discussions are in the form of asynchronous threaded discussions, not in a chat room style. Each student must write an answer to the posted question and then, following all other student answers, must also contribute substantive comments in response to their peers' postings. Postings of "I agree" or "I disagree" must also include facts and arguments from the textbook or life experience. Due to the fact that Thomas Edison State College's students are adults, real world experiences often supplement the textbook teachings in these discussions. Grading of the discussion questions generally falls into one of two extremes; namely, the student participated fully (100%) or did not participate (0%). Some mentors may include scores of 50% to indicate minimal participation.

Both the midterm examination and the final examination are constructed similarly with 4 short answer questions and 3 essay questions. Each exam is worth 100 points and together they comprise 50% of the student's course grade. As discussed above, the exam questions are closely related to the written assignment and online discussion questions. The examination questions may be synthesized from two or more previously asked questions or they may ask students to form a connection between questions from the earlier assessments.

If students were diligent in attempting to answer all of the assignment questions (instead of just those they chose to hand in for grading), and if they contributed in a meaningful way to the threaded discussions, they should be well-prepared for each of the examinations. Part of our analysis of the students' scores will focus on this area to see if we do find consistent scores between the written/online assessments and the examination scores.

THE RESULTS

At Thomas Edison State College, a new semester begins with each new month and the semester is twelve weeks in length. The revised versions of Western Civilization I and II were initially offered in the February 2008 semester. This case study includes complete student results from post-revision semesters February 2008 through October 2008 semester. As a comparison, the pre-revision semesters include those from November 2007 through January 2008. A total N of 107 post-revision students are divided into HIS-101 (N=73) and HIS-102 (N=34). The total N of 62 pre-revision students are divided into HIS-101 (N=37) and HIS-102 (N=25).

Analyses were performed on semesters previous to the revision to set up baseline data which was then compared to data analysis for the post-revision semesters. As the stated goal of these revisions was to better align all of the assessments with the course and modular objectives, our analysis focused on two areas on interest: (1) results related to the exam scores, (a) comparing pre-revision and post-revision performance (mean scores) and (b) comparing the relationship between the exams (midterm vs. final in pre-revision and post-revision), and (2) between the "low stakes" assessments (written assignments and discussions) vs. "higher stakes" assessments (midterm and final exams). As noted earlier in this case study, there were significant issues with the alignment of the course objectives and the assessment questions, especially within the examinations, or the "higher stakes" assessments. We hypothesized that with the improved alignment between the objectives and assessment questions, we would also see an improved relationship between the "lower stakes" and the "higher stakes" assessments. In addition, we hoped to find a better relationship between the examinations (midterm v. final exam), but we were unsure of how the increased amount of writing would impact the mean scores of the examinations.

The first step in our analysis was to put all of the scores onto the same scale that is used for grading in the courses. The assignments, or "low stakes" assessments combine to make up 50% of the course grade. However, there are 9 total assignments that go into this 50%. Written assignments and discussions are weighted differently, with each written assignment counting for 6.67% of the total course grade and each discussion assignment counting for 3.33% of the total course grade. The examinations combine for the other 50% of the course grade. The midterm examination accounts for 30% of the total course grade, while the final examination accounts for the remaining 20%.

Figure 1. Scatter plot of weighted scores (pre-revision, ALL)

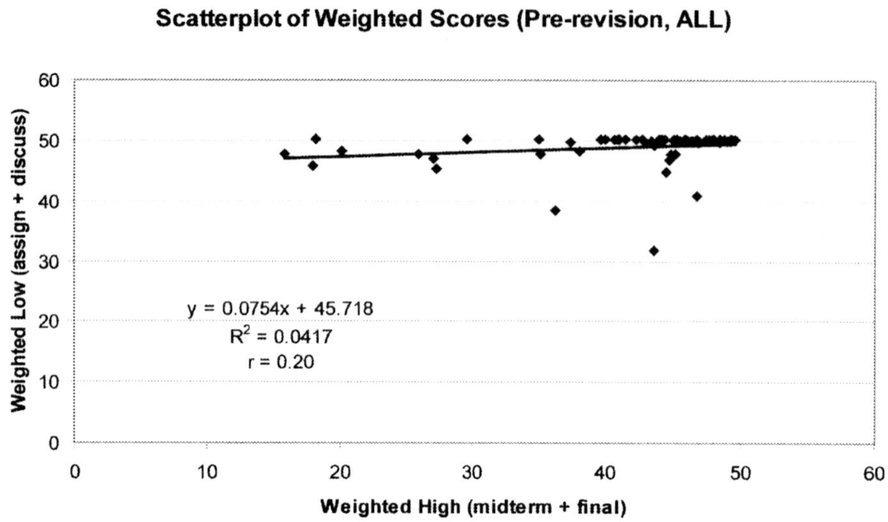

Correlation coefficients for midterm vs. final scores across all sections during the three semesters prior to the revision was moderately low at $r = 0.37$ ($n = 62$). The post-revision data analysis has so far produced a similar, although slightly higher correlation coefficient of $r = 0.42$ ($n = 108$). In analysis of the mean scores between the exams, pre- and post-revision, the overall mean of the midterm scores remained very similar, while the final scores have increased slightly in the post-revision cohort.

Highlighting the need for the revision was the correlation between the "low stakes" and "high stakes" assessments: the correlation coefficient for all students in the pre-revision group (HIS-101 and HIS-102) was moderately small; r = 0.20 (see Figure 1). These results do show a lack of consistency between scores on the written assignments and discussion assignments when compared to the results of the examinations. There are some data points on the plot that show high scores on the assignments and very low scores on the examinations, indicative of a weak relationship. As noted previously, there were student concerns about the relationship between assignments and examinations. If students were expecting that the assignments would help them in preparation for the examinations, the data are evidence that the assignments, while scored highly, did not adequately prepare them for the exams. Figure 1 also shows evidence of a restriction in range across the "low stakes" scores, as most data points are located near the top of the graph. As we discussed earlier, most students received the full 100% on all discussion assignments and the vast majority of students also score very highly on the written assignments. The correlation coefficient, therefore must be viewed in this context, as a restriction in the variability of one variable (in this case, the "low stakes" assessments) will lower the overall correlation. Figure 1 can also be used to illustrate the change from the pre-revision relationship to the post-revision relationship, as we shall see next.

This relationship improved to a moderately large correlation of r = 0.63 (see Figure 2 below) after the revisions were made to align the objectives and assessment questions. The restriction of range in the variability of the "low stakes" assessments is reduced as there is a greater spreading of the scores on the "low stakes" assessments. This increase in score variability produces a much clearer trend

Figure 2. Scatter plot of weighted score (post-revision, ALL)

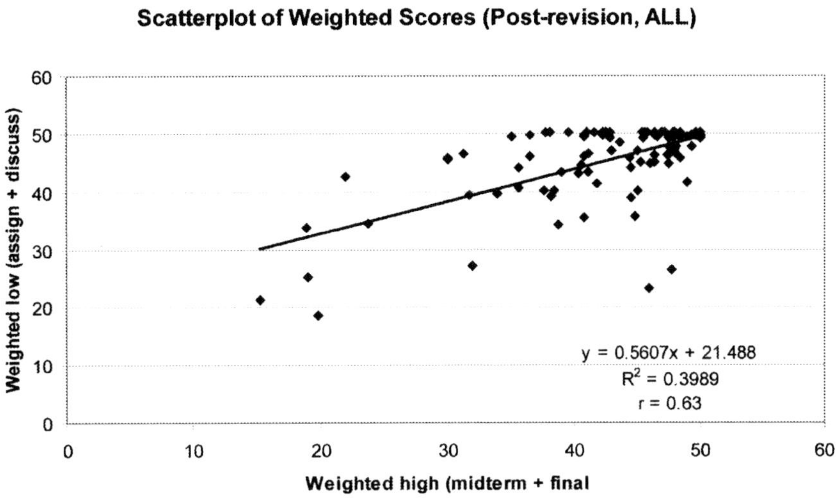

in the data, as low scorers on the "low stakes" assessments are, generally, the low scorers on the "high stakes" assessments. As can be seen, there is still a clustering of data points near the upper-right corner of the chart, as the discussion assignment grading policy of "all-or-nothing" is still grouping scores on the assignments toward the higher end, but there has been a spreading of the written assignment scores, especially toward the lower end of the scale. As seen in Table 9, the overall performance on the examinations has remained where it was prior to the revision of the Western Civilization I and II courses, evidence that the students are mastering the content at a similar level. When that information is combined with what has been seen in Figures 1 and 2, there is evidence of an improved relationship between the "low stakes" assessments and the "high stakes" assessments.

With nine semesters completed, it does seem that the revision plan that was utilized for the HIS-101 and HIS-102 Western Civilization I and II courses at Thomas Edison State College may have indeed created better alignment within the course assessments. The content specifications and the assessment plans were prepared with the goal in mind to produce in each case a high-quality course that was fully aligned from course objectives to assessment items. The data has shown that there has been an improvement

Table 9. Descriptive statistics for HIS-101 and HIS-102 examinations (pre-revision and post-revision comparison)

	Mean	**SD**
Midterm		
Pre-revision	86.57	19.97
Post-revision	85.25	19.04
Final		
Pre-revision	83.49	12.27
Post-revision	89.32	13.40

in the relationship between the assignment scores and the examination scores.

CURRENT CHALLENGES FACING THE ORGANIZATION

The main challenge facing Thomas Edison State College regarding the type of operations described in this case study is ultimately one of instituting quality control measures into the course development and revision process. A major issue affecting the outcome of this process involves striking a balance between meeting the demand for curriculum expansion, i.e. the addition of new courses, and setting realistic production targets that take into account both staffing levels and the amount of time that can be devoted to any particular project. The course development model at Thomas Edison State College involves the paring of an instructional designer and a content expert, who are together responsible for virtually all aspects of production from the time a course is first conceived until it is actually offered to students. This particular model places a tremendous amount of responsibility for quality control on the instructional designer. While many institutions have development teams involving far more personnel, each of whom is responsible for a different aspect of course production and its associated quality control issues, the model described above or some slight variation of it is much more common (Lightfoot, 2007). This typically involves two and sometimes three members of staff in addition to a content expert responsible for all aspects of course development and revision. The challenge to implement quality control measures into the course development and revision process therefore, will continue to be the responsibility of a relatively few members of staff at the majority of institutions involved in online education.

Whereas testing is certainly an area of specialization, in the course development model currently prevalent in this country not all course development teams have a member or members of staff with expertise in this area. And while those developing and revising courses might not necessarily be experts in assessment, they need to be able to differentiate between good and bad assessment items so the latter are not incorporated into learning materials and to ensure that there is close alignment between what a course claims that students should be able to do after successfully completing it and how that success is measured.

REFERENCES

Lightfoot, K. (2006). Using Bloom's taxonomy to develop learning outcomes and assessment tools. In *Step-by-step guide to an ACE evaluation, 2006* (pp. 12-14). Trenton, NJ: American Council on Education College Credit Recommendation Service.

Lightfoot, K. (2007, November). *Course development models and the role of instructional designers and course developers. How do they go about it and who does what?* Paper presented at the Annual Conference of the American Association for Collegiate Independent Study, Ogden, UT.

Palmer, R. R. Colton, Joel & Kramer, Lloyd (2007). *A history of the modern world since 1815*, 10th ed. McGraw-Hill.

Palmer, R. R., Colton, J., & Kramer, L. (2007). *A history of the modern world to 1815, 10th ed.* McGraw-Hill.

Section 2
Cases on Distance Learning Programs

Chapter 8
Great Plains Interactive Distance Education Alliance (Great Plains IDEA)

Virginia Moxley
Kansas State University, USA

Sue Maes
Kansas State University, USA

Dawn Anderson
Kansas State University, USA

EXECUTIVE SUMMARY

This chapter will examine the organizational and technological challenges encountered by the highly successful Great Plains Interactive Distance Education Alliance (Great Plains IDEA) since its members began offering multi-university online academic programs in 2000. Members include the following universities: Colorado State, Iowa State, Kansas State, Texas Tech, Michigan State, Missouri, Montana State, Nebraska, North Dakota State, Oklahoma State, and South Dakota State. Inter-institutional online academic programs are a cost-effective means of rapidly increasing access and addressing emerging educational needs. The chapter explains how the Great Plains IDEA began, operates and has evolved, as well as the mistakes made, lessons learned, and upcoming challenges. A major technological challenge was identifying a secure multi-institution enrollment system for sharing student data between enrolling and teaching institutions the award-winning ExpanSIS system. The authors hope that higher education leaders will be convinced that inter-institutional collaboration is a viable solution to many higher education challenges.

BACKGROUND

The Great Plains Interactive Distance Education Alliance (Great Plains IDEA) is a joint project of its member institutions -- 11 public universities in 11 states. The Great Plains IDEA is an academic alliance. In other words, its primary purpose is to provide access to educational opportunities by collaboratively developing and delivering high-quality, online academic programs. A secondary purpose is

DOI: 10.4018/978-1-60566-870-3.ch008

to provide a structure for member institutions to be more innovative and competitive in today's higher education marketplace by efficiently and effectively utilizing limited resources.

In 2008, the Great Plains IDEA expanded its scope beyond the Human Sciences with the affiliation of the Agriculture Interactive Distance Education Alliance (AG*IDEA). This expansion was initially driven by the interests of the agriculture deans at Iowa State University, Kansas State University, the University of Missouri and the University of Nebraska who were intrigued by the success of their human sciences counterparts and wanted to capitalize on the infrastructure that had been created to support the Great Plains IDEA programs.

The mission of the Great Plains IDEA is to enable member institutions to jointly offer educational programs with other institutions and thereby give students and professionals more access to high quality education.

Culture

"From its initiation, the Great Plains IDEA has organized itself into a culture where administrative leadership is shared, where policies are designed to facilitate academic innovation, where faculty participants provide academic leadership, and where both alliance and institutional interests govern decisions" (Great Plains IDEA Policy & Procedure Manual, Page 9). The notion of shared governance pervades this alliance. Wheeler (1999) states that the key to successful inter-institutional relationships is dedicated champions at all collaborating institutions. Wheeler also notes that since faculty members have control of content and conduct of courses, change must be championed by faculty members; however, a shared vision by top level administrators is required for institutionalization of technology-mediated inter-institutional programming.

Collaborative programs can be a cost-effective approach to new online program development; however, higher education institutions are more familiar with competing than collaborating. Implementing an inter-institutional academic program requires a new mindset, as well as new methods of operation. In order to create a successful alliance, key individuals across each partnering campus need to be on board and trust each other (Moxley, 2003). The management of trust and power is essential to building alliances. Trust and power are personal and institutional issues. The first step in building trust is to select with care the partners with whom you will work—if the people are not trustworthy, the practices will not be trusted. The most practical way to manage power is to select peers as partners. In this alliance, agreement on principles precedes agreement on policies. The Great Plains IDEA established three basic principles to guide the development of policies and procedures (Great Plains IDEA Policy & Procedures Manual, Page 10):

- **Behave as Equals**
    - Graduate schools cross-list Alliance courses.
    - Alliance courses are transcripted as institutional courses, not as transfer courses.
    - Graduate faculty status is honored by all Alliance institutions.
- **Respect Institutional Differences**
    - Alliance programs accommodate institutional differences—faculty workload, administrative and financial arrangements, and institutional culture.
    - The curriculum is the same everywhere, but course numbers and program and degree titles are unique to each institution.
- **Simplify Student Navigation**
    - Differences in procedure from one institution to another are transparent to the learners.
    - The cost to enroll in an Alliance course is the same everywhere.

- Course advising and enrollment takes place at the student's home institution.

Origins

In 1994, Dr. Joan Laughlin, in her position as associate dean of human resources and family sciences at the University of Nebraska, was organizing a master's degree to be offered to students at a distance. She and Dean Karen Craig called a meeting of the deans of human sciences colleges in the Midwest for the purpose of identifying partners – other universities that had students needing coursework at a distance and other universities having courses that might be incorporated into the University of Nebraska program. Seven institutions that took part in this initial meeting would subsequently become founding members of the Great Plains IDEA. The meeting stoked both the competitive and cooperative tendencies of the participants. A new benchmark for graduate program access had been set and no participant wanted the college s/he led to be "behind the curve." The coming together raised the bar for distance education programming and created a form of peer pressure that advanced the engagement in distance education of all participating universities.

The Great Plains IDEA has evolved from a collegial group of Human Sciences academic administrators who share a common interest in educating rural professionals through the use of distance technologies and shared courses to its current status as a premier post-baccalaureate distance education alliance that sponsors inter-institutional master's degrees and graduate certificates and develops policy and practice models for inter-institutional distance education programs (Moxley, 2007).

Conversations evolved from consideration of a videotape exchange to satellite downlinks of exemplary classes to faculty distance learning training. Major obstacles in the early years were the lack of Internet connectivity, the lack of commercially available and easy to use courseware, and the lack of distance education experience among the faculty. Early Alliance initiatives included: 1) Informing faculty about the changing marketplace for graduate education; 2) Training faculty in the use of distance technologies to promote engaged, graduate-level learning at a distance; and 3) Developing a marketplace for sharing distance education courses. It was at a faculty workshop that participants sketched out the initial concept of an inter-institutional Family Financial Planning graduate program. Program development was supported by the Cooperative State Research, Education, and Extension Service, U.S. Department of Agriculture, under Agreement Number 98-EATP-1-0403 administered by the American Distance Education Consortium (ADEC) and under Agreement Number 00-38857-0998.

Dr. Joan Laughlin - Great Plains IDEA Founding Member – Dr. Laughlin was a popular and respected leader for the Alliance and a skilled collaborator.

The notion that the alliance was about shared programs was Joan's vision from day one – it took a while for the rest of us to believe the concept was possible. Without Joan's dream of inter-institutional programs, which provided a compelling discussion topic at every meeting, I suspect the Great Plains IDEA would have failed for lack of focus.

Joan Laughlin led the alliance in its infancy – before policies and programs, before MOA's and Bylaws, before common price and guiding principles. Her administrative and leadership instincts fit the needs of this emerging alliance – and the needs of a team of higher education deans trying to figure out how to accomplish inter-institutionally what none could do institutionally and trying to figure out what systems were needed to move the work ahead. (Virginia Moxley, Letter Nominating

Joan Laughlin for the 2007 Friends of the Great Plains IDEA Award.)

Programs

The first and largest Great Plains IDEA inter-institutional graduate program is Family Financial Planning – offering both a 42-credit hour Master's degree and an 18-credit hour graduate certificate. This program is designed to prepare students to take the CFP® Certification Examination and is the first of its kind to be registered by the Certified Financial Planner Board of Standards, Inc. Family Financial Planning online courses were first offered in 2000. None of the universities that participate in this program had the faculty resources to launch such a program alone. The incentives for collaboration included the need to serve a mass market of post-baccalaureate learners seeking to become Certified Financial Planners, the need to distribute the development costs associated with deployment of a distance education program, and the desire to launch the program more quickly than would be possible at any single university given the existing workload demands on faculty. However, in spite of faculty and academic administrator good will and commitment to the program, the process of developing and beginning the delivery of the program was arduous.

Eight of the eleven Great Plains IDEA member institutions participate in the Family Financial Planning program: Iowa State University, Kansas State University, Montana State University, University of Missouri (since Summer 2007), University of Nebraska, North Dakota State University, Oklahoma State University, and South Dakota State University. The Family Financial Planning Program has generated $2,743,601 in course revenue for the Alliance and the eight participating universities from Fall 2003 through Fall 2008.

As of Fall 2008, Great Plains IDEA has sponsored eight academic programs. Graduate programs in Gerontology (six universities) and Youth Development (four universities) were added in Fall 2003. Graduate programs in Merchandising (five universities) and Community Development (five universities) were added in Fall 2005. The Family and Consumer Sciences Education graduate program (six universities including affiliate member, Central Washington State University) was added in Summer 2006. A Dietetics master's degree (nine universities including affiliate member, Kansas University Medical Center) was added in Fall 2008. AG*IDEA offers a Food Safety and Defense Graduate Certificate (four universities) and exchanges courses in the areas of Agricultural Education, Agricultural Mechanization (undergraduate) and Grasslands Management. AG*IDEA is exploring many new program ideas including Poultry Science, Swine Science and Precision Agriculture.

Prior to program implementation, a program faculty team typically meets face-to-face twice a year. It usually takes two years to develop a program and obtain program approval at each participating university. Thereafter, the program faculty usually meets face-to-face once a year and

Figure 1.

FAMILY FINANCIAL PLANNING PROGRAM GROWTH

Calendar Year	Credits Taught	% Change
2000*	72	
2001	366	
2002	681	+86%
2003	855	+26%
2004	1035	+21%
2005	1251	+21%
2006	1174	-6%
2007	1398	+19%
2008	1608	+15%

*Began Fall Semester

by conference call three times a year. The program faculty elect a faculty chair who convenes meetings and prepares the program annual report to the Board and the annual assessment report. A great deal of faculty interaction occurs through email and on restricted web pages. A Faculty Handbook is developed for each program to document programmatic decisions and orient new faculty members to the collaborative program.

Program Selection. Program selection should be a bottom-up versus a top-down decision. An alliance program is more likely to succeed when it is faculty driven (Moxley, 2005). Five criteria should guide the selection of a program for inter-institutional collaboration: 1) Demand – should be large and/or pressing; 2) Capacity – the partner institutions cannot offer the program online alone; 3) Faculty – a strong desire to participate in the program must exist; 4) Department – the academic unit must be willing to experiment in the development of an atypical program; and 5) Wisdom – based upon prior academic experience, is it sensible to select the program?

New programs are proposed, developed and delivered by faculty members and academic administrators employed by Great Plains IDEA member institutions. A program proposal briefly "describes the curriculum, the societal or industry need to be met, the intended audience, anticipated demand for the program, the minimum institutional qualifications to participate, and identifies the faculty originators of the proposal" (Great Plains IDEA Policy & Procedure Manual, Pages 77-79). All member institutions are invited to participate in the development of a new program. Institutions typically based their decision to participate upon their mission, faculty resources and the evidence of demand for the program. Those members that accept the invitation to participate identify faculty members to engage in the program development, approval and implementation.

Faculty Member Selection. This is the most critical component in successful program development. The following faculty member characteristics should be considered: teaching talent; interest in online teaching and new forms of pedagogy; area of specialization; academic reputation; graduate faculty status; department head and colleague support; contributions as a team player; and follow-through on commitments.

Faculty members from multiple higher education institutions constitute a program faculty team that functions as a virtual department. A program faculty chair is elected from among the team members and a leadership succession plan assures that no one university dominates. An administrative liaison is assigned to provide advice and counsel to the faculty team and serve as its liaison to the academic board (Great Plains IDEA Policy and Procedure Manual, Page 98). Administrative liaisons are typically members of the academic board of directors.

Program Faculty Activities:

1. Identify the target audience
2. Assess educational needs and demand
3. Articulate student learning outcomes
4. Design a common curriculum
5. Assign teaching responsibilities
6. Create a schedule of course offerings (See Figure 2.)
7. Prepare a business plan and, course and program approval documents
8. Formulate a program assessment plan
9. Share e-learning instructional strategies and techniques
10. Implement, review and revise the program as needed.

We made mistakes. Sometimes, we lacked good leadership for these faculty teams and they struggled with curriculum development - planning programs based on courses each wanted to teach rather than competencies the students needed to attain. Faculty teams, for whom collaborative programs were new, veered into discussions of administrative issues that were more easily solved in other ways. And, because we did not initially

Figure 2. Sample program course schedule – family financial planning (FFP)

	IOWA STATE UNIVERSITY	KANSAS STATE UNIVERSITY	UNIVERSITY OF MISSOURI	MONTANA STATE UNIVERSITY	UNIVERSITY OF NEBRASKA	NORTH DAKOTA STATE UNIVERSITY	OKLAHOMA STATE UNIVERSITY	SOUTH DAKOTA STATE UNIVERSITY
FALL	*Investing for the Family's Future*	*Fundamentals of FFP* *Financial Planning Case Studies*	*Personal Income Taxation*	*Family Economics*		*Family Systems*	*Family Systems*	
SPRING	*Housing & Real Estate*	*Professional Practices in FFP*		*Estate Planning for Families*	*Insurance Planning for Families*	*Financial Counseling* *Family Systems*	*Housing & Real Estate*	
SUMMER		*Financial Planning Case Studies*			*Retirement Planning, Employee Benefits & the Family*		*Housing & Real Estate*	*Family Economics*

screen for teamwork competencies, not all faculty were collaborative. Most needed help with online delivery of course work and this support was not readily available for the earliest participants.

Our most frustrating failures came in the realm of managing the communication flow to and from participants. Administrative participants continuously change. We need an effective orientation program for new participants but this has not yet been formalized. Feedback from surveys was initially provided in a 20-page document—no one read it so the utility was zero. Now we provide a slim dashboard report. New programs are required to provide business plans. Over time the Board has learned to distinguish plans justified by data from those based on over-optimistic assertions.

Governance

The founders first convened in 1994. The alliance was officially founded in 2001 and has been governed by a Board of Directors comprised of one representative from each member institution. Leadership changes biannually. Virginia Moxley, Kansas State University, Marjorie Kostelnik, University of Nebraska, and Laurie Stenberg Nichols, South Dakota State University, have served as board chairs from the time the Alliance was formalized through June 2009. The Board Chair, Vice Chair/Chair-Elect, Secretary and Treasurer are elected from among the Board members at the annual Board meeting.

Sheran L. Cramer - Great Plains IDEA Family Financial Planning Program Faculty Leader. The Family Financial Planning Program Faculty met first in 1999 under the leadership of Dr. Joan Laughlin. In one two day period the faculty agreed on guiding principles, developed a full curriculum, planned a course rotation, and assigned teaching responsibilities. This efficiency has not been matched since. The success of the meeting is a tribute to both Dr. Laughlin who convened and led the meeting and to the willingness of Dr. Sheran Cramer to assume the leadership during program implementation. Dr. Cramer led the faculty team from 2000 through 2006.

I have learned a lot in watching Sheran's leadership style. She never judges or participates in negative conversation about people or situations. She leads by focusing on the positive and by seeking optimal solutions to problems as they arise. She is calm, collected, organized, professional and kind to all. She has managed mountains of paperwork in our group as we have sought program registration from the Board of Standards for the Certified Financial Planner. In order to do her job as the faculty program leader, she requires the participation of all faculty members in the consortium, which is a frustrating and painstaking process since we are busy and we are spread across seven states. Sheran continues in her unflappable way to pull the pieces of the registration application together every time we apply. Her patience is commendable. Her attention to detail has made all of our jobs so much easier as she provides the data for the CFP® application and as she provides leadership for the faculty in accomplishing the roles of providing the coursework for the master's degree in Family Financial Planning.

Sheran is a dedicated, intelligent, hard working professional and a wonderful human being. She personifies the type of academic professional who has helped to create the great land grant universities of this nation. She is creative and practical. She is academic and yet very down to earth. She is not self-serving. Without her leadership, the Great Plains IDEA program in family financial planning would not have seen the success it has in the past six years. She has been a key player in doing all of the behind-the-scenes organizing and thinking to bring the program to its current level of success in her roles of leader, instructor, adviser, mentor, program promoter and advocate for inter-institutional graduate education. (Deborah C. Haynes, Associate Professor, Montana State University. Letter of Support for the 2007 Friend of Great Plains IDEA Award)

The governance structure will change in 2009 to accommodate AG*IDEA's affiliation with Great Plains IDEA. The original Great Plains IDEA Board of Directors is now the Human Sciences Board of Directors. A Cabinet and an Agriculture Board of Directors have been added to the governance structure. Another important part of the governance structure is the multi-institutional functional teams - finance officers, graduate deans, department heads, registrars, institutional review boards, institutional assessment directors, and continuing education administrators who make recommendations to the Cabinet and Academic Boards regarding Great Plains IDEA policies, procedures and activities. A full complement of supportive university administrators has been a key to Great Plains IDEA's success.

Cabinet. The Cabinet will convene for the first time in July 2009. It will be responsible for the overall policy and direction of the Great Plains IDEA, for sustaining the egalitarian culture of the Alliance, for resolving disputes among academic boards and programs, for maintaining relationships with the functional teams that support the Alliance and for Alliance communications. The Cabinet also approves new members, contracts with a member institution for management services, oversees fiscal affairs for the Alliance, and approves policies and procedures that will be used by the academic boards (currently human sciences and agriculture), to govern their own programs. The Cabinet is composed of two representatives from each Academic Board, two representatives from the human sciences founding members, a finance officer, a graduate dean, and a member-at-large.

Academic Boards. Fairness, simplicity and clarity are necessary as the Great Plains IDEA expands to include new disciplines. The Human Sciences Academic Board has met face-to-face twice a year for an annual and semi-annual meeting and by conference call ten times per year. This may change to one face-to-face meeting per year with the new governance structure. Each Academic

Board is charged with administrative oversight for the programs it sponsors. Thus, human sciences academic administrators are responsible for human sciences programs and agriculture administrators are responsible for agriculture programs. The new Cabinet will take care of many the issues previously addressed by the Human Sciences Board. An interesting dilemma, as yet unresolved, has presented itself in regard to interdisciplinary program governance for programs that involve more than one academic discipline.

Functional Teams. Alliance building is an intellectually and emotionally intense endeavor of great complexity. Each partner is required to give a little with the hope of getting a lot. However, because of the nature of academic institutions, when academic deans form alliances, they make work for other functional areas within the institutions they represent. For instance, although deans can provide the administrative and financial support to the faculty who develop and teach inter-institutional programs, they cannot price the programs without getting buy-in from the financial officer of the university. Deans cannot unilaterally make decisions to include graduate courses taught by other universities in students' programs of study without buy-in from graduate faculty leaders and administrators. They cannot develop a process for registrars to enroll students from other universities into shared online courses.

The buy-in and commitment of key individuals are crucial. When institutional administrators meet face-to-face with their inter-institutional peers, it is astonishing what can happen. They can become creative and find workable solutions to challenges and barriers. Had key administrators from each institution not been involved, the Great Plains IDEA programs would not be successful. The teams looked for the common denominator among institutional practices, as well as best institutional practices. In a face-to-face meeting of respected colleagues committed to a common goal, seldom does someone want to say it can't be done.

Financial Agreement

In an era demanding innovation and the rapid deployment of new, highly specialized education, universities are motivated to collaborate with their peers to accomplish together what none could do alone. Dr. Tom Rawson, Vice President of Administration and Finance at Kansas State University, and the first chair of the Great Plains IDEA Finance Team stated, "There are a lot of

Figure 3. Great Plains IDEA organizational chart

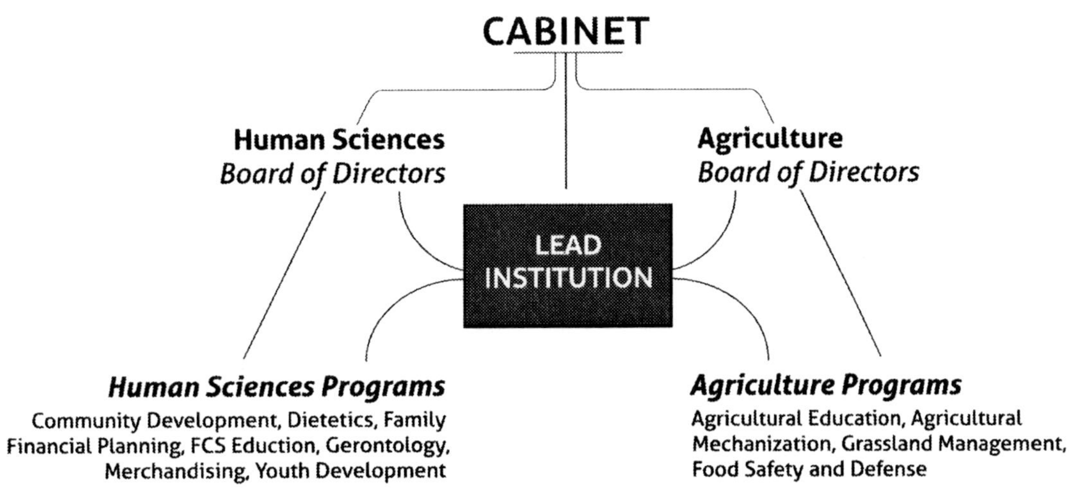

Figure 4. Great Plains IDEA revenue

GREAT PLAINS IDEA HUMAN SCIENCES PROGRAM REVENUE		
Academic Year	Income	% Increase
2003-04	$ 370,837	
2004-05	$ 768,768	107%
2005-06	$ 949,792	24%
2006-07	$ 1,162,552	22%
2007-08	$ 1,620,648	39%
TOTAL	$ 4,872,597	

start-up costs associated with academic programs. If an institution sees the need or if the state sees the need to have trained leadership in a certain area and the university does not have a program in that area, the start-up costs can be prohibitive. Through participation in an alliance, an institution can contribute courses in an area where it already has a great expertise and thereby enable the state to have trained leadership in the broader discipline without a huge investment."

A feasible business plan is critical to the success of a consortium. Institutional finance officers are experts in higher education finance. Academic administrators cannot form a stable program alliance without their support. The solution to program funding and pricing resides in the minds of the partner institutions' finance officers.

A strength of the Alliance was that by gathering individuals from across Functional Sectors at multiple institutions, people were able to make connections, to understand why certain groups approached issues differently, and to honor those differences while working for a common solution. It is also clear from the data that the participation of peer institutions played a role in forward progress and challenging long-held traditions... (FIPSE LAAP Project Evaluation Survey Response)

Knowing that annual membership fees may not be paid during financial crises, the Great Plains IDEA Finance Officers Team recommended that centralized services be funded through a portion of the students' course payment. At an impromptu gathering in a hotel lobby in Coeur d'Alene, Idaho, the Great Plains IDEA Finance Team made a bold decision that simplified the work of everyone else. They agreed to create a common price (including tuition and fees) that would neutralize the impact of "shopping around" AND they made a commitment to get their institutional governing bodies to go along with the common price (see Figure 5). They were successful. Students pay the same price no matter which member university provides the course instruction. That price is subject to annual review and adjustment and is dispersed to support teaching (75%), program administration and advising (12.5%), and Alliance central services (12.5%). This pricing model was implemented in Fall 2003 with a common price of $350 per graduate credit hour. For Academic Year 2009-2010, the common price is $430 per graduate credit and $310 per undergraduate credit.

The common price model helps ensure sustainable programs and a sustainable alliance. Prices are set at a competitive level – neither the lowest nor highest cost program provider. The Finance Officer Team reviews member institution and competitor course prices each Fall and recommends any changes in the common price to the Board of Directors for their action. Institutions are responsible for the internal distribution of the income from Alliance courses, but are encouraged to distribute the revenue where the greatest costs are incurred – to the unit paying instructor salaries (Great Plains IDEA Policy & Procedure Manual, Pages 99-101).

Graduate Deans Agreement

Like their finance officer counterparts, the Great Plains IDEA graduate deans got creative. Instead

Figure 5. Common price distribution

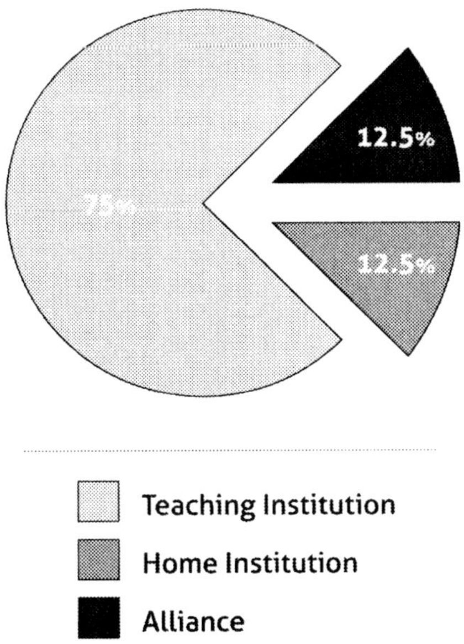

of haggling about WHY an inter-institutional graduate program couldn't work – they MADE IT WORK. It was more often a matter of reinterpreting policies than changing them. The Graduate deans contributed to enabling policy and practice environments at the partner institutions. With the support of graduate program administrators, inter-institutional programs became administratively possible—and with broader institutional faculty support they became realities. In many cases, the university-wide graduate faculty councils who had to approve the programs were skeptical that an inter-institutional web-based program was feasible and desirable. However, the graduate deans and program faculty prevailed and the programs were eventually approved.

Students seek admission at a "home institution" from which they receive course credit for all Great Plains IDEA courses and are awarded their degree or certificate. Great Plains IDEA students enroll in all Alliance courses through their home/admitting institution and pay the common price to their home institution. All Great Plains IDEA courses are listed in the course catalogs of all participating Alliance institutions regardless of whether the institution actually teaches the course, so, Alliance courses are not "transfer" courses.

Management

As a joint project supported by its institutional members, the Great Plains IDEA cannot hold assets or employ staff, therefore, one member institution provides management services, manages fiscal affairs, and employs personnel on behalf of the Alliance. This institution is designated as the "lead institution" for the Great Plains IDEA. The Great Plains IDEA strives to capitalize on institutional strengths within the Alliance and minimize the cost of central operations with a "lean" central staff. The University of Nebraska has taken the lead on several inter-institutional faculty workshops. The Board of Directors has taken bids from member institutions for items such as a marketing plan, market demand surveys and website development.

Kansas State University has served as the managing partner for the Great Plains IDEA since its inception in 2001. In 2004, Kansas State University established its Institute for Academic Alliances (IAA). Since that time, the IAA has been the unit within Kansas State University that has provided consortium management services to the Great Plains IDEA and other consortia.

The term of the lead institution is five years with satisfactory annual performance reviews and may be renewed for additional five-year terms. Member institutions interested in serving as the lead institution submit an application (Great Plains IDEA Policy & Procedure Manual, Pages 94-95). An executive coordinator and other staff, as appropriate, are employed by the lead institution. The lead institution is funded based upon a percentage (70%) of the previous year's income to the Alliance.

The Great Plains IDEA has a proven management structure and tools:

1. Lead Institution:
   a. Oversees financial management – prepares invoices, and semi- and annual financial reports
   b. Coordinates communication and arranges meetings
   c. Creates and maintains contact lists and listservs
   d. Supports governance
   e. Supports and trains campus coordinators
   f. Provides support to program initiatives and academic programs
2. Secure student data system, ExpanSIS
3. Online Course Catalog
4. Model consortium documents including bylaws, legal agreements, policies and procedures
5. Model program documents including student and faculty handbooks
6. Website for posting program information and links to each institution within the program (Great Plains IDEA Policy & Procedure Manual, Page 84)

Several positions are vital to program management: the executive coordinator, program faculty leaders and campus coordinators. Campus coordinators are professional staff members at each partner institution who serve as the information brokers for the Great Plains IDEA academic programs. The campus coordinator communicates with prospective, current, and former Great Plains IDEA students, and faculty, administrators, and other Great Plains IDEA campus coordinators to facilitate admissions, enrollments, graduations, and fund transfers (Great Plains IDEA Policy & Procedure Manual, Pages 103-106).

SETTING THE STAGE

In the beginning, what appeared to be a significant achievement—launching an inter-institutional Master's degree — had embedded chaos. The realities of institutional infrastructures limited our ability to capitalize on each other's courses. The first technological barrier was multiple online course management systems. The academic deans who organized the Alliance assumed that a common online course management system was necessary. The partner universities used various commercial products such as Blackboard and WebCT, as well as independent systems such as K-State Online. For over a year (1999-2000) deans debated which online course management system the Alliance should use, during which time several new, improved systems came into use at partner universities. A common system was neither feasible nor needed. It was decided that instructors would use the online course management system that their employing institution supported to teach their assigned courses. This has worked well. Students have not reported problems using different online course management systems and when two universities teach the same course, the instructors have been able to share course content and digital learning objects.

Each institution had a different course enrollment process, tuition and fee structure, and payment process. Students had to enroll (and sometimes be admitted) at all participating institutions and request (and pay for) transcripts to be sent back to their home institutions to document their progress. Obtaining student financial aid was difficult when students were enrolling in courses at multiple universities. Faculty members were attempting to administer the fledgling program themselves. The academic elements were in place for the delivery of a quality post-baccalaureate program, but the administrative policy, process and system elements were not in place.

Faculty members from seven Great Plains IDEA institutions had collaborated in the development of the inter-institutional online graduate program in Family Financial Planning. These institutions did not have procedures in place to effectively implement an inter-institutional post-baccalaureate program. The human sciences deans were breaking new ground in inter-institutional collaboration and believed that successful program implementation would be contingent upon achieving some fundamental policy and procedure changes at their universities. Out of necessity, they chose to seek funding to tackle the major hurdle of developing model policies and procedures to support post-baccalaureate inter-institutional distance education programs.

The authors of this case study, Dr. Virginia Moxley, Dean of Human Ecology at Kansas State University, and Dr. Sue Maes, Dean of Continuing Education at Kansas State University, wrote a U.S. Department of Education Fund for the Improvement of Post Secondary Education Learning Anytime Any Place (FIPSE LAAP) grant proposal. They envisioned a higher education environment where working inter-institutionally was no more difficult than working across academic units within a college. Where inter-institutional distance education programs were as accessible, supportive and successful as the best single-institution on-campus programs. They envisioned a policy and procedure infrastructure that would enable faculty members to rapidly initiate and sustain high quality learner-centered, inter-institutional distance education programs and enable universities to manage institutional and partnership costs, student services and faculty support in reasonable ways. Before the grant project, this environment did not exist.

In 2000, eight Great Plains IDEA universities received a three-year FIPSE LAAP grant to create a model policy and procedure infrastructure for the Great Plains IDEA that would enable: 1) Faculty to rapidly initiate and sustain high-quality, learner-centered, inter-institutional distance education academic programs; and 2) Universities to manage institutional costs, student services, and faculty support in reasonable ways.

Prior to the submission of the FIPSE LAAP project proposal, administrators within each institution agreed to work toward policies and procedures that would serve the Family Financial Planning program and that could be applied to serve regional, national, and international alliances. We were committed to developing policies and procedures that would serve not only this program but also other inter-institutional post-baccalaureate initiatives at our institutions and which would be useful and applicable to other institutions. Representatives from the National Center for Higher Education Management Systems (NCHEMS), State Higher Education Executive Officers (SHEEO), American Distance Education Consortium (ADEC), Southern Regional Education Board (SREB), Council of Graduate Schools (CGS) and University Continuing Education Association (UCEA) agreed to serve on an Advisory Board for the project.

The grant allowed the Great Plains IDEA to bring together finance officers, registrars, continuing education deans, graduate deans, academic deans, faculty members and advisory board members. This interaction resulted in significant breakthroughs as teams capitalized on ideas emerging from other teams. These collaborative achievements gave birth to the Great Plains IDEA model, a simple, academically appropriate, and cost-effective model for implementing inter-institutional distance education programs.

A chair of each inter-institutional peer team was identified with each chair from a different institution. The participants came to this project with great attitudes, and the formation of inter-institutional and institutional teams reinforced a can-do way of thinking.

Over 80 academic leaders were engaged in inter-institutional and institutional teams that revamped the policy and procedure environment for post-baccalaureate education, simplified inter-institutional partnership arrangements, and provided for academic and fiscal accountability.

CASE DESCRIPTION

After instructional technology, the biggest technological challenges for the Great Plains IDEA were the management of documents, communications, information, and data. Technology tools are necessary to fulfill the objectives of a large-scale project. The first technology tools to be utilized by the Great Plains IDEA project were obviously e-mail and listservs, shortly followed by the use of an online survey system, a web-based workspace and restricted web pages. The next technology tool upon which success hinged was the procurement of a multi-institution student information system to exchange course enrollment and grade information. The most recent technology tool has been an online program assessment reporting system.

The FIPSE LAAP project began by hiring two project coordinators, Dawn Anderson and Ruth Williams, who along with Co-Principle Investigators Maes and Moxley, formed the project staff team. The Great Plains IDEA has been fortunate to have consistent staffing since 2000. When the FIPSE LAAP grant ended in 2004, Williams stayed on as the first executive coordinator and trained Deborah Wood to take over in that capacity in 2008. Anderson, Maes, Moxley and Williams continue to assist the Alliance in various roles.

In November 2000, the project staff immediately began to plan an initial meeting of all the project participants. In preparation for that meeting, extensive research was conducted to identify pertinent issues, guidelines, questions, considerations, policies, practices, problems, challenges, suggestions, and recommendations. The results of that research were more theoretical than based upon actual experience. There were few successful collaborative models to emulate and few solid answers about how to create an infrastructure to support multi-institution academic programs. The research results were assembled into a briefing book that was organized by issues facing each functional team: faculty, human sciences deans, graduate deans, finance officers, registrars and continuing education deans. The briefing book was sent to the project participants prior to the kick-off meeting to help them prepare for the meeting.

Also in preparation for the initial meeting, the institutional teams were asked to respond to two policy and procedure surveys: one regarding faculty issues and the other regarding student issues. The surveys had multiple purposes: 1) Team Building. The survey questions generated discussion among institutional team members and facilitated mutual awareness of distance education policies and practices within each institution. 2) Baseline Project Data. The survey responses became baseline policy/practice data for the grant project. 3) Issue Identification. The survey responses were summarized and put into a matrix to simplify comparisons of policies and procedures of partner institutions. The matrices served as a starting point to allow project participants to identify institutional differences in important policies and procedures that might be significant barriers to inter-institutional, multi-state, distance education graduate academic programs. The surveys illuminated outmoded policies as well as policy disarray; policies that were not in use because: they had not been disseminated, conflicted with other policies, the developers liked the policies but the implementers disliked and ignored the policies, or the policies remained forever in draft form because no one could determine who had the authority for emerging policy areas.

Online Survey System

The initial policy and procedure surveys established a method and pattern of preparing for Great Plains IDEA meetings. Online surveys are an efficient and effective means of gathering data to assist in making informed decisions. The board decides when an online survey would be beneficial. One or more board members are usually assigned to work with the lead institution staff to develop a survey. The lead institution staff administers surveys and prepares survey reports. Faculty and administrators have responded to online surveys regarding study-abroad activities, faculty development needs, faculty compensation, marketing activities and institutional internal common price revenue distribution.

The Great Plains IDEA lead institution also uses the online survey system to administer a new student survey and exit survey to all students admitted to Great Plains IDEA programs – over 1,000 unique students as of 2008. The K-State Online Survey System is used for this purpose every semester. This is a rich data source as well as a significant decision-making tool.

A needs assessment is part of the business plan that must be prepared prior to approval of a new Great Plains IDEA program. Online surveys are often used to establish the demand or need for a particular academic program. They can also be used by the faculty team to elicit target audience or industry input to help the faculty develop the curriculum and market the program.

Communications and Document Management

Communication among and between the board, program faculty teams and functional sector teams is a challenge, but it is absolutely critical to the success of a project or consortium (Anderson, 2008). Each team had members who claimed that *they* (another team, higher education governing board, or accrediting agency) "won't let us do that." Inter-team communication usually led to the discovery that policies, even those governed by others, could be changed. We have learned that untested assumptions are the real problems and the people behind the policies, procedures, and practices are the key to solutions. When key decision-makers are committed to a project, policies, procedures, and practices can be adapted to accommodate inter-institutional programs. When key individuals are not committed to a project, virtually insurmountable barriers can present themselves. How do key people become committed? Commitment seems to occur through institutional and inter-institutional interpersonal communication, awareness of a Great IDEA (pun intended), and trust.

When the project participants go back to their campuses, their regular responsibilities preclude spending a considerable amount of time on the project. They do not have time to work on project details, but they are willing to provide feedback, thus a competent project staff with complementary strengths is critical in moving the process forward. Continuous communication with the project participants is essential.

Turnover in key positions also creates challenges. Turnover necessitates extensive communication with the new person to create a sense of commitment to the continuation of the inter-institutional programs, and an appreciation and understanding of prior work. Inter-institutional programs may be relatively fragile at some member institutions; therefore, turnover can put institutional participation at risk. A lack of communication within the institution and with key individuals may actually lead to the withdrawal of the institution from a particular program. The Human Sciences Academic Board has been discussing various methods of orienting new board members and informing key institutional decision-makers about the Great Plains IDEA history and its operations.

After the initial meeting in January 2001, the staff attempted to share research results and other documents with project participants through a QuickPlace secure website. This website contained project minutes, documents, articles, links, power point presentations, and much more. Each week an automatic e-mail was sent to project participants providing an update on what had been added to the website. However, project participants did not use the site and no one other than one project staff member posted anything, so we went back to sending email attachments – over and over and over. Thankfully, faculty and administrators found the frequent reminders more of a help than a frustration.

When the Great Plains IDEA Youth Development faculty began developing their curriculum, the project staff used K-State Online, the home grown online course management system used by Kansas State University, to create an online collaborative workspace for the faculty to exchange course and curriculum information, and other program approval information. The collaborative workspace was created like courses are created in an online course management system. However, this failed as well. It seemed that the faculty did not have the time or desire to create a user name and password at another university and learn to use a different course management system, so they went back to sending email attachments – over and over and over.

Dr. Laurie Stenberg Nichols, Dean of Family and Consumer Sciences at South Dakota State University, and at that time Treasurer of the Great Plains IDEA Board, requested that all documents be put on a restricted web page associated with the Great Plains IDEA website. By 2003, project staff were desperate to have an accessible document retrieval system for Great Plains IDEA project participants. It took a while, but eventually the project participants began to use the "Members Only" restricted web pages when they needed information. Staff just kept politely responding to requests for information by attaching the document to an email message that also reiterated the link, and common user identification and password for the Members Only web pages.

The Great Plains IDEA has grown rapidly, and with the addition of AG*IDEA, will grow even larger. Maintaining the restricted web pages for the Alliance and all the programs is extremely time-consuming for lead institution staff. We had hoped that a 2009 Great Plains IDEA website update would have "Members Only" pages that permitted all members to upload documents, but that still is not feasible. We continue to look for an online collaborative workspace that is affordable and will meet our complex needs.

Website

The Great Plains IDEA website has been continuously modified and improved. The continued success of the Alliance is dependent upon a website that works well for prospective students. But, a good website was not enough. Each member institution has to assure that the web link to their institutional information is working and easy for prospective students to use. Distinctions are made between content that is contained on the Great Plains IDEA website and content that is contained on partner institution websites. The lead institution provides the member institutions with a template and support in developing institutional web pages. Appropriate links must be maintained between the Great Plains IDEA website and the partner institution websites. Maintaining these links is an on-going challenge. The Great Plains IDEA website provides general information to prospective students and links to specific information on admissions and program requirements at partner institutions. Once a student is admitted, the home institution is generally responsible for web-based student services. The Great Plains IDEA website includes an online course catalog for those students interested in taking online courses directly from the teaching institutions. The online course catalog is also a marketing tool for member institutions to advertise their own online course offerings.

A public collaboration resource center is also housed within the Great Plains IDEA website. It provides institutions of higher education with the necessary tools to form a successful alliance.

Collaboration Resource Center Documents:

- Great Plains IDEA Model for Inter-institutional Distance Education Program Alliances
- Graduate Deans Agreement
- Bylaws of the Great Plains IDEA Alliance

- Alliance Policy and Procedure Manual
- Alliance Memorandum of Agreement
- Program Memorandum of Agreement
- Financial Agreements and Principles for Setting and Managing the Common Price
- Financial Officer's Common Price Distribution Model
- Inter-Institutional Academic Program Steps for Board Approval
- Steps for Institutional Approval and Implementation of an Inter-Institutional Program
- Teaching Institution Program Principles
- Campus Coordinator Responsibilities
- Lead Institution Responsibilities and RFP
- Administrative Program Liaison Description
- Faculty Program Chair Description
- Alliance Website and Online Course Catalog
- Alliance Student Information Database
- Program Needs Assessment/Marketing Survey
- Program Business Plan
- Enrollment Management

The greatest technological challenge for a consortium is enrollment management. Enrollment must be as simple as possible for the students. Support staff must have immediate access to the information they need to give students from multiple universities access to a particular course.

The Human Sciences deans and Family Financial Planning faculty team began offering online courses without an inter-institutional course enrollment process in place. They leapt into the fire, demonstrated that there was a demand and need for shared online courses, and then appealed to their university administrators to help them make it easier for the students. When the courses were first offered, each student had to go through the graduate admissions process at each teaching institution and pay an admissions application fee, although, admittedly some of the institutions had a "quick admit" process to take a course or two without being admitted to a program. After the students completed the course they had to request and pay for a transcript to be sent to their home institutions. Then the home institutions went through their processes of evaluating whether the course would transfer and fit into the students' programs of study. In addition, transfer grades are typically not included in the grade-point average. All of this was a burden for the students and, remember, one of the three Great Plains IDEA guiding principles is *Simplify Student Navigation*.

The first step was for the graduate dean team, finance officer team and registrar team to agree to waive admission application fees and transcript fees. We attempted to develop a common online admissions application but there was too much variation in admissions information requirements and processes among member institutions, so the next positive step was the automatic exchange of course transcripts among the registrars. This required that students sign a form acknowledging that they understand that their transcripts will be automatically sent to their home institutions. It soon became clear that a record of who had signed the form and which transcripts had been sent was necessary. The vision of an accessible centralized database was born.

The Great Plains IDEA registrar team under the leadership of Kathleen Jones, Assistant Vice President for Enrollment and Registrar, Iowa State University, began creating a process for exchanging student information in order to enroll students from multiple universities in shared online courses. The process they created was driven by the finance team's recommendation that students enroll and pay for courses at their home institution AND the graduate deans' agreement that Great Plains IDEA courses would not be transfer courses – they would be considered their own courses. Thus, the registrar team began developing an inter-institutional course enrollment process by gathering the data requirements of each member institution for enrolling students in courses.

Course enrollment system data requirements vary from institution to institution. Great Plains IDEA member institutions had both commercial registration systems and home grown registration systems. At any given time, one or more member institutions were switching systems. Therefore, there was little hope that information could be exchanged automatically or through file transfers without time-consuming, cost-prohibitive computer programming; however, there is a nationwide initiative that may resolve this problem in the future. Affordable commercial products were not available to meet consortium program needs. Recently, some commercial products have become available, but the cost may still be prohibitive.

Dr. Jones synthesized the member institutions' data requirements and created a spreadsheet for the exchange of student information. She also maintained the master spreadsheet of all Great Plains IDEA student and course information. In Fall 2002, students began enrolling in and paying for all Alliance courses through their "home" (admitting) institution. This was a great benefit to students – no admissions, transfer credit or transcript hassles when taking courses at multiple institutions! Great Plains IDEA campus coordinators facilitated this process by sending spreadsheets containing course enrollment information to the teaching institutions. The teaching institutions then added the other institutions' students to their online course management system and invoiced the home institutions. Later in the semester, the home institutions sent grade sheets to the teaching institutions for completion and return. This process worked relatively smoothly but needed to be automated immediately because emails and faxes are not secure.

The master spreadsheet was the starting point for the development of a centralized student information system. Again, Iowa State University took the lead. This time Kris Phelps, Program Manager of Off Campus Continuing Education at Iowa State University, committed the time of staff member, Diane Rupp, to develop a Microsoft Access database accessed through a Virtual Privacy Network (VPN). The registrar team identified a "no- frills" database structure. Project staff worked with the Iowa State University registrar and director of continuing education to develop the simple database. The system was demonstrated for Great Plains IDEA campus coordinators and registrars through a NetMeeting to obtain feedback and suggested improvements. The system was implemented in Fall 2003. Each partner institution began entering student course enrollment information and grades directly into the new system and was able to search for information and generate reports, such as class lists.

The Iowa State VPN system was well designed, but unreliable - there were login and authentication problems. The database got large enough to get bogged down and many users did not have proper drivers. Fortunately, the K-State Institute for Academic Alliances received a U.S. Department of Education grant that included funds to develop an inter-institutional web-based student information system. The K-State Office of Mediated Education used the Iowa State VPN system as a model for creating a secure web-based system that was named ExpanSIS. The Great Plains IDEA data were transferred from the VPN system into ExpanSIS, therefore, ExpanSIS contains all student, course and revenue information since the first program began offering courses in Fall 2000.

ExpanSIS allows partner institutions to enroll students in consortium courses, pull up class rosters and report grades. It contains institution, instructor, course, student, enrollment, program, and revenue information. Alliance institutions set up inter-institutional course offerings for each term. Partner institutions can obtain course start and end dates, textbook information for students, and other information needed to set up the course on their campus for student enrollment. Institutions then register students in the inter-institutional offering in ExpanSIS. The teaching institution can pull off course rosters and other information to set up course access for students enrolled through partner institutions.

With a consortial agreement in place, ExpanSIS meets FERPA requirements; however, students also sign a form acknowledging that their information will be shared with consortium partners as needed (Great Plains IDEA Policy & Procedure Manual, Page 111). ExpanSIS does not require special hardware or software. The application uses a secure, web-based Internet platform. The program can be accessed from any computer with Internet access. Users must have login approval from the system administrator (Great Plains IDEA executive coordinator). This platform was selected because of the ease of use for all institutions involved in the consortium. Approved users from all universities in the consortium can access needed data at any time, and any place with an Internet connection. They are not limited to one specific computer, and do not have to install any specific equipment to obtain the information needed. Secure web access is much faster and more efficient, and this system does not have size restraints we were running into with the VPN client.

All ExpanSIS users must read and agree to the terms of the ExpanSIS Acceptable Use Policy Statement prior to working with student records in the system. Each course offered by the Alliance is entered into ExpanSIS by the system administrator – in our case, the Great Plains IDEA executive coordinator. The courses are given an Alliance name and number. The course description and the program the course is associated with are entered, along with an indication of whether or not the course is required in any degree options offered within the Alliance. Each campus coordinator then maps the Alliance course to his or her university course name and number. The campus coordinators also schedule the courses taught at their own university for each semester.

Student information is contained on four screens navigated by the tabs at the top of the student record. The main screen lists the student's name, university ID, gender, former degrees and other demographic information. The contact screen contains the student's email and address information. The program screen contains information regarding the student's degree option, program, acceptance date, starting semester, program completion semester, and check boxes related to survey administration. The history screen contains a listing of courses the student has taken, the semester and the grade received.

Students enroll at the institution where they are admitted, and their course enrollment and grade information are recorded in ExpanSIS to be posted directly on their transcript without having to request a transcript to transfer courses (Figure 6). An approved campus coordinator at each university submits grades to the registrar.

Each campus coordinator can assign grades as directed by the course instructor, to all students enrolled in a course taught from their university. The coordinator may also view grades for their students in courses taught by other universities. The grade screen shows the number of students enrolled in the course, and the number of grades that have been assigned.

ExpanSIS reports are available in a variety of formats. They can be run as an Adobe Acrobat document for printing, in HTML to display on your screen, and exported to excel in CSV or raw data format (raw data contains all records in the query, and is not filtered).

ExpanSIS Reports:

- Course Enrollment – By Teaching Institution
- Course Enrollment – By Home Institution
- Course Roster – By Course
- Course Roster - By Teaching Institution
- Course Roster - By Home Institution
- Course Summary – By Term
- Enrollment Counts – Summary
- Grade List – By Home Institution
- Grade List – By Teaching Institution
- Financial Detail – Home Institution
- Financial Summary – Home Institution

Figure 6. ExpanSIS enrollment screen

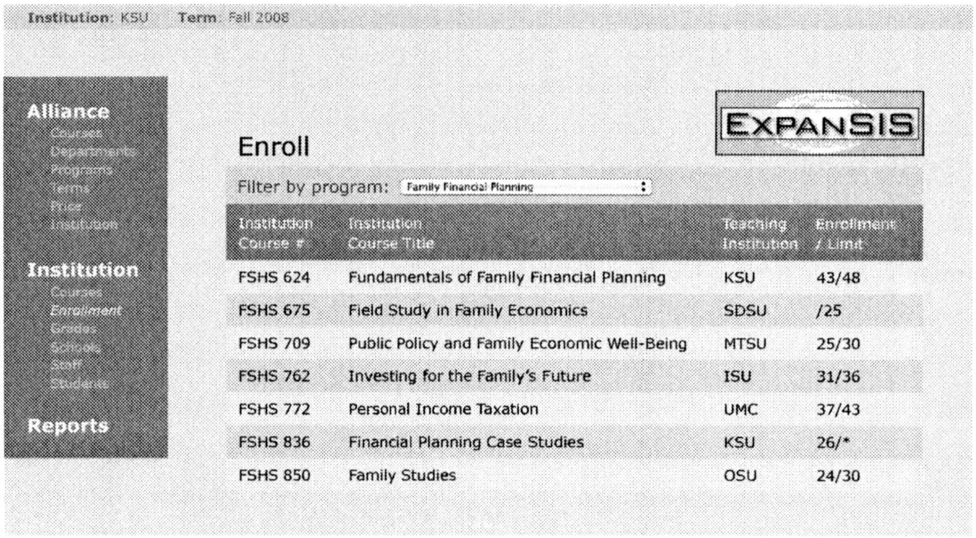

- Financial Detail – Teaching Institution
- Financial Summary – Teaching Institution
- Student Counts – Summary By State of Residence
- Student Credit Hours Earned
- Attempted / Completed CFP
- Student Post Program Employment
- Demographic Summary
- Student Survey Status

K-State's ExpanSIS system has received several awards: 2007 Innovations and Contributions Regional Award, University Continuing Education Association; 2007 Outstanding Work (WOW) Award, WCET; and 2007 Creative Use of Technology Award, Association for Continuing Higher Education. ExpanSIS is an affordable option for higher education consortia and is available through the K-State Institute for Academic Alliances.

Colorado State University Assessment Reporting Database

One of the benefits of consortium membership is sharing innovative practices. In 2004, the Great Plains IDEA Board of Directors Annual Meeting was held on the campus of Colorado State University. The Board had been struggling with how they might collaborate to assist individual institutions in assessing Great Plains IDEA programs. The program assessment technology tool adopted by Colorado State University and demonstrated by Kim Bender during the Annual Meeting became an outstanding solution for collaborative input into program assessment.

Great Plains IDEA programs are accredited separately through each partner institution. Great Plains IDEA member institutions are accredited by three U.S. regional accrediting bodies. Nine member institutions are accredited by the North Central Association of Colleges and Schools - Higher Learning Commission. Montana State University is accredited by the Northwest Commission on Colleges and Universities. Texas Tech University is accredited by the Southern Association of Colleges and Schools. The Higher Learning Commission has approved the Colorado State University assessment system.

All Great Plains IDEA programs now report program assessment plans using the Colorado State University on-line reporting site. Each program develops and reports data on at least three

Figure 7. Program assessment timeline example

SEPT 2007 – OCT 2007	NOV 2007 – DEC 2007	JAN 2008 – OCT 2008	SEPT 2008 – OCT 2008	NOV 2009 – DEC 2009	JAN 2009 – OCT 2009	AFTER OCT 2009
Revised Plan for New Cycle	Peer-Review Comments on Plan	Respond to Peer-Review Feedback	Report Results	Peer-Review Comments on Results	Respond to Peer-Review Feedback	Plan Complete

student-learning outcomes. The online reporting site contains guidelines for completing the reporting forms. The site is easy to use - information can be copied and pasted into the online database. The reporting system follows a straightforward timeline for peers to provide comments, feedback, and responses to the program assessment plan and results. The current stage in the program assessment cycle is highlighted in any given program timeline and clicking on it links the user to the stated activity.

CURRENT CHALLENGES FACING THE ORGANIZATION

The Great Plains IDEA seeks to become a more visible and prominent alliance. This vision can be met most effectively by the judicious addition of other disciplines within the Great Plains IDEA and the creation of a program inventory that includes Great Plains IDEA Human Sciences, Great Plains IDEA Agriculture, Great Plains IDEA Interdisciplinary Programs, etc. – the collective entity to be known as the Great Plains IDEA. Through affiliation with the Great Plains IDEA, other disciplines do not have to create new agreements, policies and procedures. They do not have to create a new infrastructure to support their program offerings. Higher education institutions interested in Great Plains IDEA membership should review the Great Plains IDEA Policy and Procedure Manual for information regarding membership requirements and the application process.

ACKNOWLEDGMENT

AG*IDEA: Affiliate of Great Plains IDEA.

In Memory: Dr. Paul Vaughn, Associate Dean of Agriculture, University of Missouri, envisioned a national agriculture alliance that would expand the depth and breadth of and access to agriculture education. He convinced his peers to create such an alliance following the Great Plains IDEA model. On March 6, 2008, at a national agriculture meeting, those in attendance voted to formalize AG*IDEA. Member institutions have a representative on the AG*IDEA Board of Directors and approve programs for collaborative development. He died on June 22, 2008.

REFERENCES

Anderson, D., Moxley, V., Maes, S., & Reinert, D. (2008). Multi-institution academic programs: Dealmakers and dealbreakers. *Continuing Higher Educational Review, 72*, 103–119.

Great Plains IDEA policy and procedure manual. (2003). Kansas State University on behalf of the Great Plains IDEA. Retrieved on November 25, 2008, from http://www.gpidea.org/alliance/ResourceCenter/P&PManual.pdf

Moxley, V., & Maes, S. (2003). The Great Plains Interactive Distance Education Alliance. *Continuing Higher . Educational Review, 67*, 141–154.

Moxley, V., & Maes, S. (2007). The Great Plains IDEA story. In S. Johnstone (Ed.), *Advancing campus efficiencies: A companion for campus leaders in the digital era* (pp. 161-165). San Francisco: Jossey-Bass.

Moxley, V. M. (2005, February). *Inter-institutional academic alliances-when, why, who, and how*. Paper presented at the Twenty-Second Annual Conference on Academic Chairpersons: The Chair's Role in Empowering Change, Orlando, FL.

Wheeler, B. C., Valacich, J. S., Alavi, M., & Vogel, D. (1995). A framework for technology-mediated inter-institutional telelearning relationships. *Journal of Computer-Mediated Education, 1*(1). Retrieved on December 2, 2008, from http://jcmc.indiana.edu/vol1/issue1/wheeler/essay.html

Chapter 9
The Case of e-Tutorials for Test Preparation for New Teachers Transitioning and Transforming into the Education Profession

Maria Hruby Moore
The Ohio State University, USA

Belinda G. Gimbert
The Ohio State University, USA

EXECUTIVE SUMMARY

This chapter describes the Ohio Transition to Teaching Project, which assisted adult learners pursuing an alternative license to teaching in Ohio with preparation support for the Praxis II: Principles of Learning and Teaching test. It addresses the challenges these nontraditional teachers face in becoming both "content" and "pedagogically qualified." The case describes the rationale and process for the development of an interactive online learning community that provided electronic test preparation, virtual collaboration with peers, e-coaching, and resources. The authors present the advantages of a hybrid or blended approach to instructional design, which combines the best features of both face-to-face and online formats to enable self-paced learning and appropriate levels of interaction. The case concludes with discussion of a new initiative, Project KNOTtT, which is expanding the Ohio Transition to Teaching model to Kansas, Nevada, and Texas.

BACKGROUND

Provide background on the history of the organization, type of business, products/services provided, management structure, financial status (including annual sales), strategic planning, organizational culture, economic climate and any other issues that you feel are necessary to provide the reader with a comprehensive understanding of the background of the case.

The Ohio Transition to Teaching (TtT) Project was a 5-year project funded by the U.S. Department of Education from 2002-2007. The mission of Ohio TtT was to enhance Ohio's alternative licensure pathway by supporting recruitment, professional

DOI: 10.4018/978-1-60566-870-3.ch009

development, placement, and retention of teachers, specifically in its high-need, hard-to-staff rural and urban public school districts. Statewide partners included mostly large urban school districts such as Cleveland, Cincinnati, and Columbus, working in partnership with the Ohio State University (OSU) and the Ohio Department of Education (ODE) to support ODE's alternative licensure route to teaching. The project provided access and support through an Online Resource Center to qualified applicants as newly hired educators. The Ohio TtT website, www.ohiottt.com, was referred to as an interactive online learning community providing participants with a one-stop resource for (1) online test preparation for the *Praxis II: Principles of Learning and Teaching test*; (2) online collaboration with other Ohio TtT alternative licensure teachers across the state; (3) sharing expertise from both ODE licensure specialists and master teachers for professional advice; and (4) online resources, links, and lesson plans to get new teachers started in teaching their subject area content, primarily in the shortage areas of math and science. As the Ohio Transition to Teaching Project was funded by the USDE, Office of Innovation and Improvement, this project was unique and allowed for flexible use of resources.

Management of the project resided at the Ohio State University to direct financial resources to support project goals identified in the grant proposal. Because this was a nonprofit venture, there were no sales, only the investment of approximately $2.5 million over 5 years to support 218 Ohio alternative route teachers with stipends as well as project staff of OSU and ODE with outside contractors, namely the technology company for website and e-tool development and production. Strategic planning was led by the Principal Investigator within the organizational culture and structure of OSU's Research Foundation and the College of Education and Human Ecology's Center on Education and Training for Employment. The economic climate was characterized by the state's teacher shortages, primarily in Ohio's large urban districts. Recruitment focused on potential entrants to education from among three groups: (1) recent college graduates (primarily with degrees in science or math to meet shortages in these subject/content areas); (2) career switchers from business, retirees, including military personnel (Troops to Teachers), who had bachelor degrees and practice in their fields; and (3) local paraprofessional staff, currently working in the high-need schools who wanted to become teachers and teach in their current school districts.

The issue or opportunity of this project was to expand the teacher workforce in Ohio. Recruiting was required to attract a new pool of teachers – especially for high-need, hard-to-staff school districts – primarily in the content areas of math and science. The needs of these school districts were unique, and the combination of the rate of retirements of educators, the high attrition rates of new teachers, and the continued growth in student enrollments, especially in minority populations, is was putting pressure on providing an increased supply of highly qualified teachers (U.S. Department of Education, 2004). At the same time, the career literature suggests that adults are faced with changing realities in both their work and life. Whether it is a spouse being transferred, early retirement, or changing work interests, many adults will hold 5 to 10 different jobs over a lifetime. Schlossberg, Waters, and Goodman (1995) define a transition as "life events entailing change" (p. 18) and describe it further as an event that results in new relationships, routines, assumptions, or roles. Some adults are interested in growing and learning whereas others may be forced to grow and learn due to life changes requiring them to seek new work opportunities. The goal of the Ohio TtT project was to make matches between high-need, hard-to-staff school district openings and the supply of new teachers wanting to enter the profession in Ohio. A problem is that, although these adults coming into the education profession often had past work experiences with their content area that transferred to the teaching job, they still required pedagogical training. Learning how to teach the content to students while on an alternative route required Ohio TtT to support them at a distance in their transition to the teaching profession.

Regardless, these adults were interested in transitioning to the field of K-12 education and needed to learn a new set of skills or transfer content knowledge to a new profession. Changing careers to become a new teacher requires certain knowledge, skills, abilities, and motivations. Today, more teachers need to be trained in nontraditional routes to teacher certification to fill the gap. In Ohio, several high-need, hard-to-staff schools – primarily in urban districts – received support for their alternatively licensed teachers to meet shortages in specific content areas. With funding from the U.S. Department of Education, Office of Innovation and Improvement, under Ohio's Transition to Teaching Program, this nontraditional group received the support they needed in both pedagogical course content and in passing the Praxis II: Principles of Learning and Teaching Test (7-12).

The ***context*** for this case is specific to Ohio. In 1996, legislation was adopted to allow for an alternative educator license (AEL), which the Ohio Department of Education (ODE) made effective in 1998. The AEL was viewed as a win-win solution in that it met both the needs of those interested in transitioning to the teaching profession in a shorter time frame with the immediate need to be employed, and the needs of Ohio school districts experiencing a shortage of teachers, especially in math and science. ODE uses the same licensure selection tests, the Praxis II Series, for both those graduating from traditional pathways via Ohio's institutes of higher education and for the AEL teacher's selection procedure for initial licensure. The subject matter Praxis test is used as an entrance criterion; continuation into the teaching profession on a provisional license is secured by passing the Principles of Learning and Teaching test. The National Council on Teacher Quality (2005) suggests the most significant criterion for policy makers is a teacher's academic performance because it consistently has a measurable effect on student achievement. Ohio does make sure an AEL teacher is "content qualified" by using a candidate's previous performance on academic content (transcript in subject area to be taught) and score on the Praxis subject matter test to determine the candidate's entry as a teacher of record in an Ohio classroom. The AEL teacher then has 2 years to obtain the pedagogical content knowledge to become "teaching qualified" through performance on pedagogical coursework, passing of the Praxis II: Principles of Learning and Teaching test, and actual teaching performance in the classroom.

The problem was that many on the AEL pathway were not passing the test. Becoming "teaching qualified" according to licensure requirements required these teachers to pass the Praxis II: PLT. Educational Testing Service's (ETS) Praxis II: PLT (7-12) is a test of pedagogical knowledge for grade levels 7 through 12. **Pedagogy** is defined by Woolfolk (2005) as the process of teaching and learning for children. Grossman (2005) provides two major categories of pedagogy as actual preparation for professional practice: (1) classroom instruction and interaction with students and (2) tasks and assignments. It is from these two major areas of instructional strategies that the "blended approach" results. Support for passing the Praxis II became one service offered by the Ohio Transition to Teaching Project.

This study focused on the formative evaluation data using a case study, phenomenological approach. The phenomenon, a test preparation process for understanding pedagogical content knowledge, evolved from a face-to-face version to an online, or e-format, and currently has a hybrid or "blended approach" as a result of the evaluation feedback. Qualitative data sources were collected and analyzed as part of a continuous process improvement. Participants' paper evaluations collected after face-to-face sessions, along with e-mails, threaded discussion data archived from forums posted in the Ohio TtT online learning community at www.ohiottt.com, and video cases of *Praxis Survivors* sharing their digital stories were rich sources of information. Examples of these qualitative sources are embedded within this paper, demonstrating how they were used to improve and inform future instructional design, development, and implementation of the Praxis II: Principles of Teaching and Learning test preparation for Ohio TtT participants to address adult learner needs.

This case study describes how an instructional designer with a subject matter expert analyzed and compared quantitative ratings and qualitative comments as sources from program evaluations to determine how best to support adult learners in their test preparation process. While working with a technology company, the team improved the original face-to-face (F2F) workshop and converted it into a second format option. This second format for delivery was an online, self-paced "electronic" tutorial (referred to as e-tutorial for the rest of this case). This e-option now allowed the participants, who where dispersed all over the state and would have had to travel some distance, to be able to learn via the test preparation program and practice for the test individually in their own time and space. The Ohio Transition to Teaching project used both formative and summative evaluation of its face-to-face *Praxis II: Principles of Learning and Teaching (PLT) Test Preparation workshop* and its e-tutorials (referred to as the e-PLT for electronic Praxis II: Principles of Learning and Teaching test) to inform yet a third option – a hybrid or blended approach. The blended approach has been defined as combining online and face-to-face instruction (Reay, 2001; Rooney, 2003; Sands, 2002; Ward & LaBranche, 2003; Young, 2002). Bershin (2004) expanded the definition to include a combination of different media (such as technologies, activities, and types of events) to create a program for a specific audience. The emphasis in blended learning programs is that the e-learning is complimented with an instructor lead and other live formats. The concept of blending learning basically combines two models of instruction – a traditional face-to-face learning system and distributed learning systems – using computer-based technologies as the medium for these two learning systems coming together. At one time, distributed learning environments emphasized learner-material interactions, whereas face-to-face emphasized the teacher-learner human interactions (Graham, 2006; Moore, 1989). Wagner (1994) suggested a definition for interactions "as reciprocal events that require at least two objects and two actions. Interactions occur when the objects and events mutually influence one another" (p.45). The interactive technologies available today allow for creating a "blend" of the effective aspects of these two systems – where one focuses on the content to be learned and the other on the learner's understanding of it. The challenge is in balancing the needs of the instructor with that of the learner.

The traditional face-to-face approach gives the instructor and learner the chance to physically meet as well as provides for immediate feedback. However, as Brandsford, Brown, and Cocking (1999) proposed, new technologies provide opportunities for creating new learning environments that extend the possibilities of the old technologies such as books, blackboards, and linear, one-way communications media, such as radio and television shows, as well as offering new possibilities-such as online interactive technologies to enhance that initial face-to-face meeting. Technologies today provide a powerful vehicle for extending student capabilities and offer an active, social context for supporting learning that was not before possible. For example, "Preliminary research has shown that computer-mediated communication supports a range of instrumental, informational, social, and emotional exchanges in work and leisure contexts" (Koku & Wellman, 2002, p. 3). This consideration is important in the design of adult learning systems where the student requires, simultaneously, just-in-time knowledge, flexibility, and social/emotional support. Technology provides for communication and the ability to connect learners.

It is not surprising that distance education, which is usually defined (Moore, 1972, 1989) by a physical separation (time and geography) between the teacher and the learner in an educational process, is associated with adult education literature. Correspondence courses have their history in primarily paper-based, self-instructional systems. In providing instruction at a distance, the variables of course structure, dialogue, and learner autonomy have been identified in the theory of transactional distance (Moore, 1972). Moore and Kearsley (1996) further describe transactional distance as the relationship with interactions between the learner to instructor, learner to content, and learner to learner. Interaction appears to be the variable with the most interest regarding the quality of educational experience when using technology in an online learning system.

Course design for adults addresses the following: how the course is structured regarding the approach — the amount of dialogue, communication or interaction between instructor, content, and learners during instruction — and how autonomous, motivated, or self-directed the learner is to engage in the instructional experience. All facets of course design affect the quality and outcomes of an online learning experience. Distance education systems have historically had appeal to adults in their quest for self-directedness. Thus, the interactive technologies of today have been able to improve distance learning by making access to information easier. As adults advance into the information age, it is not surprising to find that technology has revolutionized their learning. Interactive technologies have become important instruments in learning for today's students. Many adults use technology as part of their learning solution. Interpersonal interactions through e-mail, multiple exchanges of material through threaded discussions, emotional support through blogs, and collegial trust also help to characterize the learning communities that result (Koku & Wellman, 2002). Their research directed designers of online learning to examine the social networks of community members and how their use of media facilitates or constrains learning. Several studies have shown that interactive technologies allow the instructor/facilitator to engage with learners by building a learning community that transcends time and space to meet the needs of busy adults (Rovai & Jordan, 2004).

Berge (1995) has suggested it is the responsibility of those creating the learning system, as providers of a blended or hybrid approach, to address the managerial, pedagogical, technical, and social needs of the learners. A blended approach that uses technology now allows for a better way to address all four of these issues. First, management of the learning is enhanced by a blended approach because of the multiple opportunities for feedback from the participants – both face to face and online. Second, from a pedagogical perspective, a blended approach allows for the lower levels of cognitive objectives to be addressed more by the technology whereas the higher order objectives might be better observed and assessed in a face-to-face session. Third, to address the technical issues, a blended approach also allows for the program to extend and distribute the learning resources to adults during times and places that are more beneficial to meet their individual needs, but allowing for consistency in delivery of the information. Finally, to address the social benefits of a blended approach, the facilitator can begin to develop a learning community. Because learning is essentially a social experience, we will discuss how a blended approach supports interaction among peers to begin to enhance an adult learning experience by creating a learning community for both emotional and social support. For example, Edelstein and Edwards (2002) said that one of the chief tenets for a successful and engaging online course is the development of an effective system that provides ongoing student interaction. In order to accomplish the appropriate level of interaction, the facilitator often uses threaded discussions to promote interaction if needed – especially when face-to-face interactions are not possible, such as in an online setting. With a facilitator able to design and manage additional support for participants learning the intended learning objectives, threaded discussion provides the opportunity for both the e-facilitator (or e-Coach) to review issues or problems with the subject matter content.

Learning Communities

Many professions have communities of practice that support their participants with learning opportunities in a particular field, such as medicine, teaching, social work, etc. These groups of adults learning together support the notion of social learning and suggest that continuous education involves "interactions" among learners on some level – whether face-to-face or mediated. According to Swan and Shea (2005), three common themes cross over many of the cited social learning theories: cognition, which is situated in particular social contexts; knowing, which is distributed across groups; and learning, which takes place in communities. It is these three themes taken together that suggest that learning requires human interaction in a community. Unfortunately, the literature is not clear as to a definitive definition of learning community, but according to Swan and Shea (2005), learning communities are rooted in the observation that knowledge and learning stem essentially from culture – which consists of shared values, beliefs, languages, and ways of doing things. One definition from Yarnit (2000) is relevant to this study as it addresses how learning translates to the lives – both learning and earning situations - of the adults involved:

A learning community addresses the learning needs of its locality though partnership. It uses the strengths of social and institutional relationships to bring about cultural shifts in perception of the value of learning. Learning communities explicitly use learning as a way of promoting social cohesion, regeneration and economic development, which involve all parts of the community. (Yarnit, 2000, p. 11)

Other definitions support the notion of social bonds built through cohesive learning experiences and sharing knowledge to promote social improvement (Kearns, McDonald, Candy, Knights, & Papadopoulos, 1999). Kilpatrick, Barrett, and Jones (2003) suggest that, to be a learning community, the following components would be observed: shared purpose, collaboration/partnership for learning, respect for diversity, and enhance potential and outcomes.

In education terms, a learning community has been defined in a number of ways. Hord (1997) includes it as extending classroom practice to the community; bringing community personnel into the school to enhance the curriculum and learning tasks of students, or engaging students, teachers, and administrators simultaneously in learning. Another definition of learning communities relevant to adult learning systems is:

A learning community is any one of a variety of curricular structures that link together several existing courses – or actually restructure the curricular material entirely – so that students have opportunities for deeper understanding of and integration of the materials they are learning, and more interaction with one another and their teachers as fellow participants in the learning enterprise. (Gabelnich, MacGregor, Matthews, & Smith, 1990, p. 19)

Because of learning communities, researchers have suggested that knowledge is socially constructed within a community and cannot be separated from practice (Wenger, 1997). Specifically, the idea of a "community of practice," according to Wenger (1997), is characterized by mutual engagement, joint enterprise, shared repertoire, and negotiated meaning. This notion of a community of practice is that all learning is situated in practice and that all practice is social in nature.

According to Palloff and Pratt (1999), "it is the relationships and interactions among people through which knowledge is primarily generated. The learning community takes on new proportions in this environment and consequently must be nurtured and developed so as to be an effective vehicle for education." (p. 15). Opportunities exist to create learning communities because of the computer-mediated technologies available today to connect learners. A blended system supported by e-mail, instant messaging, and threaded discussions can facilitate participants' being resources for each other as well as connect the learner to the facilitator – thus enhancing any face-to-face meeting the learners would have together, particularly when the adult learners complete a program as a cohort. Imel (2002) has defined cohorts as "groups of students who enroll at the same time and complete a program by taking the same courses at the same time, a process that is sometimes referred to as lock-step" (p. 1).

Essentially, we took the best of the two previous formats to deliver the test preparation content in part face-to-face using technology and integrating them into yet a third format option with the support of the website's interactive tools, such as threaded discussions, live chats, and videoconferencing capabilities.

Currently, Project KNOTtT (Kansas, Nevada, Ohio and Texas Transition to Teaching) is continuing the work started by Ohio TtT when the U.S. Department of Education funded another round of programs from 2007-2012. We are now expanding the development of the online learning community with e-test preparation tutorials to support learners in their review of the subject matter area content tests, specifically math, science, special education, English as second language, and foreign language. These content tests are also required by many states for licensure as part of a teacher pursuing a nontraditional route to the education profession. All of these e-tutorial tools are supported with e-Coaching using social networking tools and Google applications in a Moodle environment.

SETTING THE STAGE

Describe the technology utilization, advancements, management practices and philosophies, etc., of the organization prior to initiation of the project/experience described in the case, as well as the players involved.

An overview of this case must begin with some knowledge of the organization setting, the Ohio TtT participant learners, and the content of the *Praxis II Exams.* First, the context of this project required serving individual teachers residing in separate geographical locations within Ohio often in the most challenging school districts and difficult teaching conditions. Ohio TtT participants often conducted their professional development in the evenings or on weekends and had access to the web in their homes. The content analysis was available from the "Praxis Tests at a Glance" developed by the Educational Testing Service (ETS). The structure and content of the test was used to assess a beginning teacher on his or her content and pedagogical knowledge with a passing score (decided by each state) in order to be licensed to teach in 20+ states, including Ohio where the cut score is 165. We found that the teachers were having difficulty passing because they had not taken a test in years and were not familiar with the content.

The primary focus of the Ohio TtT online learning community was to support these new teachers to pass a test of pedagogical knowledge during a 2-year timeframe. The alternative license in Ohio requires this hurdle to be jumped in order to receive a provisional license and remain teaching in the State of Ohio (2003).

Descriptions of Ohio TtT Praxis II: PLT Test Preparation formats

- Face-to-face instructor meeting for a 7-hour workshop using a paper-based system of materials and including a 2-hour practice test simulation and take-home, self-paced paper materials with two additional case studies and approximately 200 multiple-choice practice items. Feedback given immediately after in-class activities and after completion of the exam.
- e-PLT or online self-paced instruction of seven modules with three practice case studies and one test bank, a 2-hour simulated practice test, and five additional test banks of 24 items, each requiring up to 12 hours to complete. Face-to-face or e-coaching was available per individual need.
- Hybrid or blended approach delivered the essential face-to-face components in 3 hours to introduce the test preparation process, reduce anxiety of participants, answer questions and provide them direct feedback primarily on the case study written responses. Participants were introduced to the e-PLT to continue their test preparation and shown e-tools to engage in facilitator or peer support from the online learning community.

According to Heinze and Proctor (2004), blended learning is facilitated by the effective combination of different modes of delivery, models of teaching, and styles of learning, and it is connected through communication and technology among all parties involved with that course. Kirschner, Clark, and Sweller (2006) recommend face-to-face (F2F) guidance early in the process, "to be faded as learners gain expertise" (p. 5). Essentially, the third option for Ohio TtT AEL teachers was to have the instructor appear for several hours to deliver the essentials of the "why" and "how" and let them review the content on their own.

The technology used to connect participants in an online learning community was a website. Access to www.ohiottt.com allowed for communication through e-tools which included threaded discussions, live video conferencing in both a virtual community or one on one chat environments. The online learning community supported individual or small group formal or informal chats among staff and participants, as well as among participants across the state who may share a common content area or classroom management dilemma. The website also has the ability to host asynchronous resources, webinars, videos, and the electronic test preparation course for the content of the Praxis II: Principles of Learning and Teaching Test (PLT) referred to as the e-PLT. The website was Sharable Content Object Reference Model (SCORM) compliant with a learning management system (LMS) to track participant activity and that was customized to the grant requirements. The website at www.ohiottt.com was restricted to participant use only and was password protected. The website hosted the e-test preparation tools that were created from the face to face (F2F) format. The content and course structure were already available.

The rest of this case focuses on how the technology partner worked with OSU to co-develop an instructional design process that resulted in a series of customized, electronic instructional modules. Program design, implementation, and assessment of pedagogy had to be transformed into a new e-pathway for the delivery of the test preparation program for nontraditional teachers in the event that they could not attend the F2F workshop.

Because of technology, these new electronic and interactive modules provided a combination of graphics, voiceover, and text developed using Flash programming to make the content come alive without an instructor – thus providing an interactive, self-instructional experience.

Project management was used in the instructional design process. The ADDIE (Analyze,

Design, Develop, Implement, Evaluate) model was the overarching framework. Review of the following relevant framing concepts for this discussion may be appropriate here: distance education, adult learners, instructional design, the ADDIE Model, and continuous process improvement. Specifically, identifying the needs of adult learners as they both transition and transform into teachers is especially significant to understanding their motivation to learn.

Distance education is defined by Moore and Kearsley (1996) "as planned learning that normally occurs in a different place from teaching and as a result requires special techniques of course design, special instructional techniques, special method of communication by electronic and other technology, as well as special organizational and administrative arrangements" (p. 2).

Adult learners were the Ohio TtT participants. The actual instructional process began with Ohio TtT conducting a needs assessment that identified the diversity of these learners. They were mainly minorities, often English as a second language was a challenge, and they represented a larger than average number of males compared to the traditional teaching population. These learners were also working as beginning teachers in school districts and had already passed their Praxis content test in the subject they were teaching. However, they were often seen by their peers as less than qualified because they had not gone through the traditional teacher preparation pathway. These learners were already familiar with the Praxis multiple-choice format and were required to take a course in educational psychology as part of their licensure requirement during the 2-year period of employment. Examples include Pablo, a chemist from Sherwin William in Cleveland, who said, "I just wanted to teach chemistry after I retired"; Metrice from Columbus Metropolitan Library who said, "I want to work as a school librarian in my son's school to make sure he stays out of trouble"; and Karen, a recent college graduate majoring in math, who returned to her rural hometown of Circleville "so I can teach what I love at my old high school" These are all adults in professions who required both formal and informal education to make the change. *Andragogy* is the term used to describe how to teach adults (Knowles, 1970). Because adults like these have experience, motivation, and often use a problem-solving approach within their personal contextual situations to reflect and act to achieve a personal goal, teaching them is different than teaching children and teenagers.

Instructional design is the practice of arranging media, which includes communication technologies, with content to help instructors and learners transfer knowledge in the most effective way. Ideally, the process is informed by instructionally tested theories of learning and is measured. Instructional Systems Design (ISD) is the model used for efficiently creating instructional materials in a learning process focused on achieving a specific outcome. For this study, the ADDIE ISD Model was employed to help Ohio TtT participants achieve a 165 or greater on their Praxis II: PLT tests. Five phases with essential questions are employed in the ADDIE Model (Liu, 2008):

1. *Analyze* – What are the learner characteristics, the task to be accomplished, and the environment or contextual constraints?
2. *Design* – What are the learning objectives and the instructional approach to use?
3. *Develop* – How will we create the instructional materials?
4. *Implement* – How will we deliver and/or distribute the instruction?
5. *Evaluate* – How can we make sure the instruction is achieving the goals?

Related to ISD is the idea of **continuous process improvement** as a meta concept for the management system employed to make gradual or incremental changes to better a process by gathering data to inform the change. The term is derived from the word Kaizen, which in Japanese means "good change." Change provides the structure of ISD as a continuous learning system with evaluation data serving as feedback to inform improvement.

Our case then applied the five phases of the ADDIE model and transformed the F2F content into the e-Test Preparation format. The e-PLT team converted these phases into the following 12 week process: (1) Subject Matter Expert (SME) review and Content Collection; (2) Design Document, which includes identifying the learning objectives and course outline; (3) Storyboard Phase, which essentially is a word document that includes details of each slide within the module complete with graphic, voice, and text descriptions; (4) Production Phase, which includes an Alpha Link integrating all the graphics, text, and voiceover with the SME conducting both an Alpha and Beta Review before the last phase; and (5) Final Review or signoff where the course is evaluated upon completion of any necessary changes by the SME and, in this case, the Principal Investigator of the project. By starting with a subject matter expert (SME) and F2F content using a step-by-step process of chunking content or information and repackaging it into self-instructional modules was the ultimate outcome – with participants passing the test.

The phases included a content analysis, course design by the SME and instructional designer, with customized development being done by a technology partner outside the university to allow for flexibility with technology and programming that did not exist easily within the organizational structure. The final result was seven self-instructional e-modules that reviewed both the content of the test and the structure and requirements of the test, plus a practice test with four (4) case scenarios and 24 multiple-choice questions with an additional five (5) test banks for more practice with multiple choice. The e-module structure is identified below:

- **Praxis II: Principles of Learning and Teaching (7-12) Review Modules.**
  - Module 1: Praxis PLT Test Overview
  - Module 2: Content Area Review
  - Module 3: Case Studies
  - Module 4: Constructed-Response Questions
  - Module 5: Multiple-Choice Questions
  - Module 6: Tips For the Test
  - Module 7: PLT Practice Test
- **Praxis II: PLT (7-12) Practice Test Banks with 24 multiple choice items in each.**
  - Test Bank 1
  - Test Bank 2
  - Test Bank 3
  - Test Bank 4
  - Test Bank 5

The evaluation data provided the information to develop and offer this test preparation program in three formats: face-to-face (F2F), online as the e-PLT, and in a blended or hybrid model that provides an introduction with face-to-face or online coaching. The key players were three: OSU, which supplied the instructional model; the subject matter expert (SME), who was an independent contractor (former professor with expertise in pedagogical content as well as knowledge of the test competencies and performance management requirements of a beginning teacher); and to complete the system, a third-party technology partner, contracted to perform the Flash programming, create graphics, and produce the SCORM-compliant modules.

CASE DESCRIPTION

Provide a detailed overview of the project/experience described in the case in terms of technology concerns, technology components, management and organizational concerns, as well as any other

Table 1. Praxis II: PLT Descriptive Statistics for Each Ohio Teacher Group 2005-07

	Praxis II: PLT for 3 Ohio Groups					
	Ohio TtT (N = 95)		AEL (N = 260)		Traditional (N = 300)	
Age (average)	39.9		38.7		26.6	
Mean PLT Score	177.19 *		173.08		176.63	
Race	White	Non-White	White	Non-White	White	Non-White
	71.6%	28.4%	81.6%	18.4%	89.3%	10.7%
Gender	Male	Female	Male	Female	Male	Female
	45.3%	54.7%	41.8%	58.2%	49.7%	50.3%

* Note: Ohio TtT mean Praxis PLT test score was statistically significant when compared to Ohio AEL

information that would provide a comprehensive description of the project/experience to the reader. Furthermore, describe any related issues (e.g., culture, philosophies, opinions) practiced within the organization that have impacted the project/experience planning, implementation and overall management.

The State of Ohio has a law that requires that all new teachers – both those who are alternatively licensed (AEL) and those who are traditionally trained –pass the Educational Testing Service's (ETS) Praxis II: Principles of Learning and Teaching (PLT) grade 7-12 test. Both groups are required to pass this test; however, those on an AEL pathway do not **consistently** receive the pedagogical content knowledge required to achieve a passing score on this test. Because the Ohio TtT Project focused on those teachers on the AEL pathway in Ohio, a solution to passing the test was provided. The Ohio TtT Project built a test preparation process for the Praxis II: PLT as part of an online learning community (www.ohiottt.com) to support those serving in high-need, hard-to-staff school districts across the state. The program began as a face-to-face workshop and then was transformed into an electronic version. The present form is provided as a blended or hybrid model – capturing the best of face-to-face and the e-learning format. Formative evaluation data was the vehicle to improve program design, implementation, and assessment of pedagogical knowledge.

The results of a summative evaluation compared the test scores of the Ohio TtT to Ohio's other AELs and traditionally trained teachers in 2005-07. Specifically, Praxis PLT scores from Ohio TtT Cohorts 3 and 4 served as the comparison group. Data were collated from 95 participants from the population of Ohio TtT participants in Cohorts 3 and 4 who had taken the Praxis II: PLT as of September 8, 2007 to the other Ohio AELs and Ohio Traditional route Praxis scores that were available for the same time period.This study actually supported that the test preparation made a difference in the Ohio TtT participants' test scores. This was especially true of minorities (Gimbert, Moore, & Sahin, 2008). See the demographic distribution in Table 1.

Descriptions of the Various Instructional Pathways that Emerged as Solutions

A process, as defined by Merriam-Webster (2003), is "to subject to a treatment or set of procedures." The ADDIE process was used by the Ohio TtT instructional designer, so all the questions were initially answered with informal data collected and informed by theory. For example, the first cohort of Ohio TtT participants filled in a needs assessment and a form with data required by the U.S. Department of Education. The instructional designer also used learning theories from the pedagogical content to inform the workshop structure. From the F2F data, the essential workshop foundation for a completely online, self-paced tutorial was developed for the next 50+ participants in Cohort 2. Referred to as the e-PLT, the face-to-face content modules were reformatted electronically for this cohort as part of a pilot. This provided the opportunity to use new media and communication technologies and the online learning community could support some of the required interactions from the instructor and the peer support needed by the participants. Next, Cohort 3 had the opportunity to provide additional feedback after the pilot version in the spring of 2004, with the final e-PLT being rolled out at the Summer Institute (SI) 2004. Both the face-to-face workshop and the e-format were provided as an option to this group. Many who were not able to come to the SI were thrilled to have access to the online learning community – and especially this tool – because they could not travel to Columbus.

Based on the SI data from Cohort 3 – from both the paper evaluations as well as what participants had shared online in e-mails and threaded discussions – a blended or hybrid pathway that offered the best of two previous formats was developed. The essential elements of the face-to-face and online content foundation were piloted on Cohort 4. Given that the Ohio TtT accommodated 80+ participants in the last year of the program, the two test preparation formats were independently updated and offered at the last SI. A panel of "Praxis Survivors" was convened for this last session so that Cohort 4 participants could learn from their peers. The stories represented multiple perspectives and one Ohio TtT participant from each of the previous three cohorts was introduced, shared their personal experiences, and addressed questions from Cohort 4. Their stories were recorded and posted on www.ohiottt.com to share with nonattendees. More important, they became yet another source of data to use to inform the test preparation program.

According to Eterovich (1980), the ancient Greeks referred to the process of praxis as an activity engaged in by free men [sic]. Specifically, Aristotle said it was one of the three basic activities of men that corresponded to three types of knowledge: *theoria* (with the goal of truth); *poiesis* (with the goal of production), and *praxis* meaning practical (with the goal of action). In education, "praxis" is often used by educators to describe the process of experiential learning – or making sense of concrete experiences (Dewey, 1933; Kolb & Kolb, 2001; Schön, 1983). Learning is described as the acquisition of knowledge, skills, and understanding and is the product of experience and the goal of education. By providing e-modules with a simulated Praxis test experience, the Ohio TtT provided a process of experiential learning for many participants wanting to transition to a career in teaching. Therefore, "praxis" is a perfect word to describe the process of responding to our Ohio TtT participants' feedback in a continuous instructional program design, implementation, and assessment of pedagogy where change and action resulted in multiple pathways for a test preparation program for nontraditional teacher candidates based on these teachers' own praxis with the Praxis! This is a case of how Ohio TtT listened continuously – and collectively transformed the instructional process to benefit those going through the transition to teaching.

As an institution of higher education leading this effort, we begin to describe our process of continuous learning and instruction, with the adult learner in mind and change as a central theme. The radical adult educator Paulo Freire (1974) wrote about praxis as change in his *Pedagogy of the Oppressed.* According to Freire, freedom will be the result of *praxis* – which he defines as "informed action." When one is informed, or has gained knowledge, then one can act when a balance between theory and practice is achieved. We believe this was the result of the Ohio TtT

Praxis Test Preparation process for the Principles of Learning and Teaching test.

This case relied heavily on formative and summative evaluation data using a case study, with a phenomenological approach – where the phenomenon was an instructional e-test preparation process for demonstrating pedagogical content knowledge on the Praxis II: PLT. The case method for addressing adult learning (Merriam, 1998) in an instructional process evolved from a face-to-face version to an online, or e-format, and has most recently emerged into a hybrid or "blended approach" as a result of these multiple sources of program feedback. Qualitative data sources and comments, from both formal and informal methods, were collected as part of a continuous process improvement. Participants' paper evaluations collected after face-to-face sessions, along with e-mails, threaded discussion data archived from forums posted in the Ohio TtT online learning community at www.ohiottt.com, and video cases of "Praxis Survivors" sharing their digital stories were a rich source of information. Qualitative data sources contributed to improved performance on the test (the outcome) and will inform future instructional design, development, and implementation of a current series of Praxis II: Subject Matter Test preparation for Project KNOT TtT participants.

CURRENT CHALLENGES FACING THE ORGANIZATION

In detail, describe some of the challenges and problems that the organization faced at the completion of the project/experience. Furthermore, it would be helpful to the reader if you describe the current status of the aforementioned challenges and problems.

The largest problem Ohio TtT faced at the completion of the project was to keep the resource available for future Ohio Alternative Certification participants. Because of the 5-year limitation on this federally funded project, sustainability is an issue. It is possible that if we had not applied for another Transition to Teaching grant, the e-PLT tool could have been marketed and sold by the organization. However, the management team put their efforts into writing another grant proposal to allow the e-tool to remain supported and extend its reach into three other states while adding capacity to build more e-tools.

The current "change" or challenge to Project KNOTtT, a new 5-year Transition to Teaching grant (2007-2012), is to address a variety of adult learner needs and individual state licensure requirements in order to pass their initial licensure test in three Praxis states (**K**ansas, **N**evada, **O**hio) and with the Texas Examination of Educator Standards (TExES) in **T**exas, which are similar as both are produced by ETS, yet subscribe to unique testing format and slight variations in content. Based on the success of the e-PLT test preparation tutorial, the Ohio State University proposed to continue building the online learning community with the e-PLT by adding both new partner state participants and new tools to expand Ohio TtT into Project KNOTtT.

Project KNOTtT is designed to support recruitment, selection, training, coaching, and mentoring to retain teachers in high-need, hard–to-staff school districts. As a national initiative, **Project KNOTtT** addresses the teacher shortages in the subject areas of math, science, English/language arts, foreign languages, English as a second language, and special education (K-12). This 5-year project serves 545 new teachers pursuing nontraditional routes to certification in four states: **K**ansas, **N**evada, **O**hio, and **T**exas. Led by the Ohio State University, these four states will collaborate with national partners to knot together three strategic strands of support for alternative certification programs at a distance:

1. Strand 1: Online Learning Community and e-Tutorials to expand the e-test preparation support new teachers

Figure 1. Project KNOTtT organizational structure - three strategic strands and partners

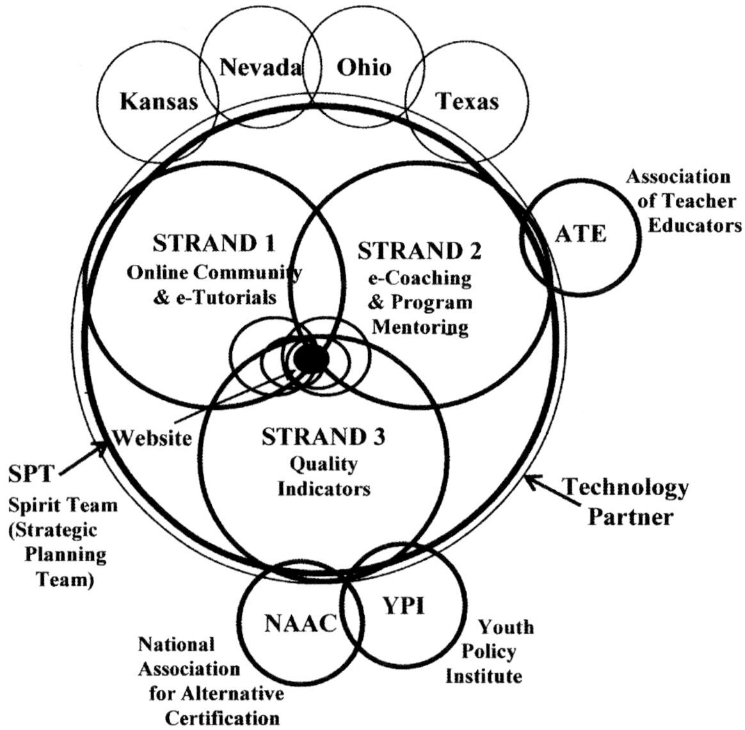

2. Strand 2: e-Coaching and Program Mentoring
3. Strand 3: Quality Indicators for Nontraditional Teacher Preparation Programs

An organizational representation of Project KNOTtT's organizational structure with the three strategic strands and its partners is presented in Figure 1.

Project KNOTtT's national partners include the National Association for Alternative Certification, Association of Teacher Educators, Youth Policy Institute, and the Thomas B. Fordham Institute. The goal of Project KNOTtT is to produce high-quality programs for nontraditional routes to licensure – many of which have "blended approaches" to their instruction and programming – through the synergy of connecting programs, participants, and partners to three strategic strands. This case of e-tutorials has now "transformed" into part of *Strategic Strand 1* which is focused on building an online learning community to host a variety of online resources including the e-test preparation tutorials to help teachers pass their state mandated subject matter tests. Strand 2, e-Coaching, will facilitate both test preparation while supporting new teachers with examples and demonstrations of good teaching practice. Strand 3 will provide a framework for assessing program quality for non-traditional teacher preparation programs. The following figure provides an organizational, conceptual image of the interdependency of these three strategic strands that are now Project KNOTtT. At the heart of Project KNOTtT where all the strategic strands come together, is the website (http://www.knottt.net) which provides access to the online learning community – representing both the opportunity *and* the challenge for communication and social networking for staff, partners, and participants as part of an community of practice.

DISCUSSION AND CONCLUSION

The ***issue*** of becoming both "content" and "pedagogically qualified" through passing state mandated tests is a challenge for many Transition to Teaching Projects. High-stakes testing prevents many potential teachers from entering the profession – although they have performed successfully within the context of a nontraditional teacher education program. This case emphasizes the essentials of how adults learn and make the transition to teaching and the support of an online learning community and its tools. How these learner teachers actually go about passing these high-stakes tests is a topic for discussion.

With all the technology available today, what is the best instructional "process" of designing and delivering e-tutorials? Do we need multiple formats or is blended instruction the best? What more can the education technology options offer us to extend instruction to their workplace? How can these teachers integrate professional development into their daily lives using the communication technologies of the online learning community? What is the role of the instructional coaches and how can they support the transfer of pedagogical practice beyond test taking? Can the community of practice of nontraditional teachers be a source of retention for school districts?

This case study was one way to examine how the Ohio TtT project and Project KNOTtT could improve its Praxis II: PLT test preparation program for nontraditional teachers as diverse learners. Examples of narrative descriptions from video case stories, feedback from threaded discussions related to Praxis test experiences captured and archived on the online learning community website, and the actual written qualitative comments from the instruction are used to further define the case of pathways for test preparation program for nontraditional teacher candidates. The answers found in the qualitative comments are from those who were both successful and not successful. Continuous process improvement provides the instructional team a way to triangulate data sources to improve the current design of a "blended approach" with the support of online e-coaches by merging face-to-face and e-approaches, as well as inform future test preparation and other program development.

We end with one example of a "Praxis Survivor" story that is the essence of the word Praxis that we addressed at the beginning of the case – which goes beyond just passing the test and gets at the essence of what being an educator is about. Teklai said his passing of the Praxis test enabled him to contribute to U.S. society as both a citizen and an educator in his own immigrant Somali community, contributing to the lifelong learning processes of his students. His story confirms that a TtT participant's journey, through a very complex process of test preparation, actually resulted in praxis as well. Praxis is defined as the process of putting theory, lessons, or skills into practice. Freire (1974) said freedom will be the result of *praxis* –which he defines as "informed action." When one is informed, or has gained the knowledge, then one can act --when a balance between theory and practice is achieved. "Liberation is Praxis: the action and reflection of men and women upon their world in order to transform it" (Freire, 1974, p. 77). We are hopeful that Teklai will transform students while we continue to support his peers as we do the same through our instructional work.

In conclusion, Wolf (2005) says, "We find ourselves in a new world with new terrain" (p. 94). Eisen (2005) recommends "blending diverse educational concepts and methods to meet the transitional needs of adults"(p. 92). To embrace the new terrain, it is especially powerful when an environment is created for a cohort or group of adults that can be supported in a community with both peer advice and mentoring from those with experience (Imel, 2002). Finally, Pappas and Jerman (2004) point out that the only certainty of the future is uncertainty. Thus, their recommendation is "Be Technological" (p. 94) and so we used a blended approach to meet the needs of Ohio TtT adult learners as they transitioned to teaching.

ACKNOWLEDGMENT

This publication is supported by the Transition To Teaching federal grant funded through the No Child Left Behind Act of 2001. The opinions expressed herein do not necessarily reflect the position of the U.S. Department of Education, and no official endorsement by the U.S Department of Education should be inferred.

REFERENCES

Berge, Z. L. (1995). Facilitating computer conferencing: Recommendations from the field. *Educational Technology, 35*(1), 22–30.

Bershin, J. (2004). *The blended book of learning*. San Francisco, CA: Pfeiffer.

Committee on Developments in the Science of Learning. (1999). Technology to support learning. In J. Brandsford, A. Brown & R. Cocking (Eds.), *How people learn: Brain, mind, experience, and school*. Washington, D.C.: National Research Council. Retrieved on May 30, 2000, from http://books.nap.edu/html/howpeople1/ch9.html

Dewey, J. (1933). *How we think.* New York: Heath.

Edelstein, S., & Edwards, J. (2002). If you build it, they will come: Building learning communities through threaded discussion. *Online Journal of Distance Learning Administration, 5*(1). Retrieved from http://www.westga.edu/~distance/ojdla/spring51/edelstein51.html

Eisen, M. J. (2005). Shifts in the landscape of learning: New challenges, new opportunities. In M. A. Wolf (Ed.), *Adulthood: New terrain. New directions for adult and continuing education* (No. 108, pp. 15-26). San Francisco, CA: Jossey-Bass.

Eterovich, F. H. (1980). *Aristotle's Nicomachean ethics: Commentary and analysis*. Washington, D.C.: University Press of America.

Freire, P. (1974). *Pedagogy of the oppressed* (M. B. Ramos, Trans.). NY: Seabury Press.

Gabelnick, F., MacGregor, J., Matthews, R. S., & Smith, B. L. (1990). Learning community models. San Francisco, CA: Jossey-Bass.*New Directions for Teaching and Learning*, (41): 19–37.

Gimbert, B., Moore, M., & Sahin, F. (2008). *A comparison of nontraditionally and traditionally prepared teacher performance on a pedagogical content test in Ohio: The impact of blended preparation.* Paper presented at the Annual Meeting of the American Educational Research Association, New York.

Graham, C. R. (2006). Blended learning systems: Definition, current trends, and future directions. In C. J. Bonk & C. R. Graham (Eds.), *The handbook of blended learning: Global perspectives, local designs* (p. 5). San Francisco, CA: Pfeiffer.

Grossman, P. (2005). Research on pedagogical approaches in teacher education. In M. Cochran-Smith & K. M. Zeichner (Eds.), *Studying teacher education: The report of the AERA panel on research and teacher education*. Mahwah, NJ: Lawrence Erlbaum Associates, Inc.

Heinze, A., & Procter, C. (2004). Reflections on the use of blended learning. *Education in a Changing Environment Conference Proceedings*. Salford, UK: University of Salford. Retrieved from http://www.ece.salford.ac.uk/proceedings/papers/ah_04.rtf

Hord, S. M. (1997). *Professional learning communities: Communities of continuous inquiry and improvement.* Austin, TX: Southwest Educational Development Laboratory.

Imel, S. (2002). *Adult learning in cohort groups. Practice application brief.* Columbus, OH: ERIC Clearinghouse on Adult, Career, and Vocational Education, Ohio State University. Retrieved from http://www.calpro-online.org/ERIC/docgen.asp?tbl=pab&ID=114

Kearns, P., McDonald, R., Candy, P., Knights, S., & Papadopoulos, G. (1999). *VET in the learning age: The challenge of lifelong learning for all.* Canberra, Australia: National Centre for Vocational Education and Training.

Kilpatrick, S., Barrett, M., & Jones, T. (2003). *Defining learning communities.* (Discussion Paper D1/2003. CRLRA Discussion Paper Series). Launceston, Australia: University of Tasmania. Retrieved from http://www.aare.edu.au/03pap/jon03441.pdf

Kirschner, P. A., Sweller, J., & Clark, R. E. (2006). Why minimal guidance during instruction does not work: An analysis of the failure of constructivist, discovery, problem-based, experiential, and inquiry based teaching. *Educational Psychologist, 41*(2), 75–86. doi:10.1207/s15326985ep4102_1

Knowles, M. S. (1970). *The modern practice of adult education: Andragogy versus pedagogy.* NY: Cambridge Books.

Koku, E. F., & Wellman, B. (2002). Scholarly networks as learning communities: The case of the TechNet. In S. Barab & R. Kling (Eds.), *Designing virtual communities in the service of learning* (pp. 299-337). Cambridge: Cambridge University Press.

Kolb, A., & Kolb, D. A. (2001). *Experiential learning theory bibliography 1971-2001.* Boston: McBer and Co.

Liu, G. Z. (2008). Innovating research topics in learning technology: Where are the new blue oceans? *British Journal of Educational Technology, 39*(4), 738–747. doi:10.1111/j.1467-8535.2008.00851.x

Merriam, S. B. (1998). *Qualitative research and case study applications in education.* San Francisco: Jossey-Bass.

Merriam-Webster Inc. (2003). *Merriam-Webster's collegiate dictionary, 11th ed.* Springfield, MA: Britannica.

Moore, M. G. (1972). Learner autonomy: The second dimension of independent learning. *Convergence, 5*(2), 76–88.

Moore, M. G. (1989). Editorial: Three types of interaction. *American Journal of Distance Education, 3*(2), 1-7. Retrieved from http://www.ajde.com/Contents/vol3_2.htm#editorial

Moore, M. G., & Kearsley, G. (1996). *Distance education: A systems view.* New York: Wadsworth Publishing Company.

National Council of Teacher Quality. (2005). *Increasing the odds: How good policies can yield better teachers.* Washington, D.C.: NCTQ. Retrieved from http://www.nctq.org/p/publications/docs/nctq_io_20071129024229.pdf

Palloff, R. M., & Pratt, K. (1999). *Building learning communities in cyberspace: Effective strategies for the online classroom.* San Francisco: Jossey-Bass.

Pappas, J. P., & Jerman, J. (Eds.). (2004). *Developing and delivering adult degree programs. New directions for adult and continuing education* (No. 103). San Francisco: Jossey-Bass.

Reay, J. (2001). Blended learning–a fusion for the future. *Knowledge Management Review, 4*(3), 6.

Rooney, J. E. (2003). Blending learning opportunities to enhance educational programming and meetings. *Association Management, 55*(5), 26–32.

Rovai, A. P., & Jordan, H. M. (2004). Blended learning and sense of community: A comparative analysis with traditionally and fully online graduate course. *International Review of Research in Open and Distance learning, 5*(2), 1-12. Retrieved from http://www.irrodl.org/index.php/irrodl/article/view/192/274

Sands, P. (2002). Inside outside, upside downside: Strategies for connecting online and face-to-face instruction in hybrid courses. *Teaching with Technology Today, 8*(6). Retrieved from http://www.uwsa.edu/ttt/articles/sands2.htm

Schlossberg, N. K., Waters, E. B., & Goodman, J. (1995). *Counseling adults in transition, 2nd ed.* New York: Springer.

Schön, D. (1983). *The reflective practitioner.* NY: Basic Books.

State of Ohio. (2003, January 1). *Teacher education and licensure standards.* Columbus, OH: State Board of Education, Ohio Department of Education.

Swan, K., & Shea, P. (2005). The development of virtual learning communities. In S. R. Hiltz & R. Goldman (Eds.), *Learning together online: Research on asynchronous learning networks* (pp. 239-260). Mahwah, NJ: Erlbaum.

U.S. Department of Education, Office of Innovation and Improvement. (2004). *Alternative routes to teacher certification.* Jessup, MD: Education Publication Center. Retrieved on December 12, 2007, from http://www.ed.gov/admins/tchrqual/recruit/altroutes/report.pdf

Wagner, E. D. (1994). In support of a functional definition of interaction. *American Journal of Distance Education, 8*(2), 6–29. doi:10.1080/08923649409526852

Ward, J., & LaBranche, G. A. (2003). Blended learning: The convergence of e-learning and meetings. *Franchising World, 35*(4), 22-23. Retrieved from http://www.accessmylibrary.com/coms2/summary_0286-23358935_ITM

Wenger, E. (1997). *Communities of practice: Learning, meaning, and identity.* NY: Cambridge University Press.

Wolf, M. A. (Ed.). (2005). *Adulthood: New terrain. New directions for adult and continuing education* (No. 108). San Francisco: Jossey-Bass.

Woolfolk, A. (2005). *Educational psychology: Active learning edition, 9th ed.* NY: Allyn Bacon.

Yarnit, M. (2000). *Towns, cities, and regions in the learning age: A survey of learning communities.* London: LGA Publications.

Young, J. R. (2002, March 22). Hybrid teaching seeks to end the divide between traditional and online instruction. *The Chronicle of Higher Education, 48*(28), A33–A34.

ADDITIONAL SOURCES

Introduction to Instructional Design and the ADDIE Model. http://www.e-learningguru.com/articles/art2_1.htm

Learning-Theories.com http://www.learning-theories.com/addie-model.html

Vavasseur, C. B., & MacGregor, S. K. (2008, Summer). Extending content-focused professional development through online communities of practice. *Journal of Research on Technology in Education, 40*(4), 517–536.

Chapter 10
Finding a Niche through an External Degree-Completion Program

Evan S. Smith
University of Missouri, USA

Terrie Nagel
University of Missouri, USA

EXECUTIVE SUMMARY

The University of Missouri began seriously investigating an external degree- completion program in 2000, working with an existing Bachelor of General Studies Degree Program (BGS). Concerns included the development of Capstone and Writing Intensive courses. The program has entailed training advisors; updating curriculum; revising student services procedures; marketing; and coordinating with other branches of Extension.

BACKGROUND

Distance learning is a competitive field, and many institutions feel compelled to provide courses, generally online (web-based) or on CD-ROM, to students. Online often connotes "degree program" to many. Yet even before online distance education came about, adult educators recognized the importance of external degrees (Houle, 1973).

Bachelor's degrees increase one's marketability (Turner & Monk-Turner, 2001); therefore, degree programs or degree-completion programs are important in distance education. To that end, the University of Missouri's Center for Distance and Independent Study (CDIS), part of University of Missouri Extension, began investigating an undergraduate external degree program in 1998-1999, and in 2001-2004, the program began to materialize. Extension's pursuit of the program included overcoming campus resistance to an external degree, drafting a proposal, and building on an existing on-campus Bachelor of General Studies (BGS) degree program. Justification of the degree-completion program included serving University of Missouri (MU) students who never completed their degrees, military personnel, and adult, non-traditional students, following the precedents of other institutions (Excelsior, Indiana University, and others); drawing on existing courses offered through CDIS; and complementing other MU programs. In seeking connections to

DOI: 10.4018/978-1-60566-870-3.ch010

MU degrees, not only would independent study enrollments and prestige increase, but the stigma of *–e* (extension, for "correspondence courses") on transcripts might disappear (Almeda, 1985); indeed it did disappear in 2002. In general, CDIS sought to follow principles such as those of the American Council on Education and Adult Higher Education Alliance (ACE & AHEA, 2000) and the North Central Association (NCA) of Colleges and Schools Higher Learning Commission (Taylor, 2000).

SETTING THE STAGE

CDIS reviewed its own print-based and online courses and their match with campus requirements, including new concerns for computer information and literacy (computer proficiency), and presented mock transcripts to administrators. The aim was to serve adult, non-traditional learners in a degree-completion program, although CDIS had contemplated a full degree program. At the time, over 78% of adults age 25 and older in Missouri did not have a bachelor's degree or better (U.S. Census Bureau, 2000).

The planning process exposed several concerns. The BGS program is housed on the Columbia campus, so one concern was using CDIS's far-campus courses (e.g., Rolla and St. Louis). Although many courses were already in place, some junior-senior-level courses needed to be commissioned or revised. CDIS also needed faculty cooperation, especially on Capstone courses (faculty-directed special projects) and Writing Intensive (WI) requirements new to CDIS and requiring special approvals. The agenda included working with, educating, and informing advisors and representatives from campus offices such as admissions; maintaining a paper trail; updating and digitizing the curriculum; marketing the program; and coordinating with other branches of Extension.

CDIS began as print-based correspondence in 1911, but by the 1970s, CDIS used computers in enrollment, some multiple-choice grading, and word-processing. By 1997, CDIS created a website and produced courses in homegrown templates for the Internet.

In 1998-1999, CDIS researched external degree programs in the US. The CDIS catalog had a page devoted to external or distance degrees but cited only other programs. In the 1960s, a Certificate in General Studies offered through MU's College of Arts and Science (A&S) and the Extension Division covered only the first two semesters of undergraduate work. Later, MU began offering a BGS through A&S with very specific requirements including residency, credits, Capstone, Writing Intensive, and Computer Proficiency. Could CDIS provide such a program with its courses? CDIS believed that it could do so, as most courses offered through CDIS were A&S and met general education requirements (see Kleiber & Zhu, 1995, for a similar plan of action).

Crucial to the BGS are "components," 45 hours in three fields of study, at least one of which must be A&S. CDIS offered multiple courses in English, philosophy, political science, and other fields to meet this requirement. After several administrative meetings, CDIS identified courses to serve the BGS population at a distance and prepared mock transcripts. These transcripts solidified the plan; top administrators took the idea more seriously. Administrators agreed that degree completion was the better option for CDIS. CDIS would need to provide a large number of higher-level courses when it had otherwise specialized in lower-level offerings.

In 2002-2003, enrollments through CDIS were given a semester classification and added to student records at the time of enrollment, rather than when completed. These changes allowed students to receive student services such as financial aid and library access. Financial aid and other student services will be discussed later; however, these changes and preparation resulted in a spring 2004 decision by the College of Arts and Science to begin offering the Bachelor of General Studies degree-completion program through CDIS during the fall semester of 2004.

CASE DESCRIPTION

Management and Organizational Concerns

Residency

Residency is much more than a physical move to a location in a city or on campus grounds. Extended degree programs often require a year of on-campus residency with courses offered by regular faculty (Olcott, 1991, 1992). Short of commuting, a distance learner cannot physically be on campus, so other arrangements or definitions must be in place. Campus residency requirements are often "reduced, waived, or eliminated" (Sullivan, Stewart, & Spille, 1997, p. 13) in non-traditional degree programs. With degree completion, students may have completed residential studies upon entry. Fall 2001 brought discussions of transcripting and its relation to residency: would courses offered through CDIS be considered resident or transfer credit? Only video courses, a small part of the total curriculum, counted as resident credit in the past (Pittman, 2001).

Residency was revisited in fall 2002, requiring action by undergraduate deans. CDIS researched how other institutions defined residency. The results went to Undergraduate Studies, who, in March 2003, defined residency as 30 hours of MU coursework enrolled in and successfully completed in one's senior year. Thus physical residence is not at issue: enrollment is. Therefore, as of spring 2003, degree-completing students must be enrolled in MU courses, but need not be in physical residence on campus. For fall 2003, residency was amended: 30 of the last 36 hours must be completed with MU courses. The path was not smooth. During 2003, this new notion of residency was challenged by faculty who felt they had been overlooked in the definition process, and the issue was resolved in fall 2003.

Credits

The BGS requires three components (fields of study) totaling 45 hours, usually in the form of 15 3-credit courses. In 2003, CDIS had almost enough courses for degree completion, but new ones are always being commissioned. In particular, CDIS has sought new or revised English, French, philosophy, political science, psychology, and sociology courses. Recent increases in course author pay rates have attracted more faculty, especially humanities, than in previous years. In winter 2004, it became apparent that CDIS needed more upper-level courses in A&S, particularly since a renumbering system in fall 2004 reduced some 200-level courses in status to 2000, or sophomore level, rather than the desired 3000, or junior level.

Capstone

A Capstone course is an independent research project under the supervision of an MU faculty member, usually resulting in a 20-30-page research paper. Service learning assignments (working with local community) will not work for distance students. Capstones presuppose that students get familiar with a faculty member that they studied under—difficult even for on-campus students to achieve. How will students get to know professors at a distance, especially since many CDIS faculty grade only one course?

Nationwide, capstone students often get feedback on their papers from not only the instructor but also a group of professionals (Wright, 1997). Sullivan et al. (1997) mention "flexible opportunities for interaction with the instructor, as well as with other students" (p. 37) as an important principle in degree programs. Through CDIS, students would interact only with the instructor.

One concern had been the component areas in which to offer Capstone courses, humanities courses being general enough to serve a large audience. Involving faculty from fields ready to serve a diverse audience resolved this concern. In November 2002, CDIS solidified a Capstone plan of finding qualified faculty by involving faculty who had already worked in the WI area; indeed some WI courses can also be Capstones, simplifying the process for CDIS. In September 2003, faculty members were recruited via a letter from Special Degree Programs (SDP) requesting faculty participation in winter 2004.

WI

In spring 2001, CDIS sent SDP fliers to the Campus Writing Program (CWP) for fall review. For peer review (Maimon, 1997), CDIS looked at precedents in Nursing, but Nursing proved to be exempted from much of this process and was semester-based. In November 2001, the Campus Writing Board (CWB) approved the BGS evening program's 16-week WI Capstone, some writing-enhanced courses, and special arrangements for portfolios. The board's concerns for CDIS included changing graders in courses (and ensuring their WI qualifications), the number of courses needed (two for BGS), and when they may be needed (academic year vs. summer). At a January 2002 meeting, the CWB asked CDIS to identify five to ten departments or emphases. From this selection, one course was targeted as a pilot WI Problems course with directed readings. Unfortunately, due to illness, the student never completed the course, so no successful pilot project was available for study.

In March 2002, A&S supported faculty working with BGS degree-completion students one-on-one (WI directed readings) rather than develop independent study courses. The September 2003 Capstone letter to faculty of September 2003 requested faculty participation in WI as well as Capstone.

It took until fall 2006 for SDP to adapt a WI template for faculty use. To qualify to teach WI, faculty attend a WI workshop. The SDP director provided updated lists of WI-qualified faculty in 2007, as more and more students hit this point in their studies. The paper trail proved challenging, with many signatures and approvals required; this was complicated by changing personnel in several offices who had to be briefed on the procedures. In 2008, the CWP expressed interest in (1) downplaying or phasing out the template that helped faculty organize their course and (2) having faculty submit proposals for WI courses through the CWP website and then sign up for payment from CDIS.

Early WI plans included a Russian culture course and an American literature course, currently pending approval by the CWB. As of 2009, several additional WI courses are planned: one covering German civilization after 1850—an adaptation of an existing course—and courses in areas such as early and middle childhood. The first WI course through CDIS opened in December 2008, enrolling 17 students as of March 2009.

Technology Concerns

The technological aspects of WI through CDIS involved adding requirements and other data to the website; e-mailing a template approved by CWB to instructors and students, and allowing direct e-mail communication between instructor and student. In the last case, CDIS departed from its norm of monitoring all student submissions: much was up to the instructor and student over the allowed 9-month enrollment period. While awaiting WI courses to be developed through CDIS, at least one instructor based her syllabus on an existing online course, making special arrangements for the student to redraft certain assignments. Other interim measures may be adopted as time goes on.

As of summer 2008, all WI courses through CDIS require faculty to submit a proposal online, to be approved by CWB, after which the student can enroll, observing rules of financial aid and census dates (semester-driven enrollment periods). Initially, delays and confusion persisted, as neither CDIS nor many faculty were accustomed to the new campus student information system (myZou), or faculty submission of course proposals online. Participants expressed confusion about where and how to start, when to expect approval, who is copied on the information, and when WI courses are designated as such in myZou.

Course Approval and Revision Cycle

As Duning (1985) notes, "Independent study courses are scrutinized [emphasis in original]" (p. 210), though this does not always improve the program's image. A&S recommended stiffening this process, beginning in fall 2001. To help foster open discussion, CDIS sent a letter to A&S chairs in early November 2001, explaining its concern for quality and faculty ownership, and describing the January 2002 course approval process. For the first time, study guides and URLs and course previews—not just a list of course titles— would be sent to chairs in all departments. University course quality was discussed, resulting in policy changes on examinations and other matters. By February 2002, most approvals were in, with no losses among MU A&S courses. By late 2009, all courses offered through CDIS are to be online, thus fulfilling the BGS Online designation and making it easier to send courses for approval—links in lieu of study guides.

Technology Components

Advising

In May 2001, Extension announced the opening of an evening college, MU in the Evening, beginning fall 2001. MU in the Evening is a collaborative effort between MU Direct, a branch of Extension, and A&S. MU Direct already offered master's degrees and some bachelor's degrees or degree completions online. The evening college's first degree offering was the BGS, with other degrees in planning. By July 2001, an advisor was hired to work within the BGS office and evening course offerings appeared on a website, with components in Business, Communication, Information, and Humanities. Only some of the courses and instructors overlap with those of CDIS and evening courses are generally in 8-week sessions.

In 2004, the BGS advisor was instrumental to the degree-completion program due to her awareness of A&S policies, transfer credit, and other matters. Further, she attended national conferences to learn more about independent study culture, even co-presenting on BGS topics. Advisors can be extremely helpful in fostering student ability to complete an independent study course, as noted by Tallman (1994).

In 2004, administrators decided that a staff member at CDIS would handle intake questions such as inquiries about the program and course enrollment questions once a student's graduation plan was created. Significant thought went into the information provided to students inquiring about the program and how best to provide that information at a distance as well as information provided after students are accepted or not. One of the BGS degree-completion admission requirements is having earned 60 or more credits from a regionally accredited institution or an associate (AA) degree from a regionally accredited Missouri institution. The College of Arts and Science handles re-admission decisions involving former MU students and admission decisions involving transfer students with 70 hours or more of transferable credit. The admissions office handles admission decisions for transfer students with 60-69 hours of transferable credit.

For the first few years of the program, A&S sent acceptance and denial letters directly to students who then queried their BGS advisor, A&S, or CDIS regarding their next steps. In order to funnel intake into CDIS and streamline this process,

A&S now sends those acceptance and denial letters to CDIS. CDIS then e-mails students about the admission decisions and their next steps and mails the letters to the students. The registrar's office also sends automated weekly BGS status reports to CDIS that also trigger communication between advisors and students.

Admission requirements for the BGS degree-completion program include a GPA requirement, at least 60 hours of transferable credit from a regionally accredited institution, and the prior successful completion of college algebra and an expository writing course. In 2006, GPA requirements for those transferring from another school increased at MU to a cumulative 2.5 GPA. The higher GPA requirement appears to have had a positive effect on the BGS program, in terms of successful admissions and course completions.

CDIS uses technology to provide critical information to students at a distance during every step of the admission and degree-completion process. Online student services play a significant role in distance learner recruitment, enrollment, and degree completion (Brigham, 2001; LaPadula, 2003). The BGS advisor was accustomed to meeting with evening students in person to create graduation plans, with follow-up discussions by phone or in person. The type of information she previously shared in this format would need to be disseminated online or via e-mail in order to meet the needs of the distance learners (LaPadula, 2003).

Providing quick, courteous, personalized service without sacrificing quality is important in building a quality program (Brigham, 2001). Careful attention is given to all of the information provided on the website and in e-mails sent when the student inquires, applies, is admitted or denied acceptance, and throughout the student's progress in the program. Guiding the student's next steps is important, whether the student is accepted or not. Information sheets about admission requirements, admission procedures, degree requirements, components, and general education course lists are shared online and by mail. Clarity is of utmost importance, and the BGS advisor and CDIS have found it useful to stress important information multiple times and in multiple formats.

While the MU admissions office and CDIS already had online admission and course registration systems in place, respectively, it was important that students receive "one-stop service" to find the information they needed to apply to and pursue the program. For example, the BGS advisor and CDIS discussed how best to develop and share the informal graduation plans created for each student. Flexibility was important since degree components were diverse, as was each student's transfer credit. CDIS programmers devised a creative solution: an online graduation plan that was e-mailed to the student after creation and available online after the student enrolled in their first course through CDIS. The graduation plan could be viewed and edited by the student's academic advisors and the student could review their progress toward their degree and contemplate their next course enrollments.

The online graduation plan has worked out very well, enhanced over the years by the inclusion of standardized comments and the ability to generate a formal graduation plan. Once students were admitted to the program, they seldom referred back to information on the website or in their information packets concerning degree requirements, so helpful standardized comments were added to the online graduation plan. The technology behind the online graduation plan was flexible enough to generate a formal graduation plan to share with the College of Arts and Science when the student is within the last 12 hours of graduating, in preparation for their degree audit. Advisors also use the graduation plan to quickly assess the student's progress and make course enrollment suggestions.

After the BGS degree-completion program was well underway, CDIS hired an academic advisor whose primary responsibility was the continued

growth of the BGS program. The CDIS advisor assists BGS students with enrolling, answers questions, updates graduation plans as students complete courses and also provides access to graduation plans online as soon as they are created, rather than e-mailing the graduation plan and providing online access with the initial enrollment. Advisors involved in the BGS degree-completion program are sensitive to student needs and creative in terms of improving communication. For example, the advisor who helps students create their graduation plans originally requested that students identify their components of interest. Because the BGS online is a degree-completion program with students often transferring in credit from other regionally accredited institutions, the advisor discovered that it worked better to ask students to provide a "top ten" list of courses for their graduation plans, rather than require them to show interest in static components that might not work as well with their transfer credit.

Students interested in the program request an information packet online; information that they provide automatically goes into the CDIS database to assist advisors and aid in marketing efforts. Packets include information about admission requirements, degree requirements, components, general education courses, university course listings, admission forms for new and returning MU students, and course enrollment forms— information also available on the CDIS website. General financial aid information is included in the packet; more detailed information goes out after the student has been admitted to the program, including contact information for their financial aid advisor. Students are also informed that the majority of their courses will include two proctored exams.

Students who are admitted to the program are provided guidance via e-mail and phone by advisors and other student services staff in areas including enrollment, financial aid, exam sites, enrollment deadlines, etc. Students may enroll online, in person, by fax, or by mail; materials are available for purchase through the bookstore at CDIS and can be purchased online or otherwise at the time of enrollment, although students are not required to purchase materials from CDIS. Access to online courses occurs as soon as the enrollments are processed and materials, if ordered, are shipped the next day. This quick, dependable service fulfills an important requirement in distance learning: access to required materials and other resources (Brigham, 2001).

Providing access to online student services goes beyond the assistance traditionally provided to distance learners in areas such as registration and admissions and contributes to the quality of their experience and their sense of community (LaPadula, 2003). Once students enroll, they receive an automated e-mail that explains how to access their courses online and provides instructor information. Course access includes links to disability services, the Online Writery (an MU cybertutorial for writing) and library services and provides contact information in case students have questions for CDIS or their instructor.

Research indicates that creating a sense of community through an array of student services increases student satisfaction, retention, and completion rates (LaPadula, 2003). Every semester, to help foster this sense of community, the advisor at CDIS tries to contact each admitted student and creates an online newsletter. The online newsletter contains information about enrollment and other important deadlines, course openings and closings, the online career center, graduation, financial aid, updates to procedures, reminders about setting up and accessing student e-mail accounts, and student profiles or awards.

Finding funds to enroll is important to students enrolled in the program. Since 2007, the CDIS advisor has provided information about a scholarship opportunity available to non-traditional students in the newsletter and in e-mails. At least 13 BGS degree-completion students have received

funds from this scholarship. Some BGS students are military veterans or currently serving in the military. The CDIS advisor is also a certifying official with the Veterans Administration, which is helpful when guiding military students through the enrollment and certification process.

Other ways to serve distance degree students include new student orientations, expanded for the entire family, allowing sharing of success stories; and recognition of supportive employers and community resources (Gibson, 1998). For general discussion of institutional mainstreaming of distance education, see Irele (2002). MU, however, does not have a virtual university or world campus such as those she describes.

Online Commencement

A reception took place in May 2008 for Evening and other BGS graduates. Only one graduating distance learner was able to participate in person, but an online commencement was also available for the first time in spring 2008 for all those completing an undergraduate or graduate degree online from MU. The online commencement website and guestbook proved very popular with distance learners and their families as it increased their sense of community and identification with other MU students.

Transfer Credit, General Education Requirements, and Mainstreaming

Mainstreaming (integration) of courses offered through CDIS facilitated the BGS but many issues remained (Darcy & Nagel, 2003). Transfer credit and general education requirements were issues in spring 2001, leading to a vote in mid-November 2001 on transferring general education credits from 2- and 4-year public institutions in Missouri. As of 2009, general education requirements are satisfied if the student transfers to MU with an Associate of Arts (AA) from a regionally accredited Missouri institution. For transfer students with a 42-hour "block" of courses but no AA, a WI course is often the only general education requirement lacking.

The BGS degree completion program differs from some that are offered nationally. Murry and Hall (1998) report that 87% of central US institutions surveyed "granted credit for prior learning when it could be documented" (p. 23). MU does not accept prior learning or experiential knowledge; generally, *courses/credits* oriented toward a baccalaureate degree from regionally accredited institutions are accepted if grades are C or better. Thus CDIS could not follow a portfolio model nor a learning-contract-based model, the latter described by Harris (2003).

On June 25, 2001, centralized transfer credit evaluation began at MU, initially with A&S—a new way to determine applicability of credit towards a degree. It was not clear how this would serve non-traditional students with special "adult" needs. The Office of Admissions took over transfer credit in October 2002. Transfer credit then allowed Missouri University of Science & Technology (S&T) and UM-St. Louis courses to be used by BGS degree-completion students.

No matter how many credit hours students have earned before entering the BGS degree program, they still need to earn a minimum of 30 credit hours from MU toward that degree. Missouri S&T and UM-St. Louis courses can serve students in six of their last 36 hours of coursework. As of fall 2004, community college courses were allowed to constitute six of the last 36 credits, but only 1000-2000-level community college courses could transfer to MU. MU accepts only 64 credits total from community colleges unless more hours are required per an articulation agreement.

Library Use by Distance Students

Prior to mainstreaming, CDIS students did not have access to MU libraries. Few courses offered through CDIS assigned library papers because of students' lack of access to MU libraries; how

could students in a small Missouri town with only a public library gain access to a scholarly journal that only some university libraries carry? What can or should the MU library provide for students off-campus? What precedents are set by MU Direct? Mainstreaming led to a breakthrough; in fall 2002, students enrolled in MU courses via CDIS were indeed MU students and thus able to use the MU library and intracampus book loans.

In summer 2004, a 4000-level sociology course helpful to the BGS program opened, using electronic reserve articles (ERes), a first for CDIS. ERes provides a central location for assigned articles. ERes was originally designed for semester use, but the library extends access to the readings in the sociology course on an annual basis. Other courses will use ERes, but not necessarily in the BGS arena.

Consortial Arrangements

In spring 2001, CDIS joined National Universities Degree Consortium (NUDC), allowing CDIS to confer with other programs, refer students, and eventually market the BGS degree-completion program. How courses from other institutions will fit into the BGS is not yet apparent, but CDIS stands to learn about other degree programs and to increase enrollments in general. Both authors have served as officers for NUDC. NUDC often brings up the issue of shared degrees. MU Direct, but not CDIS, has participated in the Great Plains IDEA, cooperatively offering financial planning courses.

Another opportunity for consortium arose in June 2001, when CDIS learned of a Missouri consortium of community colleges whose students earn associates degrees and so might provide an audience of degree completers. Further, in March 2004, CDIS learned that students in the St. Louis area were interested in a BGS without going through UM-St. Louis. Thus a new audience emerged.

General Scrutiny of CDIS

The BGS, transfer credit, approval cycle, library usage, transcripting, and other matters caused the MU campus to take much more notice of CDIS than ever before in fall 2001. Courses offered through CDIS became part of the MU mainstream in 2003, with the *–e* suffix removed from transcripts (cf. Darcy & Nagel, 2003). In general, the university approves of CDIS efforts to mainstream their courses.

CDIS is much more visible today, sending employees to campus meetings on registration, admissions, transcripting, BGS-specific topics, and key job candidate sessions. At the request of the College of Business, CDIS has pursued a Business Minor, taking advantage of the overlap of the Business Component and various courses in Accountancy, Finance, Management, and Marketing.

Procedures

In December 2003, an MU curriculum committee met to discuss how degree programs, including the BGS, should be listed and described in an upcoming catalog, inviting comment from CDIS and MU Direct. As always, certain procedures that apply on campus did not make sense for distance education. Meetings of 2003-2004 revealed that procedures of admissions, transcripting, and related matters were of primary concern. Intake procedures used by MU in the Evening would be adopted and used for the BGS degree completion website in development. Admissions and the College of Arts and Science drafted letters of acceptance and denial, with Extension input.

Conversion to Online

In late 2009, all university-level courses through independent study will be online. This entails conversion from paper and, often, copyright clearances. By 2004, "BGS Online" became the adopted marketing term, though it was only partly accurate; some courses were initially print-based or hybridized (electronic submission from a print-based course and other variations). From 2004 to 2008, many new or revised courses were commissioned at CDIS, all online. Petersen and Gearhart (2004) point out that at least some areas of the country warrant offering print-based degree programs, helping to bridge the digital divide. It remains to be seen whether prisoners—or others—will choose not to enroll in courses from CDIS that are offered only online. Meanwhile, proctored exams are only available in paper form until online exam security is perfected.

The print-to-online conversion came about to put CDIS in the mainstream, not specifically to promote a degree-completion program. Converting print courses to online is an issue only in terms of time and productivity; with many courses and agendas to fulfill, CDIS placed university courses on high priority. CDIS' system is homegrown, though it makes use of Adobe Flash animation and the like and updates templates over time. No vendor partnerships were necessary. CDIS drew on existing curricular and student-service procedures in order to go online. Over time, the template evolved, new programs were licensed, and staff changed over, providing new insights and techniques.

Lessons within courses offered through CDIS follow standard national practices in having purpose, objectives, reading assignment, commentary, study questions, and progress evaluations (submitted assignments). See Figure 1 for an example of a print course, English 2150 Popular Literature (Tolkien). See Figure 2 for an example of these elements arranged in the online version of English 2150.

Many print courses place the answers to study questions in an appendix; for online courses, CDIS converted this linearity to a list of questions with hidden answers. In some courses, students just click a link to see the answers while other courses ask students to type their own response before seeing the hidden answer. See Figure 3 for an example from The American Presidency.

Since 1997, interactive activities have become more sophisticated. Further, to humanize the course, CDIS sometimes films an instructor's introduction. In other courses, such as Gothic literature, the commentary is read aloud (see Figure 4).

Enrollment Process

Students can enroll at any time and take from 6 weeks to 9 months to complete a course. Depending on the enrollment date, a semester classification is placed on the course. For example, the cut-off period (fall census date) for fall 2002 enrollments was September 23, 2002: after that date, students were considered winter 2003 enrollees, until the winter census date, after which time students were considered summer 2003 enrollees, and so on. This pattern continued with CDIS staff adjusting to new seasonal periods of high or low enrollment. BGS degree-completion students needed to be classified as degree-seeking *and* CDIS. Academic procedures were ironed out by about spring 2003, dovetailing with catalog drafts for 2003-2004. These procedures involved areas such as dismissed students, withdrawals, graduate credit, transcripts, prerequisites, grading options, international students, financial aid, and athletic certification.

Costs

University fees went up substantially in 2003 and afterward, suggesting that lower-priced institutions might take enrollments away from CDIS in general, not to mention the BGS. Further, special fees were adopted by several schools on campus, including Business, whose course offerings through CDIS can serve as a BGS component. University fees are not expected to

Figure 1. Print course lesson, Tolkien course

Lesson 1
The Hobbit

Purpose
This lesson provides a basic introduction to J.R.R. Tolkien and his fictional worlds and creatures. Specific emphasis will be placed on reading and understanding *The Hobbit.*

Objectives
After this lesson, you should be able to

- discuss *The Hobbit* as both a work in its own right and as a "prequel" to *The Lord of the Rings.*
- discuss Gandalf's role in the adventure.
- discuss Bilbo's emerging character in the story.

Reading Assignment
The Hobbit

NOTE: I will assign no specific reading assignments in *A Tolkien Bestiary*; you should use this book with each lesson to look up the characters you encounter in your reading assignments. This will give you a broader and deeper understanding of Tolkien's mythology. Material from this book is "fair game" on exams. Students demonstrating an understanding of the material from this book in their assignments and on their exams can expect to receive a better grade for their efforts.

Commentary

The Author

John Ronald Reuel (J.R.R.) Tolkien (1892–1973) was born in South Africa. As an adult he was a philologist and a professor of medieval literature at Oxford. Philology is a discipline that has evolved into modern-day linguistics, and Tolkien's specialty was ancient northern European languages and traditions. He was a writer of both fantasy and highly respected academic works. Among his academic achievements, one of the most influential was "Beowulf: The

increase for 2009-2010, and growth in the BGS remains steady.

Marketing

The July 2004 CDIS catalog contained a description of the BGS degree-completion program, and the Mizzou Online website, devoted to online degrees at MU, went live that month. Calls came in immediately. Truluck (2003) notes that a degree program website "should include information about the program, who should apply to the program, admission guidelines, . . ." (p. 327) and other matters such as courses and their descriptions. The BGS portion of the CDIS website does just that. In May 2005, Extension marketing released a BGS brochure, with the CDIS bulletin of June 2005 shortly behind it. In summer 2005, CDIS sent letters to community colleges and Extension marketing notified former MU students about the new degree-completion program.

Marketing efforts have targeted mainly Missouri residents and former MU students. From 2005 to 2008, spring, summer and fall radio and television ads aired in Missouri providing information about online programs at the University of Missouri. CDIS sends information to Missouri community college advisors annually to share with students and at their college fairs. The BGS degree-completion program has also played a

Figure 2. Online course lesson, Tolkien course

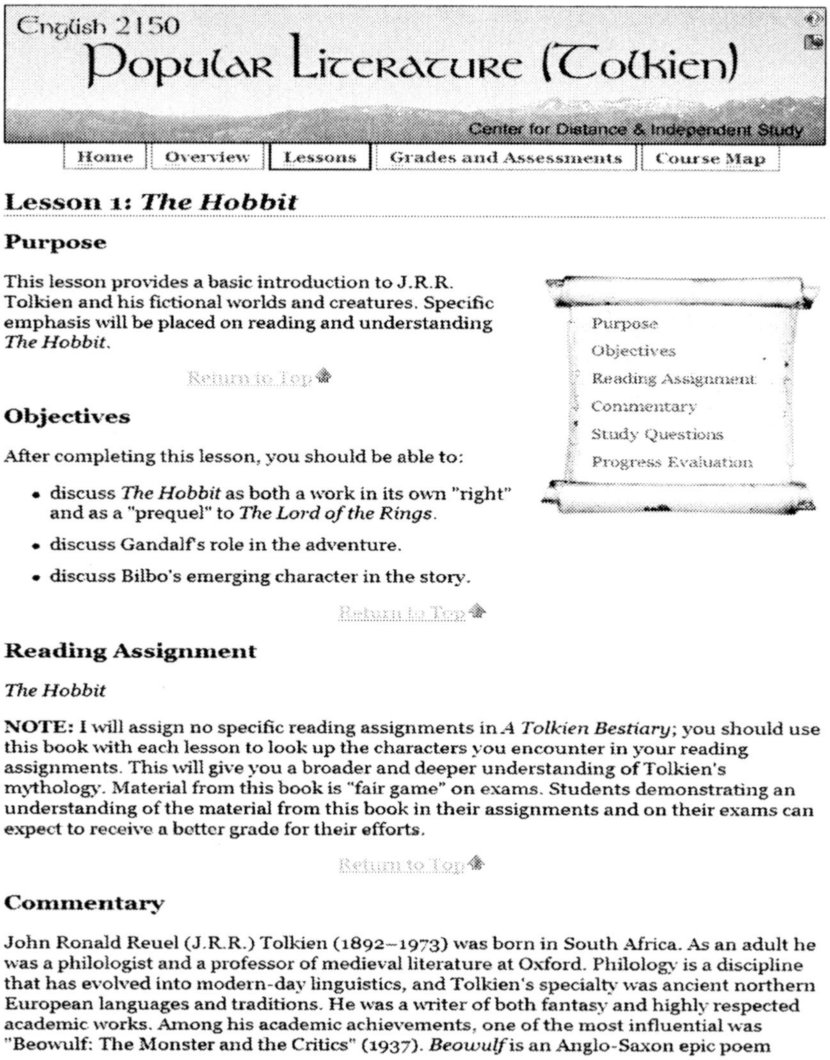

English 2150

Popular Literature (Tolkien)

Center for Distance & Independent Study

Home | Overview | Lessons | Grades and Assessments | Course Map

Lesson 1: *The Hobbit*

Purpose

This lesson provides a basic introduction to J.R.R. Tolkien and his fictional worlds and creatures. Specific emphasis will be placed on reading and understanding *The Hobbit*.

Return to Top

- Purpose
- Objectives
- Reading Assignment
- Commentary
- Study Questions
- Progress Evaluation

Objectives

After completing this lesson, you should be able to:

- discuss *The Hobbit* as both a work in its own "right" and as a "prequel" to *The Lord of the Rings*.
- discuss Gandalf's role in the adventure.
- discuss Bilbo's emerging character in the story.

Return to Top

Reading Assignment

The Hobbit

NOTE: I will assign no specific reading assignments in *A Tolkien Bestiary*; you should use this book with each lesson to look up the characters you encounter in your reading assignments. This will give you a broader and deeper understanding of Tolkien's mythology. Material from this book is "fair game" on exams. Students demonstrating an understanding of the material from this book in their assignments and on their exams can expect to receive a better grade for their efforts.

Return to Top

Commentary

John Ronald Reuel (J.R.R.) Tolkien (1892–1973) was born in South Africa. As an adult he was a philologist and a professor of medieval literature at Oxford. Philology is a discipline that has evolved into modern-day linguistics, and Tolkien's specialty was ancient northern European languages and traditions. He was a writer of both fantasy and highly respected academic works. Among his academic achievements, one of the most influential was "Beowulf: The Monster and the Critics" (1937). *Beowulf* is an Anglo-Saxon epic poem

role in the MU Recruit Back program designed to help those who left the university with a certain number of credits complete an MU degree at a distance.

The BGS brochure is updated periodically with student testimonials and photos. The updates are coordinated with CDIS website notices of new courses and student successes. PDFs appear on the CDIS site, allowing students to find courses of interest, including those in the works.

All of these efforts have paid off in terms of growth. The degree-completion program has experienced dramatic growth over the last five years resulting in 16 24 graduates to date and, in 2007-2008, growth of 40% in the number admitted to the program, as well as 31% in the number of course enrollments. One can safely assert that the College of Arts and Science and CDIS are fulfilling a previously unmet need and helping students achieve their goal, often long-awaited, of earning a degree from the University of Missouri.

Figure 3. Study question with clickable answer

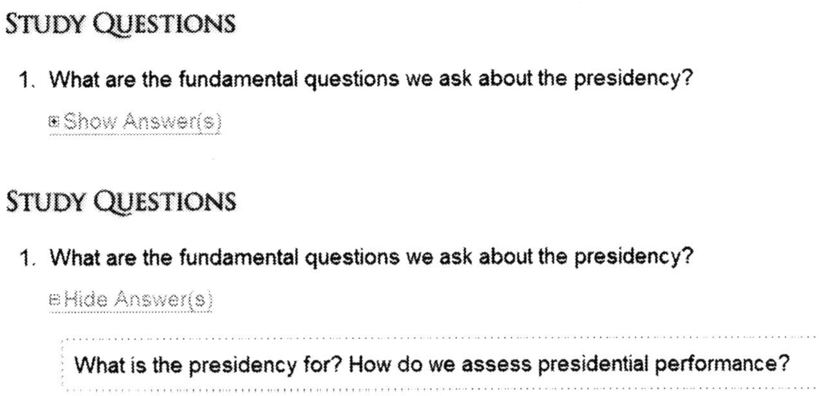

Figure 4. Audio version of course commentary

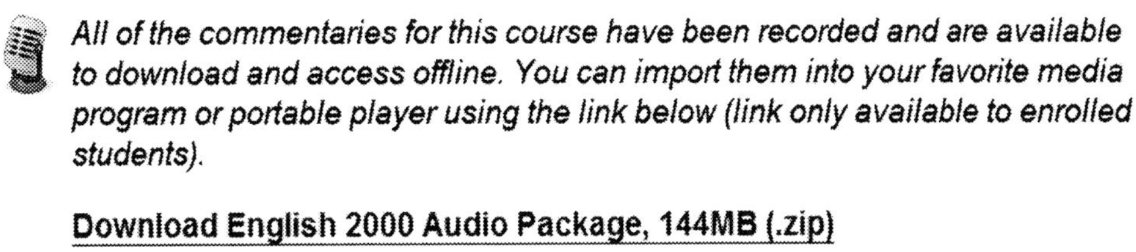

CURRENT CHALLENGES FACING THE ORGANIZATION

As stated previously, CDIS began as a print based program in 1911. Students had nine months to complete a course, with a 3-month extension possible for a fee. Students may complete their courses in as little as six weeks from the date they submit their first lesson. The original reason for the lengthy enrollment time was to allow mailing time for course materials, lessons, and exams. Now that most courses are online, and lessons are primarily submitted online or via e-mail, there is less need for such a lengthy enrollment time; thus CDIS considered eliminating the 3-month extension at the university level.

CDIS was concerned about the effects of reducing enrollment time, so a survey was sent to all active BGS students in the spring of 2008 to gauge their feelings regarding the elimination of the 3-month extension. The overwhelming response from BGS students was positive, with most students indicating that the change would cause them to procrastinate less and increase their motivation to complete courses in a timely fashion. Effective with July 1, 2008 enrollments, CDIS limited the enrollment time for a university course to 9 months, with no extension possible, except for students with major extenuating circumstances (such as a major hospitalization or military deployment). Other CDIS programs (high school, noncredit, etc.) will adopt this time limit in the following fiscal year. CDIS will compare completion data from the past three years to completion data from the first year of the new time limit to see what effect the change may have on completion rates.

Frequent communication with students also may increase completion rates. CDIS sends students several letters and e-mails throughout their course enrollment time. As of 2007, this included personalized e-mails sent in the last two months of enrollment, offering our service of setting up a course completion schedule or other assistance. In 2008, MU's Advising Council also asked CDIS to send enrolled students an e-mail each semester informing them of course completion deadlines if they intend to graduate and providing other helpful completion information. These letters and e-mails have had a positive effect on course completions and toward helping students make steady progress in their courses. This also benefits instructors, whose CDIS workload had been greatly increasing toward the end of each semester, when they were also likely to be busiest with on-campus student grading. In addition to a more equal distribution of work, instructors may also benefit financially because there is a better chance that they will grade the work within a week's time, which entitles them to a higher rate of pay.

Another positive change in student services occurred on the MU campus with the roll-out of a new student information system, myZou, effective with fall 2007 enrollments. The new myZou system replaced an outdated CICS mainframe system and includes a catalog and schedule of courses and a portal-type interface for self-service available to faculty, staff, and students. The new myZou system interfaces with the following systems: housing (new), degree audit, room scheduling (new), student loan collection and a data warehouse, providing centralized 24/7 real-time, auto-interfaced web-based access to multiple functions including admission status, registration, student accounts, and financial aid (Selman, Keely, Washington, & Nagel, 2008). Faculty and staff also have 24/7 access to electronic grade submissions, advisee lists, the ability to add advisees quickly, and an interface with Outlook for improved communication options (Selman et al., 2008).

While CDIS has its own accounting and enrollment system, it gained the capability of adding university enrollments for courses taken through CDIS to student records as they occurred, rather than reporting enrollments by campus mail to the registrar's office. This change resulted in quicker updates and helped expedite the disbursement of financial aid. When instructors notify CDIS of university course grades, grades are entered in student records at CDIS and also added electronically to student records in myZou using faculty rosters.

Students log into myZou using their MU e-mail addresses and passwords. Information is sent to students about their MU e-mail addresses and how to re-set their temporary passwords at the time of admission. Personal identification numbers (PINs) are not required.

Because multiple functions are now contained in a single student information system, myZou has made it easier for BGS students to navigate the campus functions they need, such as accepting financial aid, through their student self-service center. Given the number of benefits their university e-mail address confers, including a sense of community, BGS students are creating their passwords earlier in the admission process and checking their university e-mail more frequently. However, CDIS has noticed that online students are less likely than campus students to use these services, at least initially, without prompting from their BGS advisor and CDIS and frequent reminders in the BGS electronic newsletters. As mentioned previously, many of the online courses offered through CDIS contain links to assistance provided by the Online Writery or library resources; students use their university log-in information to access these resources, giving them many reasons to take advantage of their access and the services available.

Effective with spring 2009 semester, CDIS changed billing processes for students using financial aid. After CDIS receives notification from the financial aid office that students have aid available and those students have enrolled, cashier holds for those balances are placed in myZou, and CDIS receives payments directly from the cashier's office via electronic transfer when student financial aid is disbursed, often within just a few days of enrollment. This technological change via myZou has greatly improved the payment process. Previously, students using financial aid were expected to pay CDIS once they received their refunds but often waited months to pay their deferred balance due, especially problematic with BGS students enrolling in two or more courses at a time. Students and CDIS are grateful for the change in procedures and for the ability to view these transactions in myZou.

Other popular student self-service functions consist of reviewing transfer credit, e-mailing an unofficial transcript to the student's e-mail address, and self-certifying enrollments by running a report through the National Clearinghouse. Students are informed at the time of enrollment of the semester classification placed on their courses and this information helps them review the information found on these unofficial transcripts. The BGS advisor frequently reminds students that regardless of the 9-month enrollment time, students may count their courses only in the semester in which they enrolled, in terms of loan deferment, insurance, and full time status. CDIS also has access in MyZou to queries that show enrollments and drops/withdrawals by semester, to be compared to information in the CDIS database. This helps maintain accuracy between both systems, especially if students are dismissed from their academic programs on campus and dropped from courses or choose to withdraw from all their courses for medical or military deployment reasons, including their courses through CDIS.

CDIS must keep courses current and look for new opportunities, including new degree or certificate programs. In the latter capacity, CDIS is working with Architectural Studies and Human Development and Family Studies, both under Human Environmental Sciences.

CONCLUSION

CDIS's BGS degree-completion program will be crucial in establishing a niche in the university at large. No doubt many of the issues, e.g., periodic review of mission and program objectives, discussed by Murry and Hall (1998) will be addressed in the future, but the characteristics of BGS students at a distance are not fully apparent, beyond what is generally understood of adult learners and/or degree completers. Comparison against "best practices," such as those produced by the NCA Higher Learning Commission (2000) will be necessary. As with all distance programs, student success must be measured and evaluated, perhaps along the lines suggested in Zalenski (2000), who covers, in part, students enrolled in a University of Iowa Bachelor of Liberal Studies program that dates back to 1977 (Holstein, 2003); or Miller Brown (2002), who covers non-distance settings; or Kemp (2002), who deals with persistence in undergraduate distance education.

Further, some or all of the areas of concern noted by Sherry (1996, cited by Berge & Mrozowski, 2001) will need to be addressed, such as design issues (design, development, evaluation, revision); learner characteristics (learner styles); learner support (teacher mediation); and policy and management issues. Curriculum outcomes, as in Harris (2003), must also be assessed, as must applicability of an online bachelor's to graduate program (DeFleur & Adams, 2004).

After half a decade, the BGS degree-completion program is well underway and looks forward to greater visibility and a growing role in MU education.

REFERENCES

Almeda, M. B. (1985). Upgrading the image. *Continuum, 49*(3), 198–202.

American Council on Education & Adult Higher Education Alliance. (2000). *Principles of good practice for alternative and external degree programs for adults.* Washington, D.C.: ACE & Alliance.

Berge, Z., & Mrozowski, S. (2001). Review of research in distance education, 1990 to 1999. *American Journal of Distance Education, 15*(3), 5–19. doi:10.1080/08923640109527090

Brigham, D. (2001). Converting student support services to online delivery. *International Review of Research in Open and Distance Learning, 1*(2), 1-15. Retrieved on September 5, 2008, from http://www.irrodl.org/index.php/irrodl/article/view/23/365

Darcy, S., & Nagel, T. (2003, October). *Fait accompli: The integration of CDIS with the University of Missouri.* Paper presented at the Annual Meeting of the American Association for Collegiate Independent Study, Sioux Falls, SD. Retrieved on March 23, 2009, from http://www.aacis.org/pages/conferences/2003conf.htm

DeFleur, M. H., & Adams, J. (2004). Acceptability of online degree programs as criteria for admission to graduate programs. *Journal of Computing in Higher Education, 16*(1), 150–163. doi:10.1007/BF02960287

Duning, B. (1985). The independent study image dilemma. *Continuum, 49*(3), 209–212.

Gibson, C. C. (1998). The distance learner in context. In C. Gibson (Ed.), *Distance learners in higher education: Institutional responses for quality outcomes* (pp. 113-125). Madison, WI: Atwood Publishing.

Harris, B. A. (2003). Application of curriculum outcomes from an adult baccalaureate degree-completion program. *The Journal of Continuing Higher Education, 51*(2), 33–45.

Holstein, E. (2003). Iowa's educational legacy and vision: Widening the portal for the baccalaureate degree. *The Journal of Continuing Higher Education, 51*(3), 35–40.

Houle, C. (1973). *The external degree.* San Francisco: Jossey-Bass.

Irele, M. (2002). *Institutional mainstreaming of distance education: Guiding principles.* Unpublished doctoral dissertation, Pennsylvania State University.

Kemp, W. C. (2002). Persistence of adult learners in distance education. *American Journal of Distance Education, 16*(2), 65–81. doi:10.1207/S15389286AJDE1602_2

Kleiber, P., & Zhu, I. (1995). Planning for curriculum development in independent study. *The Journal of Continuing Higher Education, 43*(3), 25–33.

LaPadula, M. (2003). A comprehensive look at online student support services for distance learners. [from the EBSCO database.]. *American Journal of Distance Education, 17*(2), 119–128. Retrieved on September 5, 2008. doi:10.1207/S15389286AJDE1702_4

Maimon, E. P. (1997). Teaching across the curriculum. In J. G. Gaff, J. L. Ratcliff & Associates (Eds.), *Handbook of the undergraduate curriculum* (pp. 377-392). San Francisco: Jossey-Bass.

Miller Brown, S. (2002). Strategies that contribute to nontraditional/adult student development and persistence. *PAACE Journal of Lifelong Learning, 11*, 67–76.

Murry, J. W. Jr, & Hall, B. L. (1998). A systematic approach to designing and evaluating effective adult degree-completion programs. *The Journal of Continuing Higher Education, 46*(2), 19–26.

NCA Higher Learning Commission. (2000). *Best practices for electronically offered degree and certificate programs.* Retrieved on March 23, 2009, from http://www.ncahlc.org/download/Best_Pract_DEd.pdf

Olcott, D. (1991). Bridging the gap: Distance learning and academic policy. *Continuing Higher Education Review, 55*(1 & 2), 49–60.

Olcott, D. (1992). Policy issues in statewide delivery of university programs by telecommunications. *American Journal of Distance Education, 6*(1), 14–26. doi:10.1080/08923649209526771

Petersen, R., & Gearhart, D. (2004, November). *Bridging the digital divide through independent study.* Paper presented at the Annual Meeting of the American Association for Collegiate Independent Study, Bellingham, WA. Retrieved on March 23, 2009, from http://www.aacis.org/pages/conferences/2004conf.htm

Pittman, V. (2001). Distance education's scarlet letters. *The Journal of Continuing Higher Education, 49*(3), 42–43.

Selman, B., Keely, J., Washington, C., & Nagel, T. (2008, November). *myZou makeover at MU: Taking advantage of a student information system change.* Paper presented at the Annual Meeting of the American Association for Collegiate Independent Study, St. Louis, MO. Retrieved on November 17, 2008, from http://www.aacis.org/pages/conferences/2008conf.htm

Sherry, L. (1996). Issues in distance learning. *International Journal of Educational Telecommunications, 1*(4), 337–365.

Smith, E. (2004, November). *A difference in kind or a difference in degree? Developing a distance degree-completion program.* Paper presented at the Annual Meeting of the American Association for Collegiate Independent Study, Bellingham, WA. Retrieved on March 23, 2009, from http://www.aacis.org/pages/conferences/2004conf.htm

Sullivan, E., Stewart, D. W., & Spille, H. A. (1997). *External degrees in the information age: Legitimate choices.* The American Council on Education and the Oryx Press.

Tallman, F. (1994). Satisfaction and completion in correspondence study: The influence of instructional and student-support services. *American Journal of Distance Education, 8*(2), 43–57. doi:10.1080/08923649409526854

Taylor, J. A. (2000, June 22). *Adult degree-completion programs: A report to the NCA Board of Trustees from the task force on adult degree-completion programs and the award of credit for prior learning at the baccalaureate level.* ED 449847. Retrieved on March 23, 2009, from http://www.eric.ed.gov/ERICDocs/data/ericdocs2sql/content_storage_01/0000019b/80/16/d5/7d.pdf

Truluck, J. E. (2003). Questions and turning points when designing and implementing distance education degree programs. *Proceedings of the 23rd Annual Adult Higher Education Alliance Conference* (pp. 325-327), Asheville, NC, October. Retrieved on March 23, 2009, from http://www.ahea.org/conference/proceedings/2003.pdf

Turner, C. G., & Monk-Turner, E. (2001). The returns to education and degrees. *Educational Research Quarterly, 26*(3), 45–53.

U.S. Census Bureau. (2000). *American factfinder. DP-2. Profile of selected social characteristics: 2000.* Geographic area: Missouri. Retrieved on November 13, 2008, from http://censtats.census.gov/data/MO/04029.pdf

Wright, B. D. (1997). Evaluating learning in individual courses. In J. G. Gaff, J. L. Ratcliff & Associates (Eds.), *Handbook of the undergraduate curriculum* (pp. 571-590). San Francisco: Jossey-Bass.

Zalenski, A. (2000, November). *Graduation, persistence, and attrition from distance learning programs: An analysis of selected variables.* Paper presented at the Annual Meeting of the American Association for Collegiate Independent Study, Austin, TX.

Chapter 11
The Proliferation, Pitfalls, and Power of Online Education

Leah Blakey
Drury University, USA

EXECUTIVE SUMMARY

Online education has been a growing field in higher education for the last decade, and the number of students choosing online over seated classes continues to increase. The proliferation of online programs forces one to ask, "Is online education a good thing?" The resounding answer is YES, when best practices are followed. However, even the best online programs experience challenges as they deal with institutional, student, and growth issues. These struggles, their resolutions, and the knowledge gained from them are the focus of this chapter.

The proliferation of distance education programs in the last few years is unprecedented in the history of higher education. Technologies such as video conferencing and the Internet enable the delivery of educational content at a speed and level of interaction not previously possible. (Monolescu, 2004, back cover)

As a professor with a Ph.D. that focused on military history, the term proliferation conjures up very negative connotations. My seven years of graduate school were spent focusing on the proliferation of weapons of mass destruction, the proliferation of rogue governments, the proliferation of terrorists, etc. Therefore, with so many books and articles discussing the proliferation of online programs, it begs the question, "Is online education a good thing?" The resounding answer is YES, when best practices are followed.

BACKGROUND

Drury University began offering online undergraduate and graduate level courses in 1999 and 2000, respectively. Between 1999 and 2004, the growth rate of students enrolling in online classes increased by nearly 50% every year. In 2004, the

DOI: 10.4018/978-1-60566-870-3.ch011

University requested and acquired accreditation to place all of the degrees in its College of Graduate and Continuing Studies online. In their request, Drury pledged to predicate its growth on the Higher Learning Commission's (HLC) "Best Practices for Electronically Offered Degree and Certificate Programs". After careful market analysis pointed to online courses as the wave of the future for the adult learner, the program became the "technological and pedagogical hub around which CGCS configured its future growth" (HLC Change Request, 2004, 6). With this commitment to adherence to best practices and future growth, the online education wing of Drury University set off on its path for excellence.

SETTING THE STAGE

Online education has been a growing field in higher education for the last decade, and the number of students choosing online over seated classes continues to increase. This nationwide trend is incredibly strong in the Midwest, where—in many colleges and universities—the online program's growth continues to outpace all other areas of university growth. Sloan Consortium found that "The nearly 20% growth rate expected in online enrollments far exceeds the overall rate of growth for the entire higher education student population" (Sloan-C, 2002). While many of these programs are praised by their peers and customers, the online programs have experienced challenges and growing pains. These struggles, their resolutions, and the knowledge gained from them are the focus of this chapter.

The first question asked when someone mentions online classes always has to do with the students. How can you teach students you cannot see? How can you engage online students in various media of learning and experiences? How do you keep them from cheating? How can you build the rapport that small, liberal arts colleges are famous for if the students are spread out across the world? The answers to these questions are the keys to successful online teaching.

Successfully helping you deal with all of the challenges, both positive and negative, that an online program faces is the overall goal of this chapter. Online education is a rapidly growing part of higher education. The strengths it offers mean expanded opportunities for persons in all regions of the developed world, a way for students to curtail the amount of their small budgets that goes to transportation, and a way to bring great minds from diverse locations together to engage the problems of the 21st century. Mastering the art of educating people over the internet is a must for schools vying for a place in the future market, and the most successful way for them to do that is by learning from others' experiences and following guidelines for best practices in online education.

CASE DESCRIPTION INCLUDING OPPORTUNITIES AND CHALLENGES

The issues faced by a successful online program can be grouped into three categories: student, institutional, and those associated with growth. These issues are similar whether the program is at a small, Midwestern, liberal arts school; a large public university; or a community college. The successful resolution of all three of these issues is necessary for any online program to grow and prosper.

Student Issues

Reaching students in an online class is one of the issues most often debated. How can you teach students you cannot see? The truth is that a multitude of distance education programs have successfully been teaching students they could not see for decades. For example, correspondence courses have been used in colleges and universities for years, and the instructor does not see the student in a classroom setting.

The beauty of online is that now the distance can be spanned through multiple interactive means through the internet rather than through the static medium of a paper or a video distance education course. Teaching students via this global wonder is actually based on the same ideas that traditional classroom instruction has been based on for decades. These ideas take into consideration that people have different learning styles and that they need to be reached on multiple knowledge levels.

Learning Styles

Everyone learns in a way that is somewhat unique. However, most people fall into one (or perhaps two) of three learning style categories: visual, auditory, and kinesthetic (Rose, 1991).

Visual learners tend to take a lot of notes, like to see what they are learning, remember things best if they see them in vivid detail, often close their eyes to visualize or remember something, and prefer to not be distracted by sound and activity when they are learning.

Auditory learners acquire knowledge by listening to lectures, reading aloud, and talking their problems out.

Kinetic learners "remember what was done, but have difficulty recalling what was said or seen . . . rely on what they can directly experience or perform . . . enjoy field trips and tasks that involve manipulating materials . . . (and) are uncomfortable in classrooms where they lack opportunities for hands-on experience" (http://www.usd.edu/trio/tut/ts/styleres.html).

Instructors in online classes can reach visual learners with the simplest online tools. Online platforms present multiple ways of stimulating the visual learner. For example, Angel, Blackboard, eCollege, and Moodle offer places for lectures, pictures, and discussions. The key is for the instructor to find the most direct route between the student and the material the instructor wants them to learn. Often, this means short (no longer than 300 words) lectures (or a series of short lectures) coupled with visual aids such as pictures, charts, maps, etc. Even a visual learner will become lost in a lecture if it is too long or goes off on too many tangents.

When online classes first began, many auditory learners were not very successful because the online instructors did not know how to properly reach them. Now, there are many tools that can be used to stimulate auditory learners.

First, and perhaps easiest for those who might be technologically challenged is simply using what is already on the World Wide Web. Websites such as YouTube, Science Daily, www.learner.org, and www.merlot.org have a lot of free videos that can easily be worked into online curriculum. The key for instructors is to find online resources that they can use in educative ways. "Teachers in effect become curriculum makers, developing parts of the curriculum by using the internet" (Wallace, 2004, p.451). One thing to always keep in mind is if the student must download the video or if they can watch streaming video. If they must download it, then students on dial-up internet may have trouble viewing it.

In addition to these tools that already exist on the internet, there are multitudes of audio file software that can be used to deliver lectures – much of it for free. Photo Story from Microsoft, is a free tool that can be used with a Microsoft PowerPoint presentation to put sound with each of the slides. It is then streamed over the internet, which makes it easier for those students using a dial-up internet connection. Camtasia Studio and Speech-Over Studio are similar tools that can be used to add sound to a PowerPoint presentation, but these programs are not free. In the fall of 2008 Camtasia Studio 6 Student Version was selling for $169 and Speech-Over was $599, which in this budget-conscious time can possibly prohibit their use. There are also programs for adding pure audio files in an online course. The most popular of these, Wondershare PPT to iPOD, can also be downloaded and listened to on MP3 players and iPods.

The most challenging students for online instructors to reach are those who learn through kinetic or hands-on instruction. These are the students who benefit the most from laboratory experiments, field trips, and projects. Tools and procedures for reaching this type of learner vary by discipline, but every good online class incorporates a few projects for the kinetic learner. The projects normally require the student to leave the computer and complete an activity and then return to the computer to either craft a paper or share their experience via other means with their classmates. Some examples of how Drury University online instructors have reached these students include an online sociology of religion class that requires students to attend worship services outside of their faith and then report on the experience to their classmates; an online astronomy class that requires students to visit an observatory during a given week and report on the experience on the discussion page; and an online manned space flight class that requires students to watch shuttle launches and in orbit videos and discuss the events daily in the online class.

Revised Bloom's Taxonomy

Once instructors have designed elements into their classes to reach all of the various types of learners, they then must consider the level of learning they want to take place in their class. The most foundational way to consider this question is to apply the revised Bloom's Taxonomy to the list of objectives and build assignments from here. (Anderson and Krathwohl, 2001). The best way to describe this approach to learning is through a table (Table 1). (http://coe.sdsu.edu/eet/articles/bloomrev/index.htm)

This revised version of Bloom's Taxonomy combines the cognitive process (across top of table) and the kind of knowledge to be learned (down the left side of the table). By combining both what needs to be learned with the type of learning involved, instructors can better plan out their course and ensure they reach beyond the surface of learning into the deeper and more advanced levels of learning.

Cheating

Pushing students to excel and learn multiple levels of material is the goal of most classes. However, the challenge that this type of academic rigor presents often instead pushes students to cheat. Many educators feel that cheating is easiest and thus occurs more often in online classes, but most online instructors actually feel that they are better at detecting cheating online than they are in their

Table 1.

	Remember	Understand	Apply	Analyze	Evaluate	Create
Factual Knowledge						
Conceptual Knowledge						
Procedural Knowledge						
Meta-cognitive Knowledge						

face-to-face classes. "For example, in one online program an instructor who suspected plagiarism confronted the student. . . The instructor said that in the average large lecture classroom course he would have been far less likely to have identified the plagiarism because he wouldn't have been as familiar with the student's work as he was, seeing it regularly in the online discussions" (Maeroff, 2003, p.158). Unfortunately, cheating appears to be rampant in most areas of education, but instructors can work to make it more difficult for students to find the easy way out of an assignment.

First and foremost, instructors must take the time to educate their students on what constitutes cheating and how the student can best avoid it. It has been my experience that more than half of the plagiarism that occurs in my classes is "accidental;" meaning that in most cases, the student did not realize that what they were doing was wrong. Students need to be taught how to take their research findings and couple this information with their own analysis in their writings. Students should be warned against cutting and pasting and pushed to put most of their paper into their own voice and prove their own arguments. Many schools, such as the University of Maryland University College's required one-credit hour Information Literacy and Research Methods course, have been created to teach students to understand and recognize plagiarism (Maeroff, 2003).

Secondly, instructors need to construct assignments in a way that makes cheating or plagiarism more difficult. For example, if a history instructor asks students to write a five page paper on the causes of the American Civil War, the students are given an assignment that is easily plagiarized. By comparison, if a history instructor asks his or her students to write a five page paper comparing the five causes of the American Civil War discussed on the weekly discussion page and then apply the rationalist ethics theory to the two they believe are the most important, the instructor has built an assignment for a paper that is not easily downloaded from the internet. By making students pull in specific examples from class discussions and instructor lectures as well as making the question subjective instead of objective, the instructor has personalized the question to their own class and made the ease of plagiarism more difficult.

Third, there is a school of thought that pushes instructors to find new methods of assessment that are not easily plagiarized or otherwise taken advantage of by those looking for an easy way out. These instructors rely less on large exams or papers and more on smaller, interactive assignments that require multiple levels and times of involvement by the student. Wallace Pond, chief academic officer of Education America's online program argued that, "There are lots of lecture hall courses where a student might not interact with a professor at all and the final grade is based on two exams and a paper. An environment of that sort may engender poor decision making by students who turn to such activities as plagiarism. . . In a well-taught online course with an involved professor and 25 students, it would be very difficult for a student to put bits and pieces of someone else's work into his work. Everything, every word, in an online course is archived. The professor can go back and review all of the student's work" (Maeroff, 2003, p.159).

Unfortunately, there are also the cases where the person completing the exam, discussions, or paper are actually the person sitting in front of the computer screen but they are not the person enrolled in the class. There are multiple ways to attempt to combat this, but there is no perfect answer to this dilemma. Many schools require at least one proctored exam that necessitates the student to show identification and prove they are the person enrolled in the class. Many instructors check the properties on the paper they receive to verify the owner of the computer the paper was created on. None of these methods are foolproof and they all add to the time and energy an instructor or administrator must put into an online class, but they do offer a way to find a few imposters and get the word out that they are checking for such things, which is often all that is necessary to deter a student from cheating.

Finally, perhaps one of the best ways to keep students from cheating is to build a rapport with them that motivates in them a love of learning and a respect for the instructor or professor. Online instructors may go about this relationship building in different ways, but all good online instructors strive for it.

The key to successful online relationship building is timely and meaningful feedback. Good instructors talk to most of their online students at least once every day through either e-mail, asynchronous discussion, or a virtual world. They talk to all of their online students multiple times each week through one of these venues. When a student learns and comes to trust that they can reach their instructor to get questions answered, assignments explained, or lectures expanded upon; then they start to build a much deeper relationship than can be built when a student only has contact with an instructor one, two, or even three times per week. The trust that comes from this relationship will motivate students to approach the instructor with issues regarding assignments that might otherwise have been solved by cheating. This relationship also gives the instructor a perfect venue for reaching and challenging more advanced students.

Student Expectations

Most online instructors claim to have had their most academically advanced and most academically challenged students in online classes. The online environment tends to accentuate whichever one of these categories students fall into. One reason for this emphasis is the close relationship good online instructors have with their students. Another big reason is the fact that many students enroll in their first online class thinking it will be easier than the seated version of the same class. This is far from the truth. Online classes are meant to be more convenient – not easier. This is an oft made mistake. Students new to the online world quickly become overwhelmed by the larger workload of an online class. Steve Hynds, Director of Online Education at Drury University, listed this as the biggest issue he deals with regarding students. "Convenience doesn't mean easy. Convenience just means convenience . . . The bigger concern with the students I've seen is the amount of work that is required . . . then [students] feel that online is a harder or actually a more enriched program because of the challenges that they have to go through for that convenience" (S. Hynds, Interview, October 24, 2008). The increased stage on which advanced online students can perform lets stars shine and those hungry to learn truly test themselves.

Institutional Issues

While instructors and students wish to focus on the personal issues, administrators are much more concerned with institutional issues. The first hurdle the institution must cross is accreditation. After accreditation is acquired, the majority of institutional challenges will lay within the confines of the campus. These issues often involve the platform choice, staffing, scheduling, pricing and overall procedural guidelines. The biggest challenge, and often the most difficult to overcome for new programs, often deals with proving the academic quality and prestige of the program to the tenured personnel of the institution.

Accreditation

A critical point in the development of a quality online program is acquiring the proper accreditation. Drury University went through this process in 2003-2004. It was a tremendous amount of work, but the payoff was certainly worth it. The scenario is described in detail here.

In 2003 Drury University's College of Graduate and Continuing Studies (CGCS) was beginning preparation for a visit from the Higher Learning Commission (HLC) under a new dean who had extensive experience regarding accreditation. As they undertook the self-study and contemplated what they wanted to achieve from the visit, it became obvious that the ability to grant degrees online was the next step. They already had an Associate of Science (AS) degree in environmental science management online through a contract with the military. There were several packages students could put together online to construct other AS degrees that were not official nor advertised nor even considered as online degrees. There were also several Bachelor of Science (BS) degrees that offered all but one class online. In order to stay within the requirements of the accreditation body, they needed to have an HLC focused visit and request accreditation for online degree completion.

The leadership team then guided the Drury accreditation team through the arduous task of compiling all of the documents and data to demonstrate they had the necessary requirements to acquire online accreditation. The following lists highlight the areas they felt HLC would be most concerned with and were areas where Drury University could show it was ready to be granted accreditation for online degree completion.

1. **Technological improvements:** Drury had moved from initially offering online courses through Electronic Reserves System (ERes) to using WebCT as its online platform. Drury University also appointed a Technical Advisor to provide training in the use of WebCT/online pedagogy. The online office also established an initial web presence via the University server for online course information. Drury hosted its online classes on campus servers until it outgrew the servers' capability to support the program. In the fall of 2003 Drury University contracted with WebCT to host its courses at WebCT's technology center in Vancouver, British Columbia. These technological improvements gave the online program "more robust Internet access with redundant backbones and [the] larger capacity server provides evidence that Drury is committing substantial financial resources to support this online course delivery initiative" (HLC Change Request, 2004, p.30).
2. **Online tutoring:** Drury University contracted with SMARTHINKING to "provide online student support in the areas of writing, the basic sciences, and mathematics, including statistics and accounting" at no additional cost to the student (HLC Change Request, 2004, p.30).
3. **Library resources:** Drury University's Olin Library increased its ability to serve online students at remote locations by providing "online access to books, periodicals and resource materials maintained by Drury as well as access to interlibrary loan materials through Off-Campus Library Resources (OCLS) and Missouri Bibliographic Information User Service (MOBIUS)" (HLC Change Request, 2004, p.31). Olin Library also continued to offer ERes where instructors could post articles that were accessible by students via the Internet.
4. **Academic advising and registration:** Every semester the online office provided all satellite and main campus staff current details about registration, adds, drops, and advising issues. The University also designated an advisor specifically for students taking online classes, who could field an array of questions regarding online classes. Drury also added an Advising Module to help students see how specific courses "fit into their program of study. Additionally, students can access their grades at the end of each term, print an unofficial transcript, and even calculate a future G.P.A." (HLC Change Request, 2004, p.32). All registration was enhanced to allow complete web registration.

5. **Schedule coordination:** Beginning in spring 2004, the online course schedule was "developed into a rotational online schedule for the spring, summer and fall classes. The general undergraduate courses offered online are now undergoing a process of planning whereby students planning their academic strategy can rely upon specific online courses offered at planned intervals . . . . Overall, the online schedule . . . has evolved into an integral part of the CGCS institutional planning and development strategy" (HLC Change Request, 2004, p.33).
6. **Financial aid counseling:** "Beginning in fall 2003, students were notified of financial aid awards through their University e-mail accounts... This information allows students to logon to eRegistrar through the Drury University webpage to access their personal and financial information, register for online and seated courses at the appropriate times, access their academic degree history as well as a graduation audit" (HLC Change Request, 2004, pp.33-34).
7. **First point of contact:** First Point of Contact was created to "make appointments for advisors, supply information about CGCS to those who inquire, track prospective inquiries and assist in mailings, assist with student surveys and various other duties" (HLC Change Request, 2004, p.34). The goal of the student worker-staffed position was to "provide more timely feedback, to provide for a central flow of information and to relieve some of the congestion from the front liners area" (HLC Change Request, 2004, p.34). This was a key element in providing a supporting structure for online students contacting the University from remote locations who needed various student services but did not know where to find them.
8. **Science labs for the online learner:** In 2003, Online Education at Drury University was awarded a 3M Science Grant for the purpose of researching online science labs. The funding was used to "investigate and research online science labs as adjuncts to current online science courses that incorporate lab activities into the course. The investigation was undertaken as part of the continuous improvement of online science courses for non-science majors and in preparation for offering the AS in Environmental Management degree online" (HLC Change Request, 2004, p.35). Many areas of the University were included in this research and subsequent discussion including the full-time science faculty, administrators, and members of the Academic Affairs Committee. "The most important goal was to help the committee and the science faculty to develop an approach most suitable for non-science majors who cannot attend seated science courses" (HLC Change Request, 2004, p.35). The research examined the trends, options, and models used by other universities for online science lab delivery.
9. **Student technical support:** Online students needed a way to receive technical support at all hours of the day and on weekends. This was accomplished by three approaches
   a. General online information and general technical questions were listed on the online programs web pages.
   b. Specific online course questions were directed to the Online Education Office during normal business hours.
   c. Zavata, a private firm, was contracted out of Georgia that provided answers to technical computer questions 24/7 via phone.
10. **Online orientation:** Students were given 24/7 access to an online orientation website that contained basic information on all aspects of WebCT as well as practical tips.
11. **Online bookstore:** Automated enhancements were made to the Drury University bookstore's capability to serve online

students from remote areas. Students were able order their books for their online classes via the Online Education website and pay for them via a secure credit card payment method.

The amount of work that went into planning for the HLC visit was enormous and involved many parts of the University, even though the visit only focused on the College of Graduate and Continuing Studies. The payoff was substantial, though. The HLC granted Drury University approval to offer any course in the Drury catalogue online. This allowed Drury University to offer any of its Associates and Bachelor's degrees completely online.

Questions from Outside Online Education

When schools started delivering education via the internet over a decade ago, there were many questions regarding the pedagogy and quality of the instruction. Most of these questions came from the established academy. For centuries, classes had been taught the same way – with instructor and student sitting in the same room. The possibility that the same quality of teaching and learning could be accomplished with the instructor and students on different continents was difficult to imagine. However, studies such as those published by Simonson, Smaldino, Albrght, and Zvacek in 2000 have proven that done correctly, student outcomes are quite similar regardless of whether the student is face-to-face with an instructor or thousands of miles away. (Simonson, M., Smaldiino, S., Albright, M., &Zvacek, S., 2000).

In an interview with Dr. Gary Rader, who initiated a now thriving online program a decade ago, he discussed how he dealt with the doubts of the traditional academy on his campus. His response was through something he termed "chair diplomacy". In short, it took spending time with each department and sometimes even each individual in a department and showing them how online classes worked, the assessment outcomes, and final projects and statements of the students from the classes. Basically, the idea of online parity had to be marketed and sold.

And sold it was. Today online education is the fastest growing area at Drury University. Today's leaders in online education do not have to deal with "selling" the idea of online education. Instead, they deal with the varieties of options facing established programs.

Learning Management System Options

One of the first major decisions leaders in online education deal with is the choice of learning management system (LMS). There are numerous factors involved in this decision and many of the factors can be further sub-divided. The LMS must "meet the requirements, functions and features as aligning with the program's needs. Much due diligence has to be done to find out what the institution needs with input from many stakeholders" (T. Melacon, Interview, November 19, 2008). The features of the LMS must be forward thinking and able to support all of the exciting educational technology that is on the horizon. The LMS must function easily and without interruption; it must interface well with the programs the school already has in place; it must come with a strong support staff and may possibly need to offer hosting. These considerations would vary based on the institution's needs and culture, but all should be considered.

Scheduling and Staffing

Mature online programs share some of the same issues that new programs do – namely in the areas of scheduling and staffing. All types of higher education institutions – whether they be a small private institution, a community college, or a large public university – express a desire to alter the way classes are scheduled and staffed. The majority of schools still have the responsibility of staffing and scheduling falling to the department chairs or deans. This allows the experts in academia to ensure the academic qualifications of the instructors as well as the academic plan of the department by scheduling certain classes during specific semesters. Many leaders in online education would like to see changes to this, though.

In recent interviews with leaders in online education, they were asked "If you could change something about your program, what would it be and why would you change it?" Almost everyone said they sought improvement in relation to the part of the institution that develops the schedule and staffing decisions related to it. In essence, while department chairs and deans may be able to see the goals and plans for their part of the academic pie, the online director can see the whole picture for the online program. This means they can best see what students seeking a degree completely online need in order to fulfill all of the requirements. The online director also has a better feel for the pulse of the online instructors. They know which instructors generate a following and which have completed the required (and perhaps even beyond what is required) training. This appears to be the next major area of discussion between leaders in online education and traditional academia, and it may very well control the growth and profitability of the online program and ultimately the school.

Money Associated with Online Programs

Online programs are growing and outpacing the growth of all other areas of higher education. With this growth are coming large profits, which are often desperately needed by the university. The amount of money generated by online programs is both a product of the number of courses and the price of those courses. For example, at Drury University the price per credit hour for an online class is $239, while the price per credit hour for a face-to-face class is $190. The difference of $49 is one of the reasons online programs have to potential to be such large revenue makers. It is necessary for institutions to consider their student body, the culture of the university, and their competitors when setting their price schemes. This revenue potential will continue to motivate institutions of higher education to add to their online programs.

Staffing Online Departments

Most online programs are supported by an online office staffed by many individuals with significant online experience or degrees in the field. Some of these degrees include instructional design, educational technology, instructional systems technology, education with an emphasis in e-learning technology and design as well as Masters in Education degrees with emphasis in online teaching and learning. Each person in the office handles various areas of the online experience. There is normally a director, one or more technical support personnel, instructional designers, faculty liaisons or trainers, curriculum designers, marketing specialists, and others depending on the culture and makeup of the university. The jobs of these individuals are designed to complement each other to make all areas of the online program run smoothly.

The director is the highest ranking position on most online offices. This person has oversight authority and responsibility for every aspect of the program. When interviewing numerous directors of various programs, it became obvious they consider the most important aspects of their jobs to be very similar. The two most oft mentioned areas were promotion of the program both outside and inside their respective universities and maintaining or enhancing control over scheduling and staffing.

Many institutions of higher education are pushing their online departments to grow and expand in many ways because studies show that this is the most rapidly growing area of higher education. Directors of online education must market to many people in their own institution. They have to be constantly selling and reassuring faculty, chairs, deans, and trustees that the product they are delivering fits the school's mission and academic niche. They also must keep all of these constituents apprised of the changes in technology and online education so the school can evolve with the industry, remain on the cutting edge, and maintain their market share.

In order to promote their programs to potential students, many online offices are hiring marketing personnel who are well-informed and in tune with the particulars of online education in order to bring in new students. These marketing people must be able to highlight the advantages of online classes – namely, convenience and flexibility as well as address issues such as technology requirements and understanding to potential students. The marketing person must work closely with the director in order to ensure the information communicated to potential students is accurate and timely.

The director's other main task involves maintaining or enhancing control over scheduling and staffing. As mentioned previously, this is an area that most directors wish they had more control over. There is an art to scheduling the correct classes at the correct time in the correct time frame (5, 8, 12, or 16 week classes). Many students prefer to take some classes in a particular order or balance their workload by considering the intensity or time necessary to complete a class when enrolling. For example, many history majors only like to enroll in one 300 or 400 level history class at a time because they are writing-intensive classes that will require a great deal of time. The online office understands this and would like to schedule classes according to what their customer (the student) wants. The department would like for history majors to complete all of their 100 level history courses before moving onto their 200 and 300 level courses and save the 400 level courses for their senior year. They are not looking at the situation from a customer's point of view but rather the curriculum and how certain classes build upon others. These two different motivating factors push the department chairs and the director of online education in different directions regarding the scheduling of classes. This is a difficult issue to resolve. A clear cut mission statement for the online program is the only way most universities relieve the tension.

Control over staffing goes hand in hand with scheduling. The most talented lecturer with a wall full of teaching awards from the experience in the traditional face-to-face classroom may never master online teaching. Conversely, an instructor who fails to communicate well or skillfully manage their face-to-face class may shine in an online setting and become a popular and effective educator. This dichotomy is difficult to explain to chairs and deans, particularly if they have never taught online themselves. Furthermore, tenured faculty who are accustomed to having a large amount of power over their schedule balk against new and diverse systems of scheduling.

Growth Issues

Once you have a successful program up and going, you want to grow it. However, with this growth comes a whole new set of challenges. Quality control procedures must change as a program expands. Instructor training, which may have been easily managed when instructors were small in number and local, takes on a whole new dimension when growth and diversity issues push the University to hire instructors located all over the world.

Growth in an online program often creates a challenge to other parts of the university because of a loss of students from traditionally seated classes to the online environment. Successful

Figure 1.

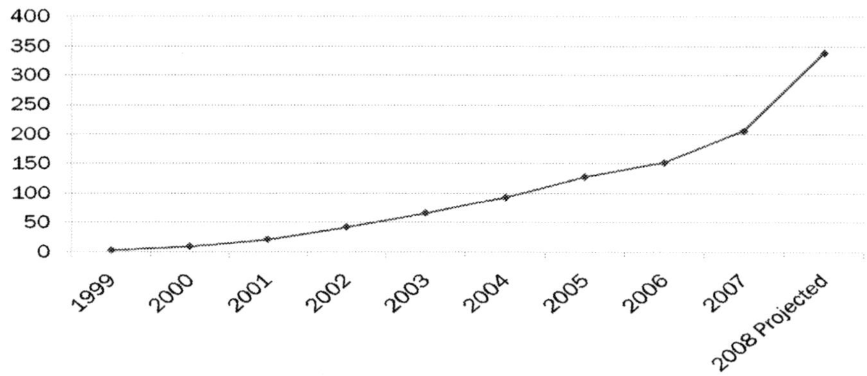

online classes logically lead to online degrees, but this is yet another challenge.

As stated at the beginning of this chapter, "The proliferation of distance education programs n the last few years is unprecedented in the history of higher education" (Monolescu, 2004, back cover). A perfect example of this is the growth in Drury University's online program. Figure 1 shows the number of online course offerings over the last nine years.

This rate of growth for online classes far surpasses the rate of growth for any other area of the university. However, this growth, while bringing a wealth of opportunity also brings it own set of challenges.

Quality Control

Assessing both an online program as well as online classes is a challenge for many schools as they grow from a small online program offering a few courses to a large program offering full online degrees. Any online education program must match its institution's mission, scope, objectives, and goals. Additionally, a strong online program must have a visionary and highly motivated director who is able to effectively communicate with multiple constituents. The director must have a highly skilled support staff that can handle the everyday workings of the program. He must also be associated with an institution that can adequately fund the program and serve all student needs outside of the actual classroom instruction.

The semester to semester, if not week to week, evaluations of specific classes are what most people examine when considering the quality of an online program. There are several tools for doing this including focus groups, in-class evaluators, in-house evaluations, or outsourcing evaluation through a program such as IDEA (Individual Development and Evaluation Assessment). The key things factors any evaluation should look for are

- Instructor to student interaction
- Student to student interaction
- Fulfillment of course objectives
- Clarity of communication
- Timeliness and quality of instructor feedback

All types of assessment tools should look to measure both faculty and student satisfaction as well as gather information that could be used to improve the program in the future.

Drury University has a multi-level concept of course evaluation. All faculty are selected through a four step approval process. First, the faculty are considered by the Director of Online Education, who evaluates their comfort level with technology and the Drury culture. Second, the department chair evaluates the academic credentials of the proposed instructor. Third, the Dean of the College of Graduate and Continuing Studies (CGCS) evaluates the proposed instructor and based on the notes from the Director of Online Education and the department chair makes a proposal to the Vice President of Academic Affairs, who ultimately makes final approval. This process is meant to ensure that no one enters the online training program without first having been vetted by all areas of the academic and administrative team.

Once the person is approved and the course assigned, they fall into a set of evaluations specifically designed for new instructors. New instructors must have all of the materials for the first two weeks of their course posted by two weeks prior to the start of the course. The course is then visited by what the online office calls a "secret student". This is a senior-level student or a recent graduate, who looks at the class from a student's point of view and give feedback as to how well the instructor did designing the class. A sample report can be found in Appendix A. This feedback is then given to the new instructor along with help to fix the necessary areas before the course begins. New instructors are also given a special mid-term evaluation, which can be viewed in Appendix B. The goal of this evaluation is to see how well the original plan has been implemented in the first half of the semester and to show administrators what (if any) help needs to be given to the instructor in order to improve the second half of the semester.

All online courses, both those of new instructors and veterans, are visited by a faculty monitor multiple times each semester. The monitor fills out a report that is then submitted to the Director of Online Education. The Director then contacts each faculty to let them know the findings. A copy of the report can be found in Appendix C. This evaluation looks at everything from the appearance of the class to the timeliness of the instructor's grading and feedback. This report is very helpful early in the semester to ensure that each class is organized in a way that students can understand exactly what is going on. It is useful throughout the semester as a measurement of instructors' timeliness and commitment to constant and quality feedback.

The final tool Drury University uses to access online classes is the Individual Development and Evaluation Assessment (IDEA) program administered through Kansas State University. This tool provides both statistical data and constructive feedback from the students. A sample of each can be found in Appendix D. These evaluations are administered towards the end of the semester. The conclusions are sent to the instructor, department chair, Director of Online Education, and the Dean of the College approximately three months later.

After receiving information back from all of these assessment tools, steps must be taken to either fix stated problems or change the expectations so that they better fit the reality of an online class. This takes a great deal of time – both for the instructor to make the changes and for the administrator charged with ensuring the changes are made. This is why it is critical that the online office team be well staffed with competent individuals who understand all aspects of the operation.

Faculty Training

Training of online faculty has evolved a great deal over the last decade. Most schools began by bringing their entire online faculty into the online office or together in a large computer lab for training sessions. For many schools, this practice has evolved to a mostly or even totally online training program. Instructors now understand that "online teaching forces a systematic rethinking of the course content in minute detail" (Little, Titarenko, and Bergelson, 2005, p.357) and are hungry for instruction in how to take the materials from a face-to-face course and apply them to an online course.

A local community college with a thriving online program has evolved its faculty training into a very impressive system. When the college's online program began in 2001, they did not have any faculty training per se. Then, they went to voluntary face-to-face training meetings. As they grew, this finally evolved into an online immersion type of training that lasted eight weeks. The instructors complained that eight weeks was too long, so it was shortened to the current system of four weeks.

The completion of this four week training program for online faculty is now required of every individual before they are allowed to teach online. The training places the instructors in the student's role. It is highly interactive and instructor facilitated. It is taught by an instructional designer and expert in online teaching. The training is broken into four modules that cover such things as managing discussion boards, completing activities in the practice course, and skilled competencies. The training began as very heavy in theory with little instruction in how to use the course management system. However, "practicing teachers, a lot of them subject experts in academic disciplines, never having taken an education class could care less about studying about pedagogy and reading the articles . . . . so we take an approach of teaching the theory at the end. . . . showing them how to do it or modeling best practices of online teaching and then relating it to theory in the back end so they can see . . . . the actual reason and philosophy behind hit" (W. Salley, Interview, October 20, 2008). Each instructor must score a minimum of 80% each week in order to continue with the training and teach at the end. They also offer optional online faculty development courses such as multi-media survival skills and using synchronous tools in the online classroom. These optional classes are very well attended because they strive to make them very valuable and not very time intensive.

Online Learning and Teaching Certificate Programs

Many other online programs are mirroring this type of program, which also has similarities with some of the Certificates in Online Teaching being offered. Colorado State University offers a Certificate in Online Learning and Teaching through their Global Campus, which they describe as being "designed for professionals in the educational, public or private training sectors who would like to develop the skills necessary to design, produce and facilitate online courses. College or university faculty, K-12 teachers, corporate trainers, program administrators, instructional designers and others who would like to enhance their skills, or prepare for a teaching or administrative career in online learning will find this program appealing" (http://www.csuglobal.org/certificate_programs.php). The four courses offered in this certificate program are

1. Instructional Theory and Design for Online Learning
2. Online Learning Technologies
3. Multimedia Technologies and Design Principles
4. Evaluation and Assessment in Online Learning

This program is similar to California State University, East Bay's certificate in Online Teaching and Learning, which they describe as meeting "the needs of university and college faculty, K-12 teachers, corporate and military trainers, educational administrators, curriculum designers, technical support staff and others who design, implement and teach online courses. Those with experience in online education and those who seek the necessary training will both find the program appealing . . . The program's online-only format makes it convenient for anyone in the world, not just those who live or work in the San Francisco East Bay Area. Students can look forward to a hands-on, rigorous curriculum in which both the instructors and fellow students provide detailed feedback on every project and should expect to spend 10-12 hours or more per week on each class. They can communicate with their instructors and advisors in a variety of ways: email, telephone, chat room office hours and/or videoconference. A high-speed Internet connection is strongly recommended" (http://www.extension.csuhayward.edu/certificate/online_teaching/index.shtml).

This program, like Colorado State University's is comprised of four courses.

1. Introduction to Online Teaching and Learning
2. Teaching Models for Online Instruction
3. Technology Tools for Online Instruction
4. Designing Curriculum for Online Instruction

Throughout this certificate program, students employ what is learned on a single class as they construct it from inception through completion. These are just two of the many great certificate programs out there that appear to be laying the ground work for institutions to follow as they set out to train their own faculty.

Purposeful Online Growth

The growth of online programs has caused many in academia to question whether the online option is pulling students out of face-to-face classes. This is a concern to administrators who are responsible for maintaining head counts in seated classes as well as the traditional academy who still feel students who can be in a seated class should be.

Students are speaking with their enrollments and telling educational institutions that they prefer online classes to face-to-face classes. The reasons have a lot to do with the hectic lives many Americans live and the economic situation they currently face. Many students are working full-time while attending school full-time. Many adult students even add family responsibilities on top of full-time jobs and school. These students want – even demand – convenience in their education. They want to go to class when it fits into their schedule, and that often means at midnight after working all day and putting their children to bed. Students are also dealing with rising gas prices that have a direct impact on their education by forcing them to consider the price of driving to campus to attend a traditional seated class. These forces are combining to push students to increasingly choose internet classes over traditional seated ones.

So, where do these students come from? Are they new to the educational institution? In some cases, yes; but in many cases, no. They are the students that had been attending face-to-face classes. Figure 2 shows how the rise in credit hours enrolled in online classes has obviously impacted the credit hours enrolled in seated classes at one university.

Rigor of Online Classes

Students are increasingly pursing the online option, however many are quite vocal in criticizing the increased rigor encountered in online classes. Here are what some history students have said regarding online and face-to-face classes:

- "The difference between online and seated classes, are many. Online classes provide the convenience of going to class at home. However, the student must be organized and disciplined as the workload is reading-writing intensive and generally double to that of a seated class." (Slagle, 2008)
- "In a seated course, if you show up, turn in your assignments and participate in the discussions, that's basically all there is to it. With online classes, your participation is based on your posts. You are required to post at least once and reply to two other students, which, if done correctly, is the

Figure 2.

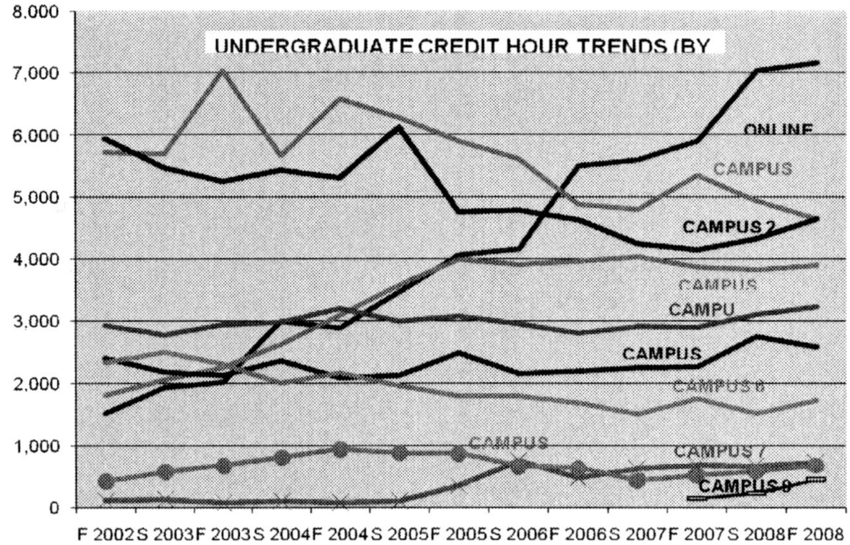

equivalent of writing a thorough, well-researched, two-page paper every week, on top of any assignments." (Bayless, 2008)

- "My 2 cents worth after 3 years of online work. I have only had 4 seated classes. . . Online classes unlike seated classes require that the student be decidedly responsible to the online program. It is much more than just one day a week but a dedicated everyday occurrence. Decisions have to be made in the beginning where your priorities are. . . This is where you must manage your time and take every opportunity to log into the classroom." (Terrey, 2008)

Of course, an online student's experience is dependent on the particular class and instructor they have. However, the evaluation statements seem to support the assertion that online classes are preferred yet require more time and commitment on the student's part.

Many faculty from traditional academia contend that the educational experience simply cannot be the same in an online class as it is in a traditional seated class. Many tenured faculty are also able to see multiple sides of the issue, such as Elizabeth Paddock, Chair of the Department of History and Political Science at Drury University, who stated, "Online classes offer students and faculty an alternative that was not available even a few years ago. The experience obviously is not the same. It often is not possible to achieve the same depth or range of discussion in online classes that academics seek in seated classes. Faculty cannot call on students in the same way, and students in online discussions miss out on the passion and thrill of intellectual discovery and debate that are such a part of the traditional classroom. However, online education offers some significant benefits that are not available otherwise.

The flexibility of online classes matches the fast-paced reality of modern society. These classes also allow a wider range of students to pursue their education, people who otherwise would not be able to fit their schedules with traditional seated courses. Finally, discussions in online classes often are more measured and less impulsive than discussions in traditional classes, and hence tend to be less passionate in some ways. At the same time, however, online discussions allow students more time to deliberate and consider their responses to questions and observations. Those students who may be intimidated in participating in regular classroom debates may find their voices more readily in the relative anonymity of the online classroom. There obviously are tradeoffs. The success of an online class ultimately rests with both faculty and students and their commitments to engagement in online assignments" (E. Paddock, Interview, November 24, 2008).

Online Degrees

Many online programs are growing past online classes into complete online degrees. It is impossible to search the internet for anything dealing with education without having ads from various online colleges and universities appearing in advertisements on the page. While many of these institutions are outside the main academic and accreditation circles, numerous well-respected institutions of higher education are venturing into the area of online degrees.

The challenges for offering completely online degrees relate back to several ideas already discussed in this chapter. First, most quality institutions will seek accreditation to deliver the online degrees. Second, when students are being drawn into a program with the promise of a degree being completely online, it is critical that the scheduling of classes for that degree be done in a calculated and well-thought out manner. As previously mentioned, online education departments strongly feel that this means they should hold the power and responsibility for scheduling. Finally, when degrees are offered totally online, it becomes necessary for the educational institution to put student services online as well.

Many educational institutions struggle with putting student services online because that reaches outside of the parameters of the online department. First and foremost, the library must offer adequate online holdings or ways to deliver research materials to students in various locals for students to complete research intensive courses. Other areas such as advising, financial aid, the bookstore, and business office must also offer students from far distances the same types of services they do for students who can walk up to their service windows. This requires staff, financial, and resource commitments from the institution.

CONCLUSION

As students demand higher education with convenience and rigor, online programs must meet this challenge. The development and growth of online programs has become a permanent mark on the landscape of higher education, and in order to perfect high quality programs a set of best practices must be followed. Divided into three categories, these can be summarized as:

Student Issues:

1. Student learning styles remain based on visual, auditory, and kinesthetic stimulation whether the student is face-to-face with an instructor or a thousand miles away. Therefore, all three learning styles must be incorporated into a high quality online class.
2. The revised version of Bloom's Taxonomy combines the cognitive process (remember, understand, apply, analyze, evaluate, and create) and the kind of knowledge to be learned (factual knowledge, conceptual knowledge, procedural knowledge, or meta-cognitive knowledge). This criteria should be used when developing objectives and assignments for an online class.

3. Instructors need to steer students away from cheating by educating their students on what constitutes cheating and how the student can best avoid it; constructing assignments in a way that makes cheating or plagiarism more difficult; finding new methods of assessment that are not easily plagiarized or otherwise taken advantage of by those looking for an easy way out; and building a rapport with students that motivates in them a love of learning and a respect for the instructor.
4. Online programs need to communicate to potential students that the convenience of an online class does not translate into it being easier than a face-to-face class. Students need to be prepared for the extra time required to complete the work and succeed in an online class.

Institutional Issues:

1. Acquiring accreditation is the first step in building a high quality online education program. The specific case study of Drury University's preparation for a focused visit from the Higher Learning Commission highlights the importance of having the infrastructure in place to support the online students and faculty.
2. It is necessary for a leader in the online education program to spend time with department chairs and deans showing them how online classes work, assessment outcomes, and final projects and statements of the students from the classes in order to garner inner institutional support for the program.
3. The choice of a learning management system is a major decision for online programs. The system needs to fit the educational institution's culture, budget, goals, and vision.
4. Scheduling and staffing need to be done with consideration given to academic qualifications and goals as well as with a vision of the online program as a whole and students seeking online degrees in particular.
5. Prices for online classes tend to be higher than prices for face-to-face classes. However, the demand for the online classes is outpacing the demand for the face-to-face classes. It is necessary for institutions to consider their student body, the culture of the university, and their competitors when setting their price schemes. In most cases, though, online classes will be a profit enhancer.
6. It is imperative that online education departments have adequate and highly qualified staff. Most offices normally employ a director, multiple technical support personnel, instructional designers, faculty liaisons or trainers, curriculum designers, marketing specialist and others depending on the culture and makeup of the university.

Growth Issues:

1. Quality control is a major issue as online programs grow to encompass several hundred classes taught by instructors located all over the world. A case study of Drury University's quality control and assessment program highlights the need for courses to be evaluated at multiple times (before, during, and after the semester) by multiple perspectives (students, peers, administrators). The evaluation must then be followed up by corrective measures and often additional guidance from the online education department.
2. Training faculty to teach online is very important and evolving process. Today's most successful strategies

include emersion in an online training course set up much like the faculty's potential online course. Most institutions of higher education have developed their own unique programs for their faculty, however several certificate programs in Online Teaching and Learning have emerged in the last few years as well.

3. Schools must accommodate student demands for flexibility and convenience in their educational experience. This often means readjusting curriculum and schedules of areas outside online education to help students fulfill their desires for education on their time table.
4. The rigor and time commitment required to achieve success in an online class are often not understand by persons who have never taken an online course. Students considering enrolling in their first online course need to have the expectations clearly communicated to them prior to the first day of class.
5. Online degree completion requires a great deal of commitment from the educational institution. The most important considerations involve accreditation, specific scheduling of classes in a calculated and well-thought out manner, and the transition of student services online.

Any online program's adherence to these best practice guidelines and examples must, of course, be tempered and made to fit the educational institution's culture and academic governance system. However, if done correctly, the proliferation of online programs is a very positive development for students seeking a rigorous educational experience highlighted by flexibility and convenience.

REFERENCES

Anderson, L.W., & Krathwohl (Eds.). (2001). *A Taxonomy for Learning, Teaching, and Assessing: A Revision of Bloom's Taxonomy of Educational Objectives*. New York: Longman.

Bourne, J., & Moore, J. (Eds.). (2002). *Elements of Quality Online Education*. Sloan Center for Online Education.

HLC Report. (2003). [*A Classroom of One: How Online Learning Is Changing Our Schools and Colleges*. New York: Palgrave Macmillan.]. *Gene*, I.

Little, C. B., Titarenko, L., & Bergelson, M. (2005). Creating a Successful International Distance-Learning Classroom. [from http://www.jstor.org/]. *Teaching Sociology*, *33*(4), 355–370. Retrieved October 18, 2008.

Monolescu, D., Schifter, C., & Greenwood, L. (2004). *The Distance Education Evolution: Issues and Case Studies*. Hershey, PA: Information Science Publishing.

Rose, C. (1991). *Accelerated Learning*. Aylesbury, Bucks.: Accelerated Learning Systems Ltd.

Simonson, M., Smaldiino, S., Albright, M., & Zvacek, S. (2000). *Teaching and learning at a distance: Foundations of distance education*. Upper Saddle River, NJ: Prentice Hall.

Wallace, Raven McCrory (2004). A Framework for Understanding Teaching with the Internet. *American Educational Research Journal* Vol.41, No.2, 447-488. Retrieved October 18, 2008, from http://www.jstor.org/

APPENDIX A

Figure 3. Secret student evaluation of new online instructor's course

Secret Student Evaluation
May 17, 2007
Course: MGMT 422
Corporate Policy and Ethics
Instructor:

I really like the homepage, very inviting!

I love that the calendar is so detailed and organized!! This is great! The students will appreciate this, so that they are able to stay on track and see what is ahead. Way to Go! I love that the instructor also adds links within the calendar function to allow students to view such things as the syllabus, assignments, etc! What a great idea!

I feel that it is noteworthy to add my opinion of group work. When I was in a course that required group work (and there was more than one class) it was a HORRIBLE experience. It usually ended up that the group would not get together, and to save my grade, at the last minute, I would have to complete the entire assignment. I was NEVER a big fan of group work.

I really like that the instructor lays out the entire semester so that students know what is ahead in this fast paced summer course.

"Discussion of the Discussion Board" I was unable to view the examples, they appeared as white text blocks with a small (x) in the upper right hand corner.

I saw in the syllabus that there is an assignment for Week Zero. Under the assignments, there are no instructions other than the lecture. I believe that it would be helpful for students if the assignment for Week Zero were listed under the assignment section.

I did not see where it was laid out when assignments were to be posted. I did see it is the syllabus, but thought it might be helpful if it was also added in with the assignment. I also saw it on the assignment drop box, but again I thought it might be helpful if it was listed within the assignment.

APPENDIX B

Table 2. New instructor mid-term evaluation form

Mid-Term Evaluation Form

Please complete this Evaluation Form. Your answers to the following questions will be relayed to the instructors in such a fashion as to ensure your confidentiality. Do not put your name on this form. These do not become part of the instructor's file, but instead are designed to give the instructor important feedback from the student at the midpoint in the semester.

Circle your response to the following questions:

A. This course is well-organized and carefully planned.

1. Strongly Agree 2. Agree 3. Disagree 4. Strongly Disagree

COMMENTS:

B. The instructor has good teaching skills; he or she produces steady interest in the subject, creates real desire to learn, keeps things moving.

1. Strongly Agree 2. Agree 3. Disagree 4. Strongly Disagree

COMMENTS:

C. Assignments are explained well and students clearly understand the tasks of each new assignment.

1. Strongly Agree 2. Agree 3. Disagree 4. Strongly Disagree

COMMENTS:

D. Class discussions are handled well; the questions are challenging and demand sound thinking, and the discussions are interesting and stimulating.

1. Strongly Agree 2. Agree 3. Disagree 4. Strongly Disagree

COMMENTS:

E. Examination questions are thought-provoking, carefully selected, and clear. Results are given to students quickly.

1. Strongly Agree 2. Agree 3. Disagree 4. Strongly Disagree

COMMENTS

F. The instructor has an excellent mastery of subject material.

1. Strongly Agree 2. Agree 3. Disagree 4. Strongly Disagree

COMMENTS:

APPENDIX C

Table 3. Course visitation report sheet

Monitor's Name:			
Course Number:			
Instructor:			
Course Duration:			
Date of Visit:			
Text Block Usage:	Yes	No	
Color Choices:	Excellent	Good	Poor
Course Navigation/Structure:	Easy	Challenging	If challenging, what are your suggestions to make it easier?
Syllabus Clearly Visible:	Yes	No	
Schedule Posted:	Yes	No	
Grading Scale:	Yes	No	
Calendar Available and Up To Date:	Yes	No	
Grading Rubric:	Clear	Neutral	Unclear
Course Materials Current:	Yes	No	
Discussion Board Read:	24 Hours	48 Hours	72 Hours
E-mails Answered:	24 Hours	48 Hours	72 Hours

APPENDIX D

Figure 4. Idea evaluation feedback form

Faculty Name: Blakey, Leah
Course: Liberal Arts & Sciences, General Studies & Humanities GSTU 493

Term: Spring 2006–2007
Page 3

Section II. Student Ratings of Progress on Relevant Objectives

This graph shows student progress ratings on the objectives you chose as ***Essential*** (Part A) and those you chose as ***Important*** (Part B). To the degree that students make progress on the objectives you emphasize, your teaching has been effective.

Part A. Essential Objectives	T Score Unadj. **Adj.**	2% of all classes	28% of all classes	40% of all classes (Avg. range)	28% of all classes	2% of all classes	Your Average * (5-Point Scale) Raw	Adjusted	IDEA Average
1. Factual knowledge	55 **52**						4.3	4.1	4.0
8. Oral and written communication skills	57 **54**						4.3	4.1	3.8
Part B. Important Objectives									
6. Creative capacities	51 **47**						3.9	3.6	3.9
7. Intellectual/cultural understanding	58 **54**						4.3	4.0	3.7

20 30 40 45 50 55 60 70 80

T Score—Comparison with the IDEA Database where the objective was selected as "Essential" or "Important" **

├✕┤ Unadjusted T Score ± one standard error of measurement.
├◆┤ Adjusted T Score ± one standard error of measurement: adjusted for student work habits (item #13); student desire to take the course regardless of who taught it (item #15); and instructor reported class size.

Similar to Section I, you may wish to assign ratings to categories. Simply assign T Scores to categories as follows: **Low** (lowest 10%) = T Score below 37; **Low Average** (next 20%) = T Score 37–44; **Average** (middle 40%) = T Score 45–55; **High Average** (next 20%) = T Score 56–63; and **High** (highest 10%) = T Score above 63.

These graphs are intended to help you identify a focus for improving your instructional effectiveness. If student progress ratings on ***Important*** or ***Essential*** objectives are disappointing, you are encouraged to discuss improvement strategies with your department head, the campus faculty development specialist, or a colleague. Such strategies could focus on matters such as teaching methods/styles, class activities and assignments, the text and other readings, assessment/feedback, and the need for course pre–requisites. You might also consider using the IDEA Diagnostic Form the next time you solicit student ratings, since it is designed to help identify specific teaching methods to use in improvement efforts.

Note: Students in your class also rated their progress on the objectives that you classified as being of ***Minor or No Importance.*** These ratings are considered irrelevant in judging your teaching effectiveness. However, a review of student ratings on these objectives, found in **Section IV** (Statistical Detail), may provide you with insights about some "unintended" or "additional" effects of your instruction.

* Statistically, adjustments can exceed 5.0 on the 5–point scale. If this occurs, "Your Average," reported in the table above, will be rounded to 5.0. However, the T Score reported will reflect the actual adjusted score, which may exceed 5.0. Therefore, identical adjusted scores of 5.0 may have different adjusted T Scores.

** Normative information (T Scores) and the process for adjusting scores were recently updated. See page 1 for "A Few Words of Caution."

Figure 5. Idea evaluation feedback form (continued)

IDEA Student Ratings of Instruction

IDEA CENTER

Blakey, Leah
GSTU 493
Senior Seminar
online
DRURY UNIVERSITY
04-23-2007 -05-04-2007
Local Code: 4279

Comments: Use the space provided in the text area below for your comments.

- Dr. Blakey is an extremely passionate teacher and she really gets involved. You can tell she puts alot of effort into it and she makes us excited to learn.
- I enjoyed taking this class.
- I thought this class was great, I was concerned when it first began, but I was able to do everything required. I ran into a problem with my internet connection, since I had dial-up, it was difficult for me to view some of the Power Point presentations unless I went to another computer. I live in a very rural area and had to drive 50 miles to find one with the program I needed. People should be warned. I also think putting a face to the names would be nice. I met some online classmates at a recent banquet, maybe a place to post pictures during our introduction process would make the class more tied together. Just some thoughts, thanks!
- Enjoyed this class alot!
- This class could have been better if the instructor had picked out books that are intersting to modern day students. Reading about Africa and meat packing industries does not leave an excitment that I can think of. Most part this class was very boring and bland. The writing for answers could have been lowered a little bit.

Chapter 12
Managing Growth of an Online Program though Policy and Mainstreaming: A Case Study

Deb Gearhart
Troy University, USA

EXECUTIVE SUMMARY

At a time when higher education is struggling to provide educational services to students on traditional campuses, many institutions are experiencing growth in the online degree programs that they provide. This case study looks at one institution's effort to deal with rapid growth in online education while maintaining academic integrity and quality in program delivery; managing program growth through the development and use of policy and procedures.

BACKGROUND

Troy University is a multi campus university established in 1887. The decade of the 1950s developed the University's long relationship with the United States Military, as extension courses were offered on nearby bases, first at Fort Rucker, near Dothan, and later at Maxwell Air Force Base in Montgomery. A separate Troy State College teaching center was established at Fort Rucker in 1961, which evolved into the present-day Dothan Campus. A similar center, begun at Maxwell Air Force Base in 1965, led to the creation of the present-day Montgomery Campus. These programs were the forerunners of the Troy University's Global Campus now with more than 60 sites and programs in 14 U.S. states and 25 countries. Distance education programs started developing on the Alabama campuses and within the regions of Global campus in 1998. When Troy State University merged into Troy University in 2005 one of the strategic decisions was to establish the eCampus to merge the online programs under one unit.

Troy University's eCampus has experienced a 47% enrollment increase since being stood up in Fall of 2006. It is critical to manage program growth to ensure the quality of education the students receive. Growing too fast without improving procedures, developing appropriate policy, and managing quality of the program, students educational success cannot

DOI: 10.4018/978-1-60566-870-3.ch012

be assured. An administrator must be aware of the issues and environment surrounding distance education program. This chapter reviews the circumstances which provide for managed growth of a distance education program, based on policy and supported by literature.

SETTING THE STAGE

This case is about how one University has worked to manage the growth of a large online program. The Troy University eCampus offers 21 online degree programs and in the three years since its establishment, eCampus has integrated all the online program delivery across Troy University and worked to develop student services and quality assurance of the program.

eCampus is based on five functional units: student services, academics, educational technology, instructional design and educational resources, and quality assurance. Within these functional units is where all best practices are met.

Student services assists students from admission through graduation and within the unit students have access to professional counselors, administrative counselors and faculty advisors to work with throughout their programs. Once admitted to Troy University online learner student services are decentralized throughout Global Campus. There are over 60 sites throughout the United States and internationally where students can go to directly to see student services support staff, if desired. Students are assigned to a home location by their address and then assigned to a Troy University Global Campus site based on that home location. This provides a more student-centered and personal touch to student services.

Academics is the unit which works with the Colleges of the university and the faculty. The Director of Academics works as a liaison, facilitator, mediator, ombudsman and initiator with the academic units of University. As developed in best practices, the academic units have control over the programs offered online to make sure the breadth, depth, and rigor of the online programs match that of the programs in the traditional classroom. Another key function is working with the academic units to certify faculty for teaching in the online programs. Other functions within Academics include workflow for scheduling, textbooks, and testing.

Educational technology includes the planning, implementation, and contract oversight for all educational technology initiatives with workflows in Blackboard and Securexam remote proctor systems. The unit is also responsible for web page design and online student service forms. All student and faculty technical support comes from Educational Technology staff and for faculty from staff within Instructional Design and Educational Resources.

Instructional Design and Educational Resources (ID&ER) is where technology and training meet. Faculty development training is conducted within this unit. All faculty members teaching online must go through training for the LMS and basic instruction design, known as TOP I (Teaching Online Proficiency). There is other training available to faculty, TOP II and specialized software training. The faculty development coordinator sets up training modules online for students to learn how to use the LMS and software. ID&ER is responsible for coordinating the eCourse redesign project where courses are redesigned for online delivery.

Quality Assurance is an ongoing process responsible for review, analysis, and implementation for improving the quality of online education meeting best practices and standards. Quality works across all units and the quality review of courses is now conducted on a regular basis.

How did eCampus work to establish a quality program that can manage growth? Through the development of policies and procedures; this case is based on a review of literature to support the work that eCampus has completed over three years.

The research literature reviewed supporting this case study looked at the issues and environment a distance education program administrator must be aware of to be an effective manager of program growth. The literature primarily reviews policy and its implementation into a distance education program.

The institution needs to determine the type of program it wants to establish. Paolucci and Gambescia (2007) have developed a categorization of distance education options that are commonly used by institutions of higher education. The six categorizations are both internal and external to the institution. In the planning process the institution needs to determine what type of distance education organization it needs and wants. The categorizations are shown in Table 1.

Once the institution determines the type of program it will develop, administrators need to be informed of the fundamental issues and enrollment facing institutions when planning for distance education programs and for the growth that comes with them. Howell, et. al. (2003) developed the following 32 trends which are still relevant today and are a good starting point for administrators in their planning process.

1. The current higher education infrastructure cannot accommodate the growing college-aged population and enrollments, making more distance education programs necessary.
2. Students are shopping for courses that meet their schedules and circumstances.
3. Higher education learner profiles, including online, information age, and adult learners are changing.
4. The percentage of adult, female, and minority learners are increasing.
5. Retention rates concern administrators and faculty members.

Table 1. Types of distance education organizations

Internal		External	
Academic Department	the academic department granting the degree has the lead in most, if not all, of the curriculum and key administrative duties in offering the online degree	**Consortium**	The university has joined other similar universities to cooperate in the offering of online degrees. The nature of cooperation for the administration and operations of the degree offerings may vary, but there exists a strong, formal relationship among the members
Continuing Education/Professional Studies Unit	a previously existing continuing education/professional studies unit within the university has the lead in many, if not all, of the curriculum and key administrative duties in offering the online degree	**Alliance**	The university has joined other universities that may or may not be similar to gain some type and level of administrative or operational service through an entity that is the focal point for the alliance of schools
Distance Education Unit	the university has set up a "separate" or adjunct distance education unit to take the lead in most, if not all, of the curriculum and key administrative duties in offering the online degree	**Outsource**	The university has turned over much of the administrative and operational duties to an outside entity that is really managing the offering of the online degree program

Source: Paolucci & Gambescia, 2007, pp. 6-7.

6. Traditional faculty roles are shifting or "unbundling".
7. The need for faculty development, support, and training is growing.
8. Faculty tenure is being challenged, allowing for more non-traditional faculty roles in distance education.
9. Some faculty members are resisting technological course delivery.
10. Faculty members who participate in distance education courses develop better attitudes toward distance education and technology.
11. Instructors of distance courses can feel isolated.
12. Faculty members demand reduced workload and increased compensation for distance courses.
13. Knowledge and information are growing exponentially.
14. The institutional landscape of higher education is changing; traditional campuses are declining, for-profit institutions are growing, and public and private institutions are merging.
15. There is a shift in organizational structure toward decentralization.
16. Instruction is becoming more learner-centered, non-linear, and self-directed.
17. There is a growing emphasis on academic accountability.
18. Academic emphasis is shifting from course-completion to competency.
19. Education is becoming more seamless between high school, college, and further studies.
20. Higher education outsourcing and partnerships are increasing.
21. Some advocate standardizing content in learning objects.
22. Technological devices are becoming more versatile and ubiquitous.
23. There is a huge growth in Internet usage.
24. Technology fluency is becoming a graduation requirement.
25. With the economy in recession, there are fewer resources for higher education and higher education initiatives such as distance education.
26. Funding challenges are the top IT concern for many.
27. Lifelong learning is becoming a competitive necessity
28. More courses, degrees, and universities are becoming available through distance education programs.
29. The Internet is becoming dominant among other distance education media.
30. The distinction between distance and local education is disappearing.
31. The need for effective course management systems and Web services is growing.
32. There is an increasing need for learning and teaching strategies that exploit the capabilities of technology (Howell, et. al., 2003, pp. 2-13).

Five of the above trends were a primary focus in the managing of growth in eCampus:

2. Students are shopping for courses that meet their schedules and circumstances. eCampus posts the syllabi for all the course being offered in the next term. It is a common faculty complaint that students shop for courses by reviewing the syllabi to find the course section they want to take. This is a service we provide to students but run into a dichotomy with faculty.
5. Retention rates concern administrators and faculty members. Retention has long been a concern for distance education program administrators. Early on in the field, research on student success demonstrated that getting a student started into the distance course right away helped retain students. In the online environment, programs use student orientations to help foster success and retention. In an online environment having

quality courses also fosters student success and retention.

7. The need for faculty development, support, and training is growing. This is an area that eCampus has expended a lot of resources in these initial years.
17. There is a growing emphasis on academic accountability. Quality assurance of the courses and training for the faculty have been key areas in the implementation of eCampus and an emphasis in providing academic accountability. The eCampus staff has worked hard to promote the academic oversight of the programs offered.
25. With the economy in recession, there are fewer resources for higher education and higher education initiatives such as distance education. The economy is becoming more of a factor since the Howell, et. al. article was written. eCampus, like other programs, is facing budget constraints and is working to maintain all current levels of service.

As stated when Troy University merged from Troy State University with its separately accredited campuses a strategic decision was made to create the Troy University eCampus. Many of the questions addressed by Magjuka et. al. were part of the discussion and are still being reviewed as Troy University moves through SACS re-accreditation this year. Magjuka et. al. (2005) discussed eight issues, from their experiences at the Indiana University's Kelley Direct Program, which administrators need to consider when looking at distance education program growth:

1. Which student group is to be served?
2. Will the distance education program be treated differently than the campus program? For example will Distance education courses be noted differently on the student transcript?
3. To what extent will the program provide a blended experience or will all of the program be at a distance?
4. Will faculty be allowed to teach in-load, overload or a mixture?
5. Should an institution put resources to already designed courses or train the faculty in online pedagogy and course design?
6. To what extent should there be a standard template for teaching in a distance education program?
7. To what extent should courses yield interactivity on the basis of the course design or should we rely on faculty to lead interactivity?
8. Should the program consider a learning management system, open source or create its own?

Pisel (2008) discussed the importance of strategic planning for higher education institutions, particularly when it comes to planning for distance education. Distance education programs are more successful if there is a systematic strategic planning process. However, Pisel noted that there are three levels to successful planning, strategic, operational and tactical, with strategic planning being the highest level. Pisel developed a strategic planning model that incorporated all three levels in 10 different phases:

1. Planning initiative – internal and external factors help determine the decision to start, identification of assets; assignment of a team to conduct the planning.
2. Planning guidance and scheduling – two functions in this phase: leadership intent is articulates and a planning schedule is established.
3. Analyses – includes the environmental analysis (SWOT) and the needs/gap analysis.
4. Mission Refinement – describes how distance education is important to and aligned with the core mission and future vision of the institution. Must include an organizational mission which is for internal consumption

designed to provide direction and inspiration to the institution.

5. Assumptions – are used to fill a gap in knowledge so planning can continue. A valid assumption is logical, realistic and essential and is a hypothesis on the current situation or on the future course of events.
6. Strategy development and courses of action – this phase moves the assumptions to courses of action based on the SWOT analysis.
7. Functional Analyses – the purpose of this step is to take the courses of action through a functional analysis and makes a recommendation to leadership on the course of action to take.
8. Implementation – moves away from the planning to determine who has authority and responsibility for implementing of the program including resources and personnel.
9. Assessment – often neglected, assessment is the continuation of the process and provides data to bring the plan and program in line with the mission and vision of the institution.
10. Periodic review - is required to identify and assess the impact of changes on the plan and the program. (Pisel, 2008, pp. 2-7)

It doesn't matter the size of a distance program. Leary and Berge (2007) looked at challenges of distance education for a small organization. Does using distance education in a small organization make sense? Will there be a return on investment? Small organizations face many barriers to successfully using distance education, including the lack of infrastructure and the lack of training. Leary and Berge provided three ingredients for successful distance education for small organizations: 1. combine the distance education culture with the strategic goals and activities of the organization; 2. encourage a web savvy staff; and 3. hire, maintain, or train a good training professional.

Leary and Berge provided strategies for successful distance education. In a small organization multipurpose distance education serves as a training method, a communication network, a knowledge management system, and education system, among other useful tools.

From this literature a distance education program administrators need to consider, as done with this case, to:

- Determine who the program will serve
- Analyze the environment and plan for the internal and external factors needed to make the distance education program a success
- Implementation of the program and assessment of the program.

This is appropriate no matter what the size of the distance education program.

Case Literature on Developing Policy

In setting the stage the literature looked at the type of distance organization to be developed and the issues and environment administrators need to review in making the determination on how the distance education program will be developed. Moving on to the actual growth that eCampus has experienced in three years the case looks at the how the mission has driven development and continue to look at how policy has affected development, supported by the following articles.

Mission and Mainstreaming

Managing growth of a distance education program depends highly on whether the program is part of an institution's mission and is mainstreamed into the institution.

The Troy University mission is:

Troy University is a public institution comprised of a network of campuses throughout Alabama and worldwide. International in scope, Troy University provides a variety of educational programs at the undergraduate and graduate levels for a diverse student body in traditional, nontraditional and emerging electronic formats. Academic programs are supported by a variety of student services which promote the welfare of the individual student. Troy University's dedicated faculty and staff promote discovery and exploration of knowledge and its application to life-long success through effective teaching, service, creative partnerships, scholarship and research. (http://www.troy.edu/mission.htm)

The eCampus mission statement expands upon the University mission by concentrating on the "nontraditional and emerging electronic formats".

Troy University's eCampus serves traditional and adult learners in the state of Alabama, as well as those located across the nation and globally, by providing online graduate and undergraduate degree programs that afford them opportunities to gain knowledge, skills, and competencies needed to live and work in the 21st century using technology and a learning environment that overcome time and place constraints associate with the traditional campus setting. The eCampus provides access to critical technology, resources, and tools allowing students and faculty members to access and use a rich and flexible learning environment, along with high quality support services, which enable online learning in a student-oriented environment.

Through its mission statement eCampus mainstreams the distance program in the University. Mainstreaming is said to reflect the process of seamlessly integrating distance education into conventional education (Irele, 2005, p. 2). According to Irele (2005) there are three dimensions to the mainstreaming of distance education: 1) the achievement of parity of esteem between traditional and distance education; 2) the convergence of the means of instructional delivery, student groups, and types of institutions; and 3) integration of distance education's constituent elements, outreach, use of technologies and teaching strategies, into traditional education, especially into institutional policy. However, incorporation into institutional policy can be a double-edged sword in that what is necessary for maintaining a distance education program can be disruptive to the traditional institution.

When mainstreaming distance education into the institution the institution needs to determine if the integration of distance education will be centralized or decentralized. Centralization is defined as the location of decision-making authority in a single office within the organization. Decentralization is defined as the transfer of decision-making authority from supervisory to functional units, or between units (Irele, 2005, pg. 3). The development of the Troy University eCampus established a centralized structure in the institution. According to Irele (2005), components of include an institution-wide strategic plan incorporating distance education, a team-based approach to course design and development, a central office for recruiting and registering distance learners, central funding for distance education, and centralized marketing of distance learning courses and programs.

- An institution-wide strategic plan incorporating distance education

The Troy University strategic plan for 2010 includes the development of the eCampus - 2.1 Establish an *eCampus* whereby quality academic programs and support services can be delivered *via* digital processes by fall 2006.

- A team-based approach to course design and development

eCampus has the only instructional design unit at the University which includes the training of faculty for online teaching, the eCourse redesign project where courses are redesign for distance delivery, and the course quality assurance reviews with the academic units.

Table 2. To make mainstreaming and growth work develop policies in the following areas

Faculty	Rewards (e.g. stipends, promotion and tenure merit increases, etc.); Support (e.g., student help, technical assistance, training, etc.); Opportunities to learn about technology and new applications (e.g., release time, training, etc.); Intellectual property (e.g., ownership of materials, copyright, etc.).
Students/Participants	Support (e.g., access to technology, Library resources, registration, advising, financial aid, etc.); Requirements and records (e.g., residency requirements, acceptance of courses from other places, transfer of credit, continuing education, etc.).
Management and Organization	Tuition and fee structure; Funding formula; Collaboration (e.g., with other departments, units, institutions, consortia, intra- and inter-institutional, service areas, etc.); Resources (e.g., financial resources to support distance education, equipment, new technologies, etc.); Curriculum/individual courses (e.g., delivery modes, course/program selection, plans to develop, individual sequences, course development, entire program delivery, interactivity requirements, test requirements, contact hour definitions, etc.).

Source: Irele, M. E. 2005, pg. 7

- A central office for recruiting and registering distance learners

The recruitment and admissions process into the online degree programs offered by eCampus is done centrally through the eCampus staff. Since Troy University is a global institution the administration of student services is done regionally. This allows for students to have face-to-face, hands on services if they desire. When students are admitted centrally they are notified of their regional contact for advising and services.

- Central funding for distance education

eCampus is part of Global Campus at Troy University and is part of the revenue generating arm of the University. As part of the revenue generating arm of the University the program receives no state funding but it totally self-supporting with a budget that adequately meets the programs needs.

- Centralized marketing of distance learning courses and programs.

Troy University uses the services of a marketing firm for the entire University. Advertising and media plans are developed annually and tracked carefully. Within Global campus advertising is often a cross over for the on site programs and the online programs. Many students cross over course/class delivery modes in doing their Troy University programs.

- Policy for distance education programs

Irele's research demonstrates that strategic plans and institutional documents of distance education programs provide three basic goals for the distance education programs: the intent to increase access for students; intent to increase revenue to the institution and intent to enhance the institution's position as a national and international leader in distance education (Irele, 2005).

However, policies at many institutions do not support these goals. There are often restrictions on who can participate in the distance education programs; some programs are still not considered equal quality educational programs.

The Irele study demonstrates that policies are weak in Faculty and Students/Participants and strong in Management and Organization. Administrators suggest that written policies indicated the awareness of the issues and more acceptance of distance education into the mainstream of the distance education program. Although implementation of the policies are problematic; especially in gaining faculty acceptance of policy related to distance education. There seems to be an air of mistrust of administrators by faulty. At Troy

University the Director of eCampus' role has been to build a trusting relationship with faculty to improve acceptance of eCampus. One common area of mistrust administrators by faculty is the "cash cow" syndrome. Faculty want to avoid administrators who are looking at distance education for financial gain only and bypass complete documentation of student credentials to ensure students' success in their chosen programs. The perception is that administrators would forego academic quality of the distance education programs for making money.

According to Irele (2005), it is evident that distance education teaching and learning strategies can and do indeed co-exist well within higher education; however, it is the institutionalization of distance learning that is the concern. Implementation of policies is a barrier to systematic integration of distance education into institutions. Irele paper suggests that mainstreaming is a term overused and in most cases not fully complete.

Move to Academic Quality Control - Getting Buy In on Using the Technology to Developing Quality Courses

McGrath (2006) discussed distance education policy at the departmental level which can also be applied at any level within the institution. An appropriate distance education policy should respond to the unique needs of online faculty and learners in a way that supports and protects faculty. Open communication during the development of the policy with a properly established committee of administrators and faculty with online teaching experience. The policies developed must be sustainable, tied to assessment efforts, and informed by available disciplinary research.

Schauer, et. al. (2005) comment that the two groups most impacted by the high demand for technology based distance education are the department chairs and faculty. Department chairs are the first-line academic leader for adopting distance education and are often asked to provide vision, encouragement, direction, guidance, resources and coordination. Technology integration in teaching is one of the biggest issues facing department chairs. Faculty are required to create, develop, and deliver distance education courses and programs. The Schauer, et. al. study found eight issues related to faculty, technology and distance education programs.

The Troy University eCampus has primarily addressed the academic issues described by Schauer in Table 3 through the formation of the eCampus Academic Communication Committee (EACC). This committee is the primary communication means between eCampus and the academic colleges and units (see Appendix A). Representation on the committee is comprised of the primary academic stakeholders and areas of discussion for the Committee might include, but not be limited to: implementing programs and curricula; the development and implementation of guidelines and procedures for teaching effectively through eCampus; managing change; faculty qualifications, development and training to teach online; implementing the Troy University textbook policy; scheduling of courses and faculty; and the quality assurance process in eCampus.

Other areas of the above table are addressed in the Troy University Academic Operating Procedures, such as the workload policy and the proctoring policy found at http://www.troy.edu/academics/aop/index.html. Student issues are resolved the same way for the University no matter how or where the academic program is offered and the Oracle, the student handbook, provides the students all information there.

CURRENT CHALLENGES FACING THE INSTITUTION

Once a program is established and policy developed, implementation of policy into the everyday management of an online program is the challenge to an institution, as it was in the case with Troy

Table 3. Issues to developing academic adoption of distance education

Issue	Theme	Concerns related to each issue
Faculty	Obtaining faculty buy-in and their involvement in implementing distance education.	• Workload issues related to development and delivery • Faculty incentives to integrate distance technology • Compensation for course development and delivery • Faculty acceptance of distance education • Access to appropriate professional development for faculty • Faculty awareness of new delivery alternatives and technologies • Pedagogical shift from teacher-centered to learner-centered • Recognition of distance education teaching in tenure decisions
Technology	Appropriate media supplied for transmitting instruction and the availability of various technologies.	• Technology support for faculty and students • Appropriateness of technology for program and pedagogy • Reliability of technology supporting distance education courses • Faculty and student access to technology • Student technology literacy • Security of courses
Setting distance education direction	Making the decision to implement distance education, determining who leads the implementation process, and then following the chain of command.	• Shared vision (buy-in) by faculty of need for distance education • Leadership within the department to pursue distance education • Development of appropriate department plans for distance education • Pace of implementation of distance education courses • Marketing and promotion of courses and programs • Justification of need for using distance education • Inclusion of distance education in institutional strategic plan • Institutional administrative acceptance of distance education.
Finance	Integrating funds into the budget for distance education, therefore, having minimal long-term reliance on grants and contracts for supporting distance education.	• Cost of implementation and delivery of distance education • Treatment of distance education in institutional budget structure • Sufficient departmental budget to develop courses • Availability of internal and external grants • Recognition of distance education in funding formula(s) • Competition from other educational providers • Source of additional revenue for the department • Competition with on-campus courses for enrollment

continued on the following page

Table 3. continued

Issue	Theme	Concerns related to each issue
Student issues	Provision and availability of resources and materials for distance classes. Providing an interactive environment between students, university faculty, and support staff and services.	• Encouraging faculty – student interaction; encouraging students – student interaction • Development of a sense of learning community in students • Access to library and instructional materials • Course and program services (admissions, registration) • Help desk support for students (course and program information) • Availability of support services (advising, counseling, & tutoring) • Availability of financial aid
Quality and effectiveness	Course development to include planning for quality courses and evaluation of their effectiveness. The offering of distance courses as a complete thesis process and not just a course here or there provided.	• Course standards (including technology tools) • Academic integrity of student work • Measuring the effectiveness of the learning experience • Program or course credibility outside of department • Assessment of course effectiveness by department • Student evaluation of distance education courses and programs • Evaluation of student performance (testing & grading) • Course enrollments limits (faculty/student ratios)
Policy and governance	Policies related to implementing and maintaining distance education courses and delivery are responsibilities of higher administration. The governance of courses and overall logistics related to distance education implementation are higher administration's responsibilities.	1. Intellectual property (course ownership) 2. Acceptance of distance education courses (transfer and articulation) 3. Academic control over course content and delivery 4. Appropriate academic calendar for distance education courses 5. Provisions for consortia, partnership and commercial providers 6. Presence of an institutional "acceptable use policy" 7. Compliance with ADA regulations 8. Collective bargaining agreements
Regulatory and legal	Regulating the offering of distance education in regard to copyright, crossing state boundaries, etc. Maintaining the rigor of distance courses and programs.	Copyright and 'fair use" Establishment and participation in consortia and partnerships Recognition and treatment by accrediting agencies Regulations imposed by state/system higher education boards, councils and offices Regulations imposed by taxing authorities (legislatures) Regulations imposed by the federal government

Source: Schauer, et. al., 2005, pp. 10-13.

University eCampus. In an ideal world, policy and procedures would be established before a distance education program begins, but in reality that does not happen. Program administrators and University officials learn as they go. Initially the two main areas of concentration for program administrators are setting up the online programs and admitting students to the programs. In eCampus it was a daunting task to pull together the degree programs offered across the University. Troy University is made up of four campuses in Alabama and Global Campus which extends nationally and internationally. The plan to bring all the Troy University online offerings into eCampus was a three year process; the Global Campus online offerings were first, then the online offerings from the Alabama campuses and finally the GoArmyEd program with the eArmyU. Challenges faced in this implementation of eCampus included developing the schedule of courses, admitting students, training instructors and developing quality courses.

Troy University eCampus runs on five nine week terms, an accelerated schedule. The eCampus scheduler needed to develop term schedules and long term schedules so students could plan when to take the courses they needed. The scheduler also needs to watch faculty load for both full-time and adjunct faculty to keep within University policy.

Admitting students to the eCampus online degree programs is done centrally in eCampus but providing for most student services is decentralized through out the Troy University regional teaching sites. This was done strategically to provide local student services with the local Troy University site.

Training instructors to use the course technology is always a challenge. Troy University uses Blackboard as its learning management system. Blackboard training was available in all of Troy University, however many were not updated to the current version of Blackboard. The eCampus instructional design and training staff developed the Teaching Online Proficiency course known as TOP. TOP not only had the required Blackboard training but also added best practices in online teaching and course design. To meet our accrediting standards all faculty were asked to participate in a TOP assessment to determine proficiency or were then placed in additional training. TOP II was developed as the next level of training. McGrath (2006) noted that training is important. Even though administrators look at technology to save but faculty are looking at it for pedagogical use not its financial value. The development of online training in short video modules has used both the technology pedagogically and cost-effectively.

Along with the training of instructors comes the development of quality courses. With the start-up of eCampus was the eCourse Redesign Project. The Deans of the University worked with eCampus instructional designers to lay out a plan for course redesign. The first year included 110 general education courses. The second year looked at the Master level program courses, and the third year included some additional course redesign with the start up of the quality assurance process. The redesign project has different challenges in designing course templates for general education courses versus master level courses, but all were based on the master course syllabi for the online programs with the approved learning outcomes. The eCampus has elected to work with the Quality Matters rubric for the quality assurance process. The Quality Matters standards were worked into the TOP II training and the quality assurance course reviews are based on the rubric.

The challenges mentioned above are still the challenges facing the program only growing exponentially. Course enrollments average between 18,000 to 20,000 a term; with an approximately 12,000 student headcount. Course offerings are up 132.5% to 411 a term, with course sections up 217.6% to 788 a term. There are approximately 700 to 750 faculty members teaching a term. Student admission, student services, faculty training and course quality assurance are challenges facing this growing program daily.

Lessons Learned

So what has all this literature demonstrated and how has it helped Troy University's eCampus? It takes blending all the separate functions within a distance education unit to make the whole of it run smoothly, commonly known as a systems approach.

Need for a Systems Approach

Traditionally distance education programs had been seen as a distinct and separate from of education (Irele, 2005). Distance education programs were often considered separate entities because of the systems approach in distance education. According to Picciano (2001) systems theory is a most appropriate framework for studying and planning distance learning in an educational organization. Distance education programs are required to provide all the services to students that they receive on the traditional campus. Distance education programs have developed a system approach to provide these services, often being viewed as a separate form of educational delivery. Figure 1 is the student services model adopted by eCampus.

Managing an online program and planning for growth in the program is what all distance education administrators strive to do. This case study has discussed how to do that through the use of planning tools and policy and procedures to concentrate on the student admissions process, student services, faculty development and a quality assurance process. What developed is a systems approach to distance education which is now becoming mainstreamed into the institution and into higher education. The goal of distance education, particularly at the Troy University eCampus, is to provide quality education experiences for the twenty-first century student in a flexible learning environment.

Develop Your Own Case for Program Growth

How would you develop your own case form program growth based on this case study?

1. Take a look at what type of distance education program do you have, based on Table 1 and review your program against the trends presented by Howell, et. al.

Figure 1. Student Services Model (Used by permission of the author, Brindley, 2008)

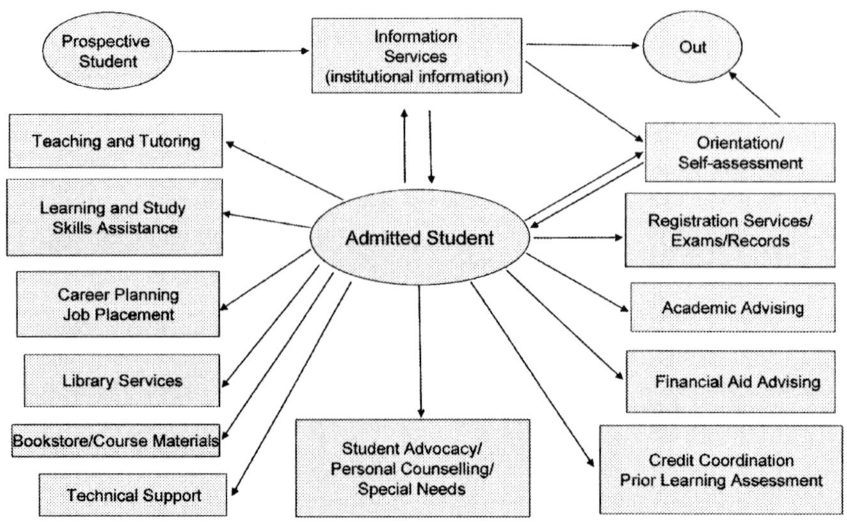

2. Next you move into the strategic planning phase of your developing your plan for program growth – complete a SWOT on your program
3. Questions you should ask – Does the mission of your institution include distance education? Is your distance education mainstreamed in your institution? Is distance education an integral part of institutional policy, especially for promotion and tenure, faculty load, etc? Does the institution provide adequate funding, personnel and facilities to support our distance education program?
4. Do you have a quality assurance process in place? Do you have academic support for course and program development, for well-trained faculty?
5. Do you have adequate student services, all available at a distance?
6. Implement you plan for growth and evaluate your success or develop you lessons learned for future program growth.

EPILOGUE

As Troy University moves into its strategic planning for 2011 to2015, the Chancellor's vision is to move all of Troy University's degree programs online. TROY's eCampus will continue this case in the strategic planning to continue to grow, continuing to mainstreaming online learning into the University through policy and academic support.

REFERENCES

Howell, S. L., Williams, P. B., & Lindsay, N. K. (2003). Thirty-two trends affecting distance education: an informed foundation for strategic planning. Retrieved 1/5/2009 from the *Online Journal of Distance Learning Administration, V. VI, No. III, Fall 2003.* http://www.westga.edu/~distance/ojdla/fall63/howell63.htm.

Irele, M. E. (2005). Can distance education be mainstreamed? Retrieved 9/19/2008 from the *Online Journal of Distance Learning Administration, V. VIII, No. II, Summer 2005.* http://www.westga.edu/~distance/ojdla/summer82/irele82.htm.

J.E. Brindley. (Personal Communication, September 20, 2008).

Leary, J., & Berge, Z. L. (2007). Challenges and strategies for sustaining eLearning in small organizations. Retrieved 9/19/2008 from the *Online Journal of Distance Learning Administration, V. X, No. III, Fall 2007.* http://www.westga.edu/~distance/ojdla/fall103/berge103.htm.

Magjuka, R. J., Shi, M., & Bonk, C. J. (2005). Critical design and administrative issues in online education. Retrieved 9/19/2008 from the *Online Journal of Distance Learning Administration, V. VIII, No. IV, Winter 2005.* http://www.westga.edu/~distance/ojdla/winter84/magjuka84.htm.

McGrath, L. (2006). Developing eLearning policies at the departmental level. *MERLOT Journal of Online Learning and Teaching. 2:3 pp. 177-186.*

Paolucci, R., & Gambescia, S. F. (2007). Current administrative structures used for online degree program offerings in higher education. Retrieved 9/19/2008 from the *Online Journal of Distance Learning Administration, V. X, No. III, Fall 2007.* http://www.westga.edu/~distance/ojdla/fall103/gambescia103.htm.

Picciano, A. G. (2001). *Distance learning making connections across virtual space and time.* Upper Saddle River, NJ: Merrill Prentice Hall.

Pisel, K. P. (2008). A strategic planning process model for distance education. Retrieved 9/19/2008 from the *Online Journal of Distance Learning Administration, V. XI, No. II, Summer 2008.* http://www.westga.edu/~distance/ojdla/summer112/pisel112.htm

Schauer, J., Rockwell, S. K., Fritz, S. M., & Marx, D. B. (2005). Implementing distance education: issues impacting administration. Retrieved 9/19/2008 from the *Online Journal of Distance Learning Administration, V. VIII, No. III, Fall 2005.* http://www.westga.edu/~distance/ojdla/fall83/schauer83.htm

Troy University Mission Statement. (2008). Retrieved March 16, 2009 from http://www.troy.edu/mission.htm.

ADDITIONAL READING

Allen, I. E. & Seaman, J. *Online nation: Five years of growth in online learning.*

Harris, M. (2009, January). Lowendahl, & Rust, B. (2009). Predicts 2009: Education faces challenges in a difficult economy. *Gartner Industry Research ID Number, G00164595*, 28.

Retrieved from http://www.sloan-c.org/publications/survey/pdf/online_nation.pdf. 09/27/2009.

Strong faculty engagement in online learning. APLU report retrieved from http://www.sloan-c.org/APLU_reports. 09/27/2009.

APPENDIX A

eCampus Academic Communication Committee (EACC)

December 7, 2007

MEMORANDUM

TO: Dr. Hal Fulmer, Associate Provost and Dean of Undergraduate Studies
Dr. Dianne Barron, Associate Provost and Dean of the Graduate School
Dr. Don Hines, Dean, Sorrell College of Business
Dr. Lance Tatum, Dean, College of Education
Dr. Maryjo Cochran, Dean, College of Communication and Fine Arts
Dr. Edith Smith, Interim Dean, College of Health and Human Services
Dr. Bill Richardson, Interim Dean, College of Arts and Sciences
Dr. Deb Gearhart, Director of eCampus
Dr. Elizabeth Bloom, President, Faculty Council
Dr. Cherie Fretwell, Chair, Graduate Academic Council
Dr. Jeff Spurlock, Chair, Undergraduate Academic Council
Ms. Andrea Roberts, President, Troy University Student Government Association
FROM: Dr. Earl Ingram, Vice Chancellor, University College
SUBJECT: eCampus Academic Communication Committee

The Academic Steering Committee has recommended, and Dr. Ed Roach, Executive Vice Chancellor and Provost has approved, the establishment of an ad hoc committee to enhance communication between the eCampus and its academic stakeholders. The attached document outlines the purpose of the committee and its membership. Following the May I ask you to either plan to participate as a member or make a suitable appointment according to the membership guidelines in the attachment.

We plan to hold the inaugural meeting of the committee during the last week of January or the first week of February and would request all appointments be communicated to me by January 18, if at all possible.

Please let me know if you have any questions.

CC: Dr. Ed Roach, Executive Vice Chancellor and Provost
Dr. John Schmidt, Senior Vice Chancellor of Student Services
Dean Herbert Reeves, Dean of Student Services

eCampus Academic Communication Committee

Purpose

To provide a structured means to enhance two-way communication with faculty of Troy University on issues and topics related to the quality and continuous improvement of online education. The Committee would provide a forum for discussion of academic policies and topics related to effectively delivering online courses associated with Troy University programs delivered through eCampus.

More specifically, areas of discussion for the Committee might include, but not be limited to: implementing programs and curricula; the development and implementation of guidelines and procedures for teaching effectively through eCampus; managing change; faculty qualifications, development and training to teach online; implementing the Troy University textbook policy; and scheduling of courses and faculty in eCampus.

The ad hoc Committee does not establish policies related to online education in Troy University, but from time to time may develop policy recommendations that would be vetted through normal academic policy-making lines of authority.

Membership to the eCampus Academic Communication Committee

- Director of eCampus (Chair)
- Representative from the Undergraduate Academic Council (Chair or designee)
- Representative from the Graduate Academic Council (Chair or designee)
- Representative from the Faculty Council
- Faculty representatives from each College (two each College; one from Library)
- Director of Academics, eCampus
- SGA President (or SGA designee)

Implementation

The eCampus Academic Communication Committee would be established by appointment of representatives from the colleges, councils, and the SGA effective upon the approval of its establishment.

Section 3
International Cases

Chapter 13
The Appropriation of Wiki in a Higher Blended Learning Course:
A Case

Beatriz Fainholc
UNLP-CEDIPROE- Buenos Aires, Argentina

EXECUTIVE SUMMARY

This chapter introduces the description of wiki appropriation. It considers the tool inclusion in an online educational innovation, supported in student and group-centered learning approach, to improve the quality of the blended learning offered. It states that the university course of educational technology, through wiki application as an appropriated methodology, beyond its consideration as a Web 2.0 tool, gives the opportunity to enhance the student protagonism into the deepening of the conceptual field of the subject by a collaborative knowledge construction. The evidence shows that the application worth to develop the learning strategies towards the student comprehension and its social skills in universities contexts. The results shows that the transformation of reactive attitudes into creative ones is a long process of change mediated by emotions and metacognitive work. Both facilitate a change of the students' focus, perspectives, and mentalities, understandable by the help of collective learning, among diverse variables.

BACKGROUND

This case related to the incorporation of the Wiki tool as part of the Web2.0 Internet proposal in education, was/is an education-technological innovation of the Course of Educational Technology – undergraduate level- in the Education Sciences Dept, in the School of Humanities of the National University of La Plata, Buenos Aires, Argentina. This experience has taken place since 2005 and continues today. The team who worked in this experience was lead by the author of the present case, full professor of the course. Nevertheless, the same group of professionals is practicing this innovation and others in connection, at the CEDIPROE Foundation, directed by the above mentioned author. She has worked in both institutions for more than 18 years leading different programs and introducing new technologies from the educational technology field.

DOI: 10.4018/978-1-60566-870-3.ch013

Although the main objectives of both educational organizations, is the training of teachers and faculty– basically with the articulation of ICT in the curricula -, the main emphasis on their tasks lies on research (action and evaluation) as their important role.

The CEDIPROE Center, -Centro de Diseño, Producción y Evaluación de Recursos Multimediales para el Aprendizaje-, was created in 1990 (www.cediproe.org.ar).The CEDIPROE has been working in distance education programs to train future producers and evaluators of educational and cultural materials, in a proposal that was completely new and unheard of in Latin America. This is why it received two awards from the Canadian Government as a model of an innovative organization.

With the advances of the networks and the Internet, the institution, now with an NGO profile, registered in the National Secretariat of Science and Technology of the Ministry of Education, Argentina, works, teaches, and researches in ICT innovations in education – face to face and at a distance.

We will describe the theoretical pillars and the context of the project of Wiki tool as an educational-technological innovation. Wiki is a page or collection of Web pages designed to enable anyone who accesses it to contribute or modify content (excluding blocked users), using a simplified markup language. It is often used to create collaborative (Frydenberg, 2008) websites and to power community websites. Wikis is used, too in business to provide intranet and knowledge management systems. His creator, Ward Cunningham, originally described the wiki as "the simplest online database that could possibly work" simplest online database that could possibly work." http://en.wikipedia.org/wiki/Wiki

We have to remember that Wiki (from the Hawaiian which means "fast") is a collaborative and interactive website, where users read, write, produce, edit, etc. diverse information by consulting and progressive agreements, keeping a history of changes which allows to easily access the information accumulated while it is shared.

Wiki is a simple online database which works as a piece of server software that allows users, to freely create and edit web page content using any web browser. A Wiki supports hyperlinks and has a simple text syntax to create new pages and cross-link between internal pages on the fly. Wiki is unusual among group communication mechanisms in that it allows the organization of contributions to be edited in addition to the content itself (Duffy & Axel, 2006). In our case the information is referred to the concepts of the educational technology field.

Wiki is an intensely accessible and collaborative hypertext software which challenges and gives greater complexity to the traditional notions of - as well as access too, - authorship, editing, and publishing. Brian Lamb (2004) points out that in Wiki anyone can change anything, using a simplified hypertext markup. It includes Wiki Page titles which are meshed together and its content is egoless, timeless, and never finished.

It is an open structure tool, which allowed participants to add new pages, or change the content of existing pages: this was the way which was implemented in the blended course of Educational Technology. This experience still takes place today, with the introduction of new trends which provoke changes towards the improvement of the higher learning course of Education Sciences at the University with the goal of training professors.

Consequently, for the students, working and becoming socialized with the use of Wikis (Guth, 2007), represents, among diverse topics, to teach the differentiation in features and uses, of similar Internet technologies (such e.mail, forum, discussion board, blogs, etc-, in comparison to wiki.

This means to teach different skills related to information management, to carry out a correct navigation/interaction in Internet with a comprehensive and critical reading and evaluation of this environment. Also, the students consider they are prepared for the use of Internet tools. However, they perceive that in these study-work situations, they begin to elaborate a conceptual deepening of the central repertoire of the discipline in question, articulated with an interdisciplinary approach (5) that enriches it. It would be very important if they could achieve the conceptual discrimination and ranking through the exchange and discussion among class (real and virtual) mates within wikis.

Setting the Stage

Educational Concerns of the Introduction of Wikis into a Course as an Innovation

The features and the context of the project of Wiki tool in education as a educational-technological innovation, establish to remember that innovation means not to continue doing things at the same way we have always done them, for instance using the Internet to communicate and share questions, through email among students. An educational and technological innovation seeks to improve the quality of teaching and learning and satisfy new demands, among these: motivation, achievement of competencies, time for learning, reduction of costs, the development of higher thought functions, to provoke the cognitive residue and change to other areas in order to optimize awareness, to reflect within the pedagogical practices, etc. According to this, using email or/and Power Point slides in class or uploading them onto a site are not "innovations". They are technological adoptions but hardly relevant for a real change in the educational-technological model of knowledge construction with authentic protagonism of the students and the faculty guide, which would mean a significant difference. In other words: more steps and new experiences are needed to really include ICT in education, today.

In order to have success with progressive innovative application – in this case virtual methodological ones, it is important to introduce them so users may incorporate their use in a friendly way.

This is why we first implemented the blended learning proposal for the course, which involved a mediated interaction with a mix between diverse ICT (by Moodle software:see: Fainholc, B. Blended learning. Report UNLP. www.cediproe.org.ar) and face to face interactivity as different possibilities to study, communicate with professors and classmates, evaluate with e- portfolios, carry out action-research, and produce a collaborative WebQuest, among many ICT innovations that were included.

The course of Educational Technology introduced students to basic educational technology concepts, which ensured their gaining of basic competency in understanding and applying them with the use of the Internet. The course builds individual and group skills in problem-solving related to the information society and the professor's role in it, learning to do a critical reading of websites through a reflective navigation and creating WebQuests using many resources, not only electronic ones.

Creating a course in which the use of a Wiki was a key component, it encouraged students to embrace the technology, apply it in educational contexts, through the main concepts of educational technology, which as future professors they were required to interact with it on an almost daily basis.

In our opinion, the professors who teach educational technology and related fields, like "research methodology", or others, should teach how to overcome the isolation and the rupture from interaction to transform the student groups into knowledge or learning communities, emphasizing the value of the exchange in networks, with solidary attitudes, and collaborative tasks. This proves that the use of telematic tools genuinely, permeate the teaching task because ICT are integrated into the disciplinary content of the field.

Another important input in this case with the goal to improve the quality of the virtual education offered, was to strengthen the on line educational technological innovation supporting it in the learning student and group-centered conceptions.

This means that the learner and his/her team are very important in an independent way, to decide when, what, and how they will learn by interacting and discussing, most of the time in an asynchronous communication, with professors, related to diverse main concepts of a specific content.

In this context the teaching by learning based on the model of development of skills and capacities, establishes that guidance from the professor, the personal and group advice and practice, and immediate feedback are critical.

The significant point is that both professors and learners have the opportunity to discuss the process and be aware that they are practicing socio cognitive and metacognitive skills with feedback from classmates who are observers and from the instructor, besides dedicating time and effort to deepening the definition of the concepts in the Wiki encyclopedia.

To be successful in the improvement of these mechanisms, the students and professors need to stay current with online technology, even if they are not programmers or Web designers. They need to be self confident and effective on line with a fundamental knowledge of the various tools and resources available on the Internet: in this case with the Wikis, internalizing its methodology and spirit of collaboration and adopting reciprocal monitoring attitudes among the users.

Some Contextual Figures

It is useful to know that more than half of the university students use the Internet and other tools as a resource to study and as a personal communication tool. At the same time, 40% uses Internet generally to complete university tasks and 20% are just beginning to use it. This data is very relevant for the design of bimodal courses and to introduce ICT innovations in the university. There is a great tendency (one third of the students) who use the Internet to be up to date on information.

60% of the students do not access the Internet from their homes. The university provides free access to the Internet but not everyone takes advantage of this opportunity. The access of the students, at least four times a week is done from public computers and cybercafés, as the National Institute of Statistic - INDEC, www.indec.gov.ar shows for Argentina. So, for the students, working and becoming socialized with the use of Wikis, represents to recognize pros and cons in comparison to any other Internet tool. It would be very important if they could achieve the conceptual discrimination and ranking of the Educational Technology field, which would be amplified by the exchange and real/ virtual discussion for a knowledge construction.

Regarding to the professors and their possibility of ICT application, although we may observe a very favourable climate towards their application, they advise its use basically to access information (61,7%), and they recognize its importance in the labour market with a development of expertise in Internet abilities through its tools (93.5%) as the Monitor of ICT Policies for Latin America and Caribbean, says. http://lac.derechos.apc.org/index.shtml

OBJECTIVES

In this context the objectives of the experience of the Wiki Study Case were and are:

1. to identify the aspects leading to a technological appropriation of Wikis in a university level virtual learning,
2. to propose an adjustment of the Wiki tool to the educational technology field to deepen this area so as to facilitate and re-create the conceptual knowledge of the discipline, and

3. to train and support students of Education, as future professors and faculty, so they may internalize new collaborative mentalities in virtual approaches using Wiki tools, towards a reflective transfer to the pedagogical practice.

Besides this, we must mention that the experience promotes the elimination of reluctance in Faculty attitudes and improve their behavior in relation to ICT, -even more so in countries in the south of the world-, where it is usual to find an absence of systematic training proposals for university staff. This training, in this way, represents an advantage for a systematic-holistic approach to a practice of a conceptual deepening (of any kind of knowledge area), including Internet tools and an updating through collaborative work, through a community of practice.

This is very important because we have to remember while students belong to the digital generation, university professors, in general, still belong to traditional paradigms of educational practice.

Since ICT are – in general-, introduced in regular classrooms and in virtual ones by different interests (in general, trends and pressures), without a previous analysis of their characteristics, possibilities, potentialities and limitations, - and without legitimizing them as real contributors to "good teaching practices", - the present case (carried out over the past 3 years), tries to identify how a genuine contribution of Wiki tool to the pedagogical work might be achieved, not only to evaluate the possibilities of the Web2.0 in education, – where Wikis play a part-, but mainly to discriminate basic theoretical and practical categories towards its recognition, to the improvement of the university teaching in an electronic blended format.

We recognize that ICT in Web2.0 Internet, enhances the possibilities not only of access to any kind of information and facilitates communication among people, but also promotes protagonism of the students in their knowledge building and its deepening in their reflective thought through the conceptual comprehension and sharing of alternative proposals, which is hardly insignificant, within the paradigm of comprehension pedagogy.

The content of Educational Technology - E.T. – of this university course presents many topics which include concepts, ideas, principles, which have been substantively selected, put in order, and ranked to give consistence to the academic work (to read biblio and webliography, to discuss these through different practice exercises as part of the educational work).

The main challenge here is to give meaning to the information accessed through reading, to recognize the importance of concepts within the field of E.T., and to transfer ways of thinking from one field to another, in order to expand our knowledge and allow us to act increasingly more intelligently.

The conceptual comprehension processes are stimulated when professors and students, when faced with new knowledge, search for and discuss who said that, when, why, within which theoretical framework (Fink, 2003), and which methodological perspective supports it in the construction of said statements.

This contributes to a needed content validation, which is very important in general and particularly, when using information from the Internet, perhaps written for different proposals and contexts. In this way, the stimulation of the formulation of criteria and their implementation lead to the development of critical thought, undoubtedly something which is longed for in university training, and even more so when dealing with the training of faculty or teachers.

The implementation of Wikis, like any telematic tool, provides the possibility to enhance the educational proposal mentioned above or perhaps, to trivialize it, if we do not recognize that information is not the same as the content to be taught or learned, or the production of knowledge. In no time students easily begin to use the Wiki for the collaborative writing and studying of the concepts of the referred subject of the course.

Portals have a huge informational capacity in communicative networks with the potential to create and maintain virtual communities for the exchange of shared projects, proposals, focuses, and solutions related to scientific, technological, social problems, and have the possibility of being linked in Wikis.

This contributes to generating a solidary culture through collaboration among students, faculty, regions, etc., networks which allow us to adopt and adapt, in a solidary way - to link and to look for the best solutions and for the best projects. In this framework we can say that a Wiki is a moral tool, as well, because we are helping to teach this kind of values by experiencing them. This should be what characterizes any educational tool: the enhancement of pertinent and valued actions.

Besides this, without a doubt, professors see that their workload has increased. As they are faced with such quantities of information, they must guide students to differentiate, –and teach them to do it-, between trivial and relevant information. The focus lies on different tasks and building meaning depending on the projects.

The Web 2.0

It has been proven (Heidegger, 1992) that the interaction between human being and machine must include and develop more protagonism because this emphasis intervenes in the profile and aims of the educational proposals and their products, which are mediated by technology and ICT, today. Or in other words, the goal is to think of an alternative pedagogy supported on telematic tools and the Internet, within a virtual interaction.

The Web 2.0 without a specific definition points to a new educational paradigm as a change in the proposal of designing, to think of activities and use technological tools. These could be Flickr (or communities of photographers), Wikis (to manage content) and Wikipedia (or encyclopedias as the case that we are presenting now), blogs (or personal pages), podcasting (or audio files in mp3 format), the subscription to RSS feed, a program which allows us to download different things (Duffy & Axel, 2006), etc.

Basically the Web 2.0[1] has the potential to enhance collaborative learning. This type of learning is defined by the interaction process with participants who have clear objectives to learn in a planned situation. Each participant acts as a resource for the others. It requires a well-thought coordination of the interventions and the specification of objectives, tasks to be carried out, and content. It is important for the members of the group to distinguish these components, which are useful to the process (in other words the implicit factors that makes things happen), distinguishing them from the content (the explicit factors regarding learning), and the product (or end result) expected from the work.

As we may see, pedagogy[2] is undergoing an enormous revision and transformation due the transversalization of the Educational Technology field, which includes not only a revision of objectives and media concepts but mainly overcoming the technical point of view or the mystification of ICT.

Beyond believing or not in the scientific level that may be found in the Web2.0 productions, there is no reason to doubt, beforehand, about the level of skill or specialization of a Wikipedia collaborator, working within an open source software, in comparison to an expert who belongs to a scientific community or publishes academic papers. It is more useful to recognize the difference among diverse domains and work philosophy of each. What we should do is think more than just to have content, which we do have; the point is to be in contact with an innovative model of collective action and be able to extract some benefits for the future.

The WIKI Tool Appropriation towards a Wiki Pedagogy

Besides the features that we state before about the Wiki, this tool was not used for syllabus or/and project management, in this higher education case. The basic use was for collaborative writing/note taking and studying he main conceptual concepts of the subject named above. "Wiki open editing" has the possibility to acquire profound and subtle meanings and effects on its usage. To allow everyday students as users to create and edit concepts at any page in a web site, it necessary to remember some Wiki limitations or risks, due to its openness. But it is exciting; it encourages a democratic use of the web and promotes content reconstruction by users, collaborating,- in our case-, to increase the field of E.T.

Wiki proposals develops alternative communication in discussion contexts, among students towards the facilitation conceptual comprehension of the Educational technology field. In this way, it is very interesting to study and identify its main concepts which impact in student cognitive structure due to its character of mediator, in the style of facilitating a kind of special invitation to increase and deepening the knowledge area.

Using Wikis, in this way, we verify the increase of the student motivation, as they learn, allowing them to develop new competencies, and reflective functions of thought as the important points to improve the quality of education.

TO WORK COLLABORATIVELY WITH AND WITHIN WIKI

The personal and interpersonal competencies such as, adaptability, stress management, mood, and motivation, are qualities to be developed towards personal mastery and self-control of emotions: social group and networks are the natural environments to develop and achieve these by interacting among people: learning and regulating through a relationship of feedback. This is the basic environment that we have to take into account in Wikis, besides the cognitive impact and residue; that is, considering emotions, designing creatively a personal view without bias and cognitive tags which influence the way we interpret the present world, in order to react properly in social terms, is one of the great benefits to include the strategy and tool of Wiki in diverse pedagogical proposals.

Since emotional mastery give us a context to work in harmony,the creative ability to solve conflicts, a space between a fact and its judgement, an empathic listening to daily problems and the possibility to express feelings and beliefs without prejudgements nor fears, it is worthwhile to work with the collaborative environment of the Wikis.

Case Description

The incorporation of the Wiki tool in the case that we are presenting had the specific purposes of:

1. proposing an adjustment of the Internet Wiki tool to the educational technology field to deepen this area in order to facilitate and re-create the conceptual knowledge of the discipline, and
2. facilitating the collective construction of the conceptual knowledge of the discipline, and at the same time, improving the cognitive and meta-cognitive skills of the students
3. The YouTube video found at http://www.youtube.com/watch?v=gia9htchml0&feature=related helps us to understand the achievement of this goals in the students and to reflect with them about some of these particular objectives.

SOME CHALLENGES FOR THE IMPLEMENTATION OF THE PROGRAM

1. Time for the Wiki tool sensibilization, its knowledge, acceptance and transfer to a content. The use of the YouTube video http://www.youtube.com/watch?v=jIgk8v74IZg greatly helped us to clarify, the before mentioned moments.
2. To appropriate (Fainholc, 2002) the content and the design as previous steps of the education – technological innovation. Both steps were done in 2005, when we began to interact with Moodle software in the blended learning course and it was not very common in this part of the world. This anticipation permitted to create – including adjustments till today-, a reliable Wiki system useful for future courses, as well, and this was in our case.
3. To train professors in the use of Moodle, which included the Wiki proposal.
4. To train the students in the new virtual tool, basically to be aware not only of the use of the new virtual proposal, but also the collaborative work possibility to be achieved, quite different from regular e-mail and other methods. Here, it was interesting to underscore the long achievement of the work done by the professors, as well, towards the importance of preparing and producing by themselves an encyclopedia with the concepts used in the E.T. field, which they have to study, exemplify, and apply in different works.
5. To be very patient in overcoming the limitations in the use of this Wiki technology in particular with students with difficulties in terms of access and their willingness to participate in the virtual environment. Much guidance and incentives were given in face to face interaction with many activities in order to encourage the use of the virtual class, in this case through debates in the forum.
6. Another challenge in this online environment is that appropriate and useful material takes longer to be developed. The students have to be patient as they edit and review many times their work and productions in the Wiki virtual environment with the guidance of the professors.
7. The harmonization between the academic and technical staff was another issue to take into account.

Methodology/Approach Used

The methodology was a *case study* presented through an action-research of the role and use of Wikis as a real-time synchronous communication tool during the blended learning Course of Educational Technology, carried out at the National University of La Plata, Buenos Aires, Argentina, (2004- 2007) and run in a blended learning format.

Qualitative methods and ethnographic observation as specific methods within the qualitative approach were used beyond the quantitative register given by the technological system. The decision to select basically, these *qualitative methods*, is link to the belief that we can register how, step by step the students underwent a big transformation from the general reactive attitudes into committed and creative ones, through *technological mediations and interaction, sharing emotions and meta-cognitive work.*

From other methodological point of view of the design of the Wiki interactive instructional environment, we had to prepare/design in advance, many "*Advanced Organizers*" (Novak & Gowlan, 1984) to establish the conceptual ranking, looking very carefully for the supra-ordinate, coordinate, and subordinate concepts of each, in order to make the approach rigorous and its ulterior classification to be done by the students in the collaborative discussions and discursive and exchange work.

Figure 1.

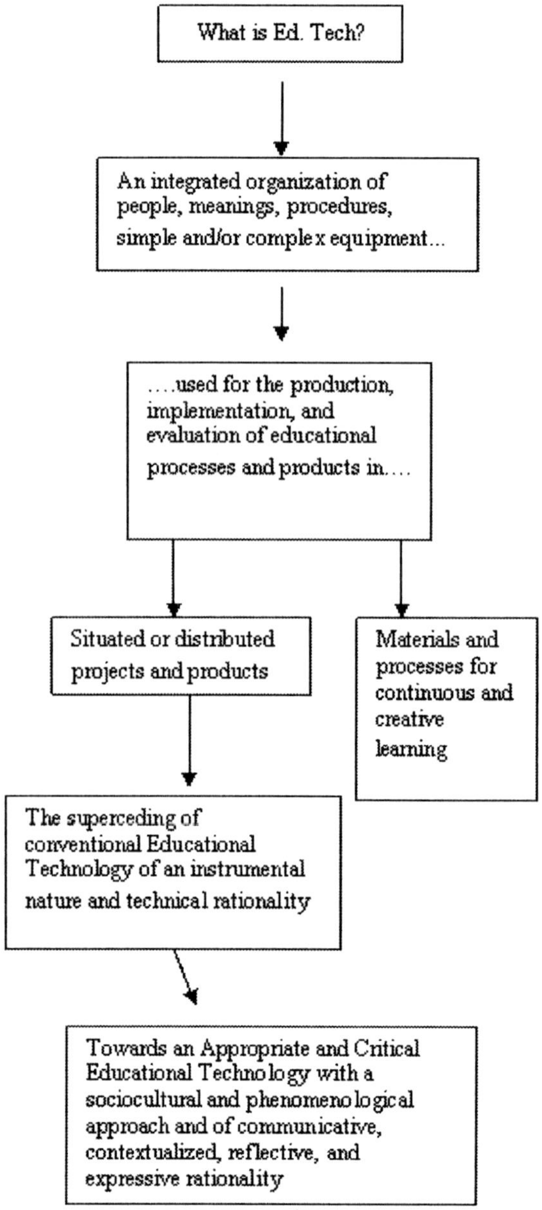

An Advanced Organizer to Work with Wiki

The Proximal Development Zone (PDZ by Vigotski) also helped to guide the students to understand why it is important to work through successive approaches in the conceptual and situated theoretical approximations (Stone Wiske, 2006), - following constructive and connectivist learning theories, - which were prepared by the professors in the preparation of the mediated interaction.

The role of facilitation, looking after the construction of negotiation of meaning processes about different concepts and their close relationship to many other concepts, within the discussions, helped to clear up many personal and group positions, to help them be respected and cultivated within a framework of democracy and solidarity, such as, discerning what a network is (in this case a socio-cognitive one), which prepares proactive attitudes for flexible mentalities and the social field for a technological network, which is what the Wiki proposes.

Special emphasis was placed on the formative evaluation during the research, with the design of several instruments for self administration, which contributed to student awareness, maturation, and meta-cognition achievements.

Besides that, the techniques used for collecting the data for the course consisted of different questionnaires, such as:

- Previous and final self- evaluation,
- Survey of the Internet learning,
- Questionnaire on the opinion on studying in a blended system,
- Survey about Internet access and interaction habits,
- Academic Evaluation of the on line course.
- Evaluation of the teaching strategies (Wikis, WebQuests, Forum,etc) and materials
- Formative continuous self – evaluation.

CAPACITIES, COMPETENCIES, AND SELF-EVALUATION PROPOSED

At the very beginning we clearly stated the competencies to be developed by students, which would become the criteria of evaluation of the Wikis, in different formats: one by e- portfolios[3] to enhance their self-evaluation and their conceptual change (Moneo Rodriguez, 1999) and meta-cognition[4] development, in order to live the comprehension pedagogy (Perkins, 1995), not only to understand topics and principles but to carry out the transfer of the theoretical study of concepts to practical issues of the educational technology field, to solve problems, to judge strengths and weaknesses of technologies and others.

It is true that the application of the mentioned processes requires from the students the availability of cognitive resources (Hura, 2008) such as motivation, attention, memory, comprehension, among many others. These cognitive resources and their activation depend on the type of material, the pedagogical strategy, and the awareness that the students have of their own cognitive skills, that is, their meta-cognition: self- regulation that they were able to deploy during their learning.

The learning strategies as cognitive resources, from trial or repetition, to the self-regulation or meta-cognition, passing through strategies of production and organization, are defined as a set of mechanisms of self-regulation which the students uses actively, to plan activities, monitor their actions, in order to reach the expected results, which are reviewed and evaluated in terms of effectiveness related to the action carried out.

The following capacities were to be developed and transformed into competencies to be shown or compared during the evaluation process with the use of numerical criteria to evaluate the achieved skills, and this way, to be transformed (by interaction processes) into cognitive resources through the use of ICT and the Wiki application.

Nevertheless many of the capacities that we will named after, have been developed not just specifically through wiki, but with a mix of ICT (such as library and other Internet resources, like discussion forums, etc).

This suggests us that we have to continue doing a lot of research in these areas to overcome confounding situations or/ and factors. For instance, social and communicate skills were deeply deployed with wiki but we have to recognize that with e-mail communication and discussion forums, blogs, etc they will take place, too. The inclusion and simultaneous deployment of other ICT, includes a brunch of variables, which make us very difficult, to determine the relationship between the use of wiki and students' affective and cognitive changes, among others, at the point which it has reached the nowadays researches.

Wiki pedagogy (Fountain, 2005) is in progress, and for this reason, it is an educational technological innovation yet, which must be researched to enhance its ontological existence and epistemological contribution to knowledge production. This makes its implications particularly important in and for the university context.

The achieved skills were/ are:

A. Referred to the production of technological knowledge:
   a) Information Management: to look for, organize, and evaluate information, to interpret and communicate through the use of ICT, to develop critical judgement. This accounts for **25%**
   b) Develop an understanding of the greatest issues of the current field of educational technology (comprehension of concepts, analysis, establish relationships to develop synthesis, rank concepts depending on the topics, extract main ideas). This accounts for **25%.**
   c) Application of the above mentioned to various situations (exemplify, transfer to practice). This accounts for **20%.**

B. Referred to interpersonal and group relationship within the collaborative work:
   a) Cooperation, debate, discussion, confronting ideas and re-signifying them to build knowledge
   b) To acquire social skills of respect recognizing the feelings of others in

groups/networks, in order to react and act properly and proactively.

c) Emotional mastery involves communication: to think before speaking (or writing); to understand and accept the point of view of others; to negotiate a solution, to try not to provoke irritating situations; to not become involved in others' feelings; and to analyze each problem with responsibility within the interaction with others. This accounts for **20%**

C. Referred to creativity:

a) Artistic development, creative imagination, management and ethical distribution of resources, written production. This accounts for **10%**

b) Analysis, selection and combination, to produce, implement, evaluate, improve and research proposals, materials and activities for diverse media, for face to face and virtual educational environments, with telematic communication, that is easily adaptable. This accounts for **20%**

Among the *cognitive skills* planned by the professor and taken into account within the student's interaction with Wikis were:

1. Information search seen in the following specific actions: to find where the information is stored, to ask questions related to these, to use the library, Internet and reference material.

The diverse skills were: to read in a thoughtful, comprehensive, and critical way, different URL, to interact with respect within the virtual environment, to access the information produced during the process, by the forum, as well by the Wikis, which were always open, in order to facilitate opportunities to access as well as communicate among everybody.

2. Information assimilation and retention in the following specific actions: to listen and to concentrate on achieving comprehension, to remember how to code and form representations, to read with understanding, to register and to control the tasks done with comprehension.

The diverse skills were, within the virtual platform: to accept guidance from classmates and professors to improve listening and concentration in order to solve different problems; to practice critical reading in Internet after selecting a website to be read.

3. Orientation skills through the specific actions of: establishing priorities, to plan time in the correct way, to deploy resources, to get things done in a timely manner or try to get the most important tasks done in time.

The diverse skills were the internal organization of concepts and sub-concepts in a semantic map; to discriminate and rank key concepts in order to solve problems; to give examples, analogies, different relationships among these, to enable the continuous conceptual building, etc. The portfolios and the WebQuest in all of the cases served to reinforce the above mentioned skills.

4. Inventive and creative skills in the following specific actions: to develop an inquisitive attitude, inductive reasoning, to generate ideas, hypothesis and predictions, to anticipate new perspectives, to use analogies, to take advantage of interesting or strange facts and points, to avoid rigid stereotypes.

The diverse strategies were forum with asynchronous contribution which enriches collaborative discussion and role assumption, to solve problems, to analyze cases, examples, etc..., to prepare WebQuests where creativity has to be increased specially looking for new hyper links and the relationship among diverse topics within the subject matter, were the main point.

5. *Analytical capacities through the specific actions of: developing critical attitudes and reaction, deductive reasoning, to evaluate ideas and hypothesis.* The diverse skills to be developed were to solve problems, to analyze webliography, to do many practical exercises of analysis, to interact in Forums.

The specific actions were: to identify alternatives and basically to carry out rational and argumentative selections of future actions to solve problems and to act. The strategies used to achieve this were in all cases to look for alternatives to a successful resolution, like the collaborative work of the Webquest.

6. Social and Communicative skills[5] seen in the following actions: to express oral and written ideas after reading them. To avoid interpersonal social conflicts, to cooperate and to receive cooperation, to compete with ideas, to motivate them, etc

The strategies were deployed in e-mail communication among classmates who formed the small work group to discuss and generate ideas based on consensus; to participate in the Forum to express in written ways different opinions on some topics of the course, so as to expand them and reach consensus, or discern by contrasting some productions to others. The Wikis encouraged all of this collaborative work.

7. Meta-cognitive and self-regulated abilities carried out through the following actions: to select pertinent strategies to solve a determined problem, to focus the attention on this, to transfer learnt principles from one situation to another, to know what is demanded in the task, to discriminate among the media to reach objective, to recognize their own capacities in order to compensate deficiencies, etc.

The strategies named above were developed not only in the Wikis, but also applied in problem-solving methodologies, - to expand, analyze, carry out comparisons with other concepts, examples, etc. -, and to compensate group difficulties when facing different resolutions.

This "declarative" work is referred to the "what" will be done, and it is accompanied by the "how" or procedural knowledge through the actions to be carried out, articulated through the conditional knowledge, which is referred to the "why", related to knowing if the selected strategy works, to know when to use each, or when not to.

In sum, emotional and socio-cognitive skills are the basic aspects of the collaborative and meta-cognitive work that concerned the course in general and the work with Wikis in particular.

More on the Implementation of Wikis in our Case

The skills and competencies named before are in development within the Wiki as an encyclopaedia strategy. This use of Wikis is new to the university context, in this country. What these democratizing tools may — and may not — enact in and across educational settings remains to be seen, because in spite of its potential, a Wiki is complex and challenging and "in-the-making".

Wiki constructivism (Holmes, Tangney, FitzGibbon, Savage & Mehan, 2001) focuses on – during its implementation-, the community rather than on the individual learner. Students and professors are not simply committed to deepening and developing the educational technology discipline with more and up to date information – such as our case-, but are actively involved in the re-creation of technological and interdisciplinary knowledge, that will benefit other students.

Wiki is a profound interactive proposal. Interactivity (Fainholc, 1999) means: "inter": among, and "activity " as the interposition of personal or/and interpersonal aspects, and mediated by technological actions within the development of social actions. When this interposition is planned by didactic /instructional activities in order to develop the comprehension of concepts, understand procedures, modify attitudes, in other words, to develop capacities to transfer them in term of actions: this is the essence of the competencies, in this case, developed by Wikis, within the constructivist paradigm.

In this way, interactivity is the main pedagogical characteristic for education in comprehension, solidarity, and commitment, within the space proposed through the collaborative learning in network, but we can implement/combine other technological methods (such as blogs, WebQuests, etc). The interactive learning mediated by ICT shows that more effectiveness may result more significant, practical, and imperative in comparison to the learning by heart or repetition.

The model centered on collaborative construction of knowledge through electronic interactivity and mediated communication is shown in Figure 2.

THE CASE

In the elaboration of the concepts for this Wiki, 85% of the students of the Course have participated, some of them with a greater commitment than others.

Regarding the content that they had to discuss in the Wiki, we must remember it was developed by the professor, and it appears in Figure 3 (http://wik.ed.uiuc.edu/espwik/index.php/Tecnologia_Educativa).

The Wiki work is the result of a community which establishes an unwritten pact of collaboration among equals: each content /definition was negotiated and re-built as a result of the group activity. Nothing is definitive, because it is submitted to constant discussion in small groups (or community) and it could be changed. The professor guides and advises, but does not direct; suggests strategies and paths to follow; motivates, stimulates curiosity, encourages searches, facilitates different contexts.

Dynamic

To promote collaboration, not more that three students per group were assigned specific dates throughout the semester to post their notes from the classes -after reading and discussing the bibliography,- on the Wiki. This was done to ensure that they had to complete their Wiki notes prior to the start of the following class, which was not always achieved, since there were delays.

Classmates then reviewed the "Wikipedia-style" notes pages, and added information that they learned but the original authors may have omitted. The instructor provided guidance and proposed spaces to be explored while accompanying and giving feed back to the students in their exploration and writing, sometimes using an "Advanced Organizer" and/or skeleton onto a Wiki website and encouraging students to create high-quality bibliographical and lecture notes about what they have understood, always emphasizing the possibility of living the experience where their work had been edited or changed, possibly by other students whom they did not know, due to the anonymity of posting on Wikipedia. In our case, the professor stimulated them to leave their names so as to have this information for the evaluation.

In general terms, we observed a good level of conceptual discussion and integration in each group, using the biblio and webliography determined by the professor. From the work dedicated to each concept, the students re-build their own discourse, in other words, they themselves re-build the concept: they agree on a post-discussion integration, they write it, and edit it on the Internet.

Figure 2.

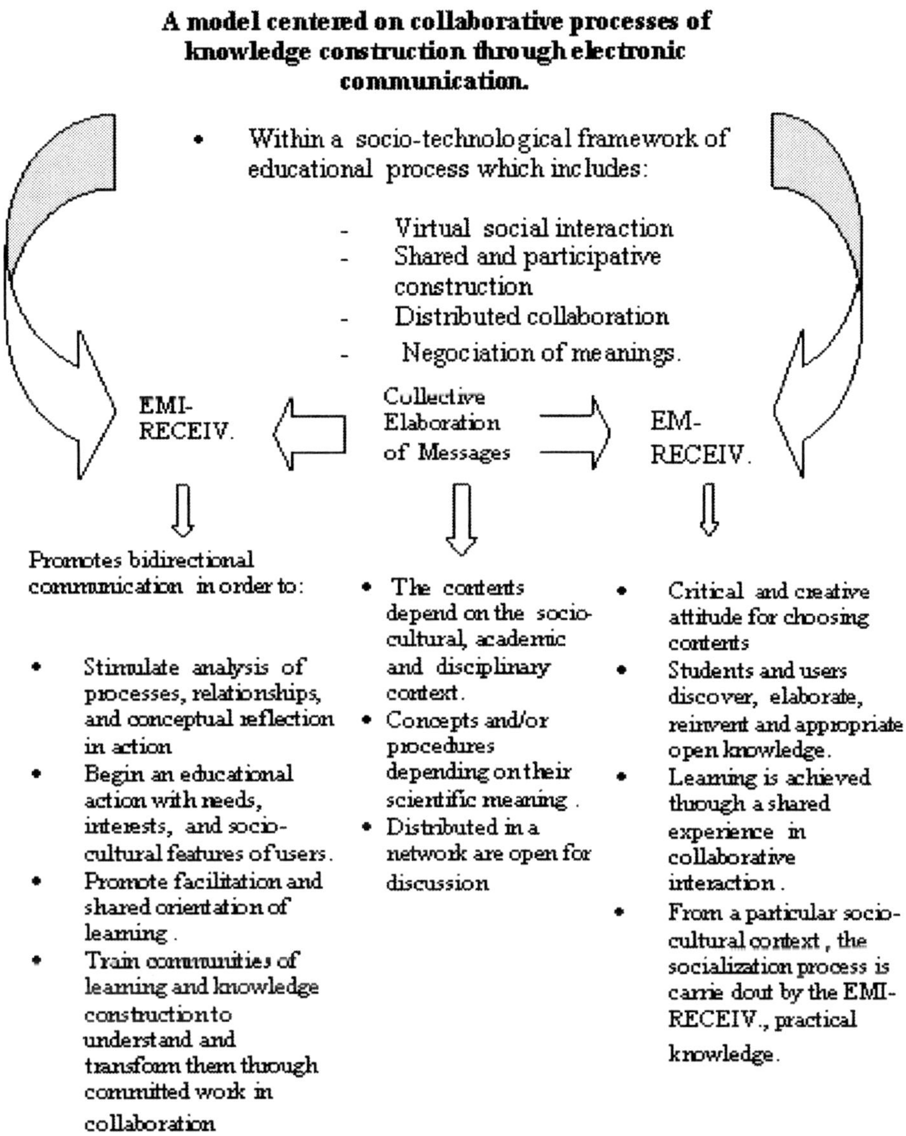

The **technological interaction was** favored by the groups that most interacted with the Wikis, since they polished the definitions that other groups had already conceptualized, in other words, they accessed the selected concepts more than once to complete and/or re-write them.

Regarding the **stimulus for cognitive development, it is possible to** highlight that the students used and complemented their definitions with concepts, that they have worked with in others disciplines and quoted, as well, including authors which the course program didn't include: it was an interesting opportunity to observe real conceptual integration. Anyway, it is important to mention there was no use of examples nor analogies in their definitions, for the most part.

Figure 3.

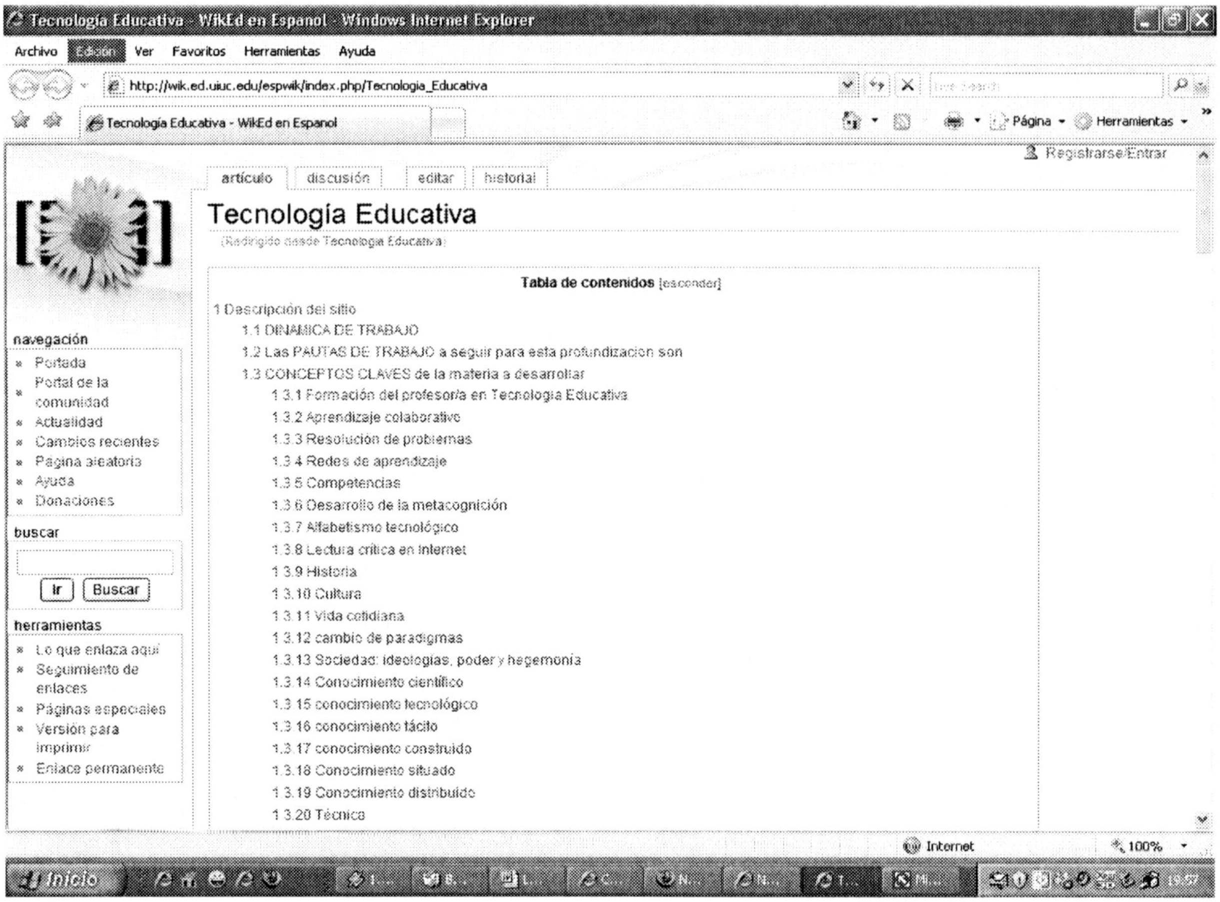

The **socio-emotional** aspect has already been mentioned, although it wasn't observed directly in the work of building Wiki definitions, but many of them, were analyzed in the Diaries. Here the students mentioned feeling engaged and motivated in the use of what was for them a new tool, although they did mention a certain reluctance in its use, since they didn't know how it worked.

Although there was high group participation in the development of the Wiki, at the very beginning, most the students, said it was important but did not participate. Participation was only increased when the professor set forth that at least seven concepts had to be developed by each group, which was not achieved as seen in the following quantitative data:

1. From a total of 46 participating students, at least 29 individuals worked on the Wikis: we know that because they left their names as proof of their participation. We must acknowledge the fact that we saw much Wiki work without student signature.

Participation in the Wikis

73% Participated, 27% Didn't participate.

2. Of the 29 students who participated, more than 85% collaborated on no more than two topics or definitions; only a few of them did so on mor ethan two.

Figure 4.

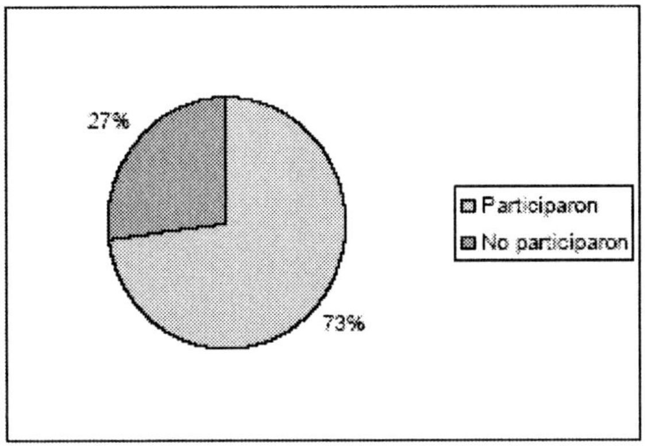

The mean and median measures shows 4 different topics students have accessed. The most was 6.

3. Although we cannot establish a direct relationship between this data and the performance in the Wiki and the evaluation test marks obtained, we can hypothesize that the average rate of the evaluation marks was lower (between 5 and 7 points, where the maximum is 10), in comparison to the students who participated in the Wiki collaborative work.
4. We can appreciate that the periodic participation in student access to the Wikis, always lay in the same students most of the time. Besides this, it was observed that the collaborative work within and outside each of the work groups, lay more on the student that represented the group, and who was responsible for said intervention, than equally among all of the members of the group.
5. In reference to **meta-cognition and self-regulation**, the students stated in their Diaries that at the very beginning, they did not understand which was the role of the Wiki within the proposal of the Course. But as time went by, they interacted in the practice, managed to plan, verify, supervise, analyze definitions, and their evolution. This is evident in the contributions that they made to the Wikis.

The Wikis of this case are available at http://wik.ed.uiuc.edu/espwik/index.php/Tecnologia_Educativa

The following image is taken from the Diaries: a complement for the Wiki work (Figure 5).

Diaries: A Complement for the Wiki Work

In this vein, the development of the Wiki innovation, shows the process and product of interactivity within a collective network with relative commitment.

EVALUATION

Within the qualitative evaluation perspective used in this case, it shows, the progressive transformation of student's general reactive attitudes into creative ones, logically located within a long process of change mediated by emotions and meta-cognitive work, encouraged by this innovation. Emotions and meta-cognitive work, both have facilitated a change in the student's focus and perspectives of facts, understandable through a collective network exchange carried out in this experience.

The quantitative data registers a total of 141 students enrolled in the E.T. course given in blended learning format from 2005 to 2008, with a high satisfaction level at 90%. They agree in emphasizing the good pedagogical experience of living the social interaction mediated by ICT in collaborative work with Wikis, among other ICT.

Some items included in the different questionnaires as techniques used for collecting the data

Figure 5.

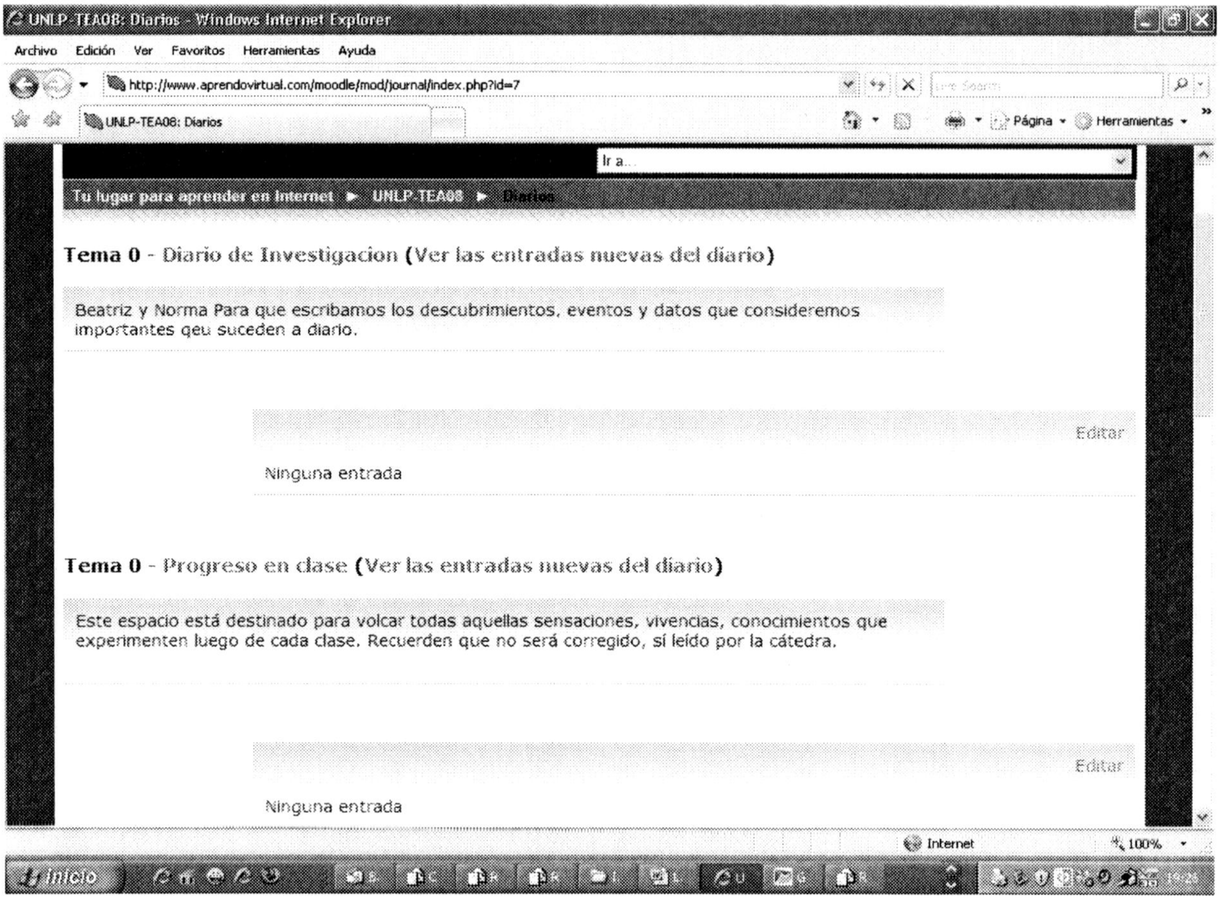

of the experience in the course, have consisted on a kind of self-reflection, like:

- Has the use of a Wiki in the virtual classes helped me in my learning?
- Could I see the value of the collaboration when I used a Wiki?
- Was it easy for me to post information on the Wiki?
- Was it hard to learn how to edit information on the Wiki?
- Were the concepts and notes posted on the Wiki helpful for me while I studied or reviewed the bibliographical material?
- Could the group help me gather the meaning of the collective work?
- Etc.

The *qualitative research,* within the interpretive paradigm, enabled us to achieve a comprehension of the real meaning that the participants have given to the project triugh the items mentioned above. This was easy to see in the "ethnographic observation", and in different interviews as well the development, along the course, of the *Diaries*, by the users.

The Evaluative Action-Research

The study within a context of an evaluative action –research[6] involved a self-reflective process, about the group and the educational situation as a whole. Within this dialogue, an awareness of the socio-cultural, emotional, cognitive and meta-cognitive mechanisms is encouraged, which mobilizes the students, to achieve comprehension in order to become skilled thinkers and problem solvers, so they may then, transfer these processes, to the real practice and to be knowledge producers.

In this context each individual (student) was and is a resource for another person. To achieve this within the Wiki, a good coordination of interventions and the specification of objectives and techniques to collect data was required.

During the discussions we may observe many times how social maturity, reliability, and self-reliance increased, when members could distinguish among the implicit factors which make things happen, from the content of the explicit factors related to the learning topic, and the product or final outcome of the work in the Educational Technology discipline.

This involves the socio-emotional and communicative development of competencies and ICT skills, which had to be learned and developed, so as to be able to exchange discourses, to engage in dialogue, to discuss, and analyze not only content but virtual meetings, etc. These allow each member, to evaluate his/her self-performance towards self-development and direction, through a meta-cognitive continued and fluid practice.

The work was carried out in teams of students, first with preparation and afterwards with search and discussion of the main concepts of our specific area. It was easy to observe/evaluate through quantitative methods (produced by the technological system) combined with qualitative approaches, how, - step by step-, they live a self monitoring of their knowledge building, within this technologically mediated exchange, located now, in their mentalities, focuses, perspectives of facts and others points of view, as a consequence of this interaction, not without setbacks.

The main point of this interaction mediated by technology in our opinion, has produced not only a conceptual change (Moneo Rodriguez, 1999) but also a cognitive residue[7], only understandable in this case, through the interactivity with the Wiki tool within the collective virtual network discussions, which – we would see throughout -, will increase motivation, deepening of comprehension and commitment.

Although 45% of the students had never heard about the possibilities offered by the use of Wikis prior to the course, - and they were afraid, reluctant or / and ambivalent in relation to its use, - 65% had heard about the encyclopedia Wikipedia. The group that never heard before the wiki use, were not very enthusiastic about the publicness of Wikis and they had difficulties in the co-interacting, discussing, and agreeing throughout the Wikis (and some of them had difficulties connecting to the Internet), but at the same time, in a contradictory way, they acknowledged the possibility to acquire skills as writers and critical readers, with a cognitive awareness of analytic and pragmatic applications during these processes. Nevertheless, we know that this is a long process.

The Diaries were the qualitative spaces where they highlighted the positive and risks points of the application in the course and particularly of each ICT used. They expressed their emotional and meta-cognitive opinions and feelings by living the mediated ICT experience. A lot of students underscore, in relation to the Wiki, its possibility of integration, which they experienced as useful after reading the bibliography, while others, who were a bit more behind in socializing with the Wiki, still do not know what it is for.

Many of the students value not only the pertinence and relevance of the collaborative work and highlight the ease with which remote exchange is possible by Wikis and Forums, - and the feedback given by the professors-, but mainly this experience as a future responsibility of education as professors which involves commitment.

Practical implications taking into consideration the strengths and risks/ weaknesses of the Wiki experience in virtual higher education

Strengths

1. It was a positive experience regarding the genre (Fainholc, 2008) variable, if we take into account that most of the students are women, with little or no technological socialization. This practical training was strategic, critical and meant to overcome bias and stereotypes referred to the traditional relationship between "woman and technology", as something which is not expected culturally, yet today.
2. The Wiki facilitates the development of conditional knowledge by interacting within it, which means that the students can be more aware of their decisions by evaluating the conditions to guide them towards effective operations which would regulate the whole process.
3. The public nature of the Wiki inspires students to "gain a greater sense of collective ownership" (Guth, 2007), over the shared content and process to which they contributed.
4. The Wiki strategy shows a good teaching proposal because the students can access to their productions – which are edited, registered, and returned with recommendations given by the professors, which at the same time, allows the growing of other constructions (Holmes, Tangney, FitzGibbon, Savage & Mehan, 2001) or reconstructions of knowledge during the students's work, and to continue with their self- learning. Although the ways that students submitted their assignments and the professor gave feedback, are very much like standard methods of digital assignment dropbox, in most course management system, and it not mean any innovation in the evaluation circuit, we select them because in our opinion, they were/ are effective to the learning control – individual and group - of the students.
5. The Wiki is a technological methodology and a resource which facilitates the cognitive self-monitoring[8] during the socio-pedagogical face to face and virtual, synchronous and asynchronous interaction, observed during the same.
6. The Wiki provides users, the possibility to acquire awareness and critical understanding about how to work with the resource: this objective not only facilitates the interaction analysis but also gives the students the chance to produce their own hypertexts, with the use of the Moodle software, which is characterized by its usability and flexibility.
7. The Wiki enhances the collaborative (Frydenberg, 2008) habits of study, where diverse topics like conflicts and disagreements, and so on, stimulate the confrontation of ideas in different argumentations, to internalize respect. This "self- explanation" produces a verbalized knowledge, so it may be communicated to others, provokes social self-regulation through self-correction, during the presentations of alternative views, etc. These strategies have the chance to be developed, at the same time, with the strengthening of the individual independence through the student pedagogical organization because they have to manage their agenda, and virtual schedule for the resolution of activities, and others.
8. In this way, it increases interactivity, commitment, and enthusiasm within the interaction through social on line encounters, as well as off line, and this develops respect for diversity, solidarity, and socio – emotional skills. The learning mediated by Wiki was / is relevant, because it has enhanced not only the exchange possibilities through telematic communication and interaction, but also, to implement socio-psychological mechanisms which facilitate the changes in personal cognitive, attitude and competencies mediated by this tool.

Risks / Weaknesses

1. The openness of Wikis is a problem because anyone can change anything. Although this did not take place in our case, it is worthwhile to mention its negative aspects, as well as the positive ones.
2. The Wiki provides a space where "knowledge becomes networked but remains ephemeral: it changes, and can be changed and mediated by the community."(Duffy & Axel, 2006)
3. Students do not always work on the Wiki: they are sometimes distracted and use the computer to check their e-mail in class, which makes it necessary for us to repeat several times the same work instructions.
4. In many situations students don't discuss but cut/copy and paste concepts from different resources, which proves that the professor and the textbook remain invested in the traditional authority. This leads us to explore or acknowledge new spaces of authority, although we already have the Web2.0, of social cooperation, which in turn calls for a revision of the university curricula which in general continues to be traditional, and within which the Wiki innovation was implemented.
5. At the same time and contrary to this, professors don't assume their new on line roles, which involve a great change, and return to the old and traditional proposal of lecturing.
6. Sometimes technical problems hindered access and interaction with the Wiki which frustrated the collaborative experience.

Findings

We are able to identify the worthy aspects of a technological appropriation of Wikis in a university virtual blended learning course, which contributes to:

1. deepen the positive and negative implications of the Web2.0 (Fainholc, 2007);
2. transform the students's general *reactive attitudes into (re)creative* ones within a change process mediated by emotions and a meta-cognitive work,
3. diminish the focus on students (developing facts, concepts, and perceptions individually), in contrast with the new conception of people working in a collective network with commitment.
4. The Wikis are working not only towards the collective spirit in this context but also towards producing "meta- knowledge " (awareness of knowledge which is generated mentally), and "metacognition" (or awareness of the processes which are involved in knowing)[9] results.
5. Contrary to the multiple research done on the "no significant difference" phenomenon (Russel, 2001) through the inclusion of ICT in education and in consequence of the Wiki, our results are: www.nosignificantdifference.org (Figure 6)

No Significant Difference Phenomenon

1. The shift and transformation from reactive student behavior to a proactive and creative one takes place as a result of (a long) process with setbacks covered and mediated not only by Wiki and other ICT interaction, but mainly by emotions and a reflective work. This indicated that both of these domains, emotional and socio-cognitive, helps people (or not) to make conceptual changes of perspectives and to begin to accept with respect different points of view / interpretations of facts, settings, and opinions in order to make the best personal decision, but always involving others.

Figure 6.

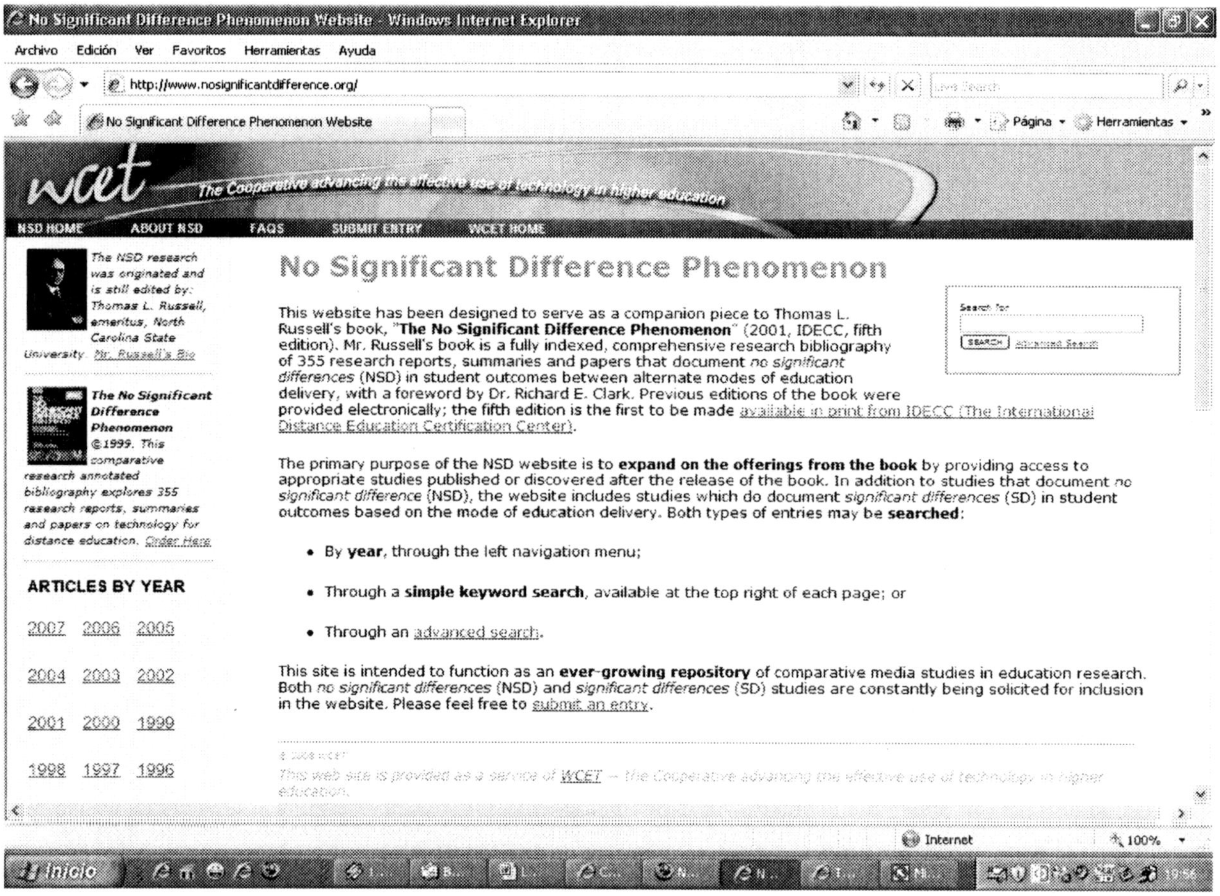

Emotional and socio-cognitive skills, involve knowing self-emotions, recognizing different reactions from others and beginning a self-monitoring process towards a real transformation to a conscious, proactive, and aware improvement.

2. Wiki collaborative work shows an improvement of the interpersonal skills needed for the building of knowledge through the management of stress, giving feedback and the stimulation of motivation towards a successful relationship of people as builders.
3. In consequence, educational interactivity and collaboration through the use of ICT and Wiki specially in higher education, increases the socio-constructivist paradigm, the ethical commitment approach, and the interactive model through connectivism (Siemmens, 2004)

Learning theory within the epistemological line of building knowledge, acting through successive aproximations, in an experiential, active, and interactive way for a contextualized and distributed problem-solving methodology, by a variety of means and towards the inhibition of isolation, individualism, and loneliness, which could be some distance learning risks.

REFERENCES

Duffy, P., & Axel, B. (2006). The use of blogs, wikis, and RSS in education: A conversation of possibilities. *QUT Online Learning & Teaching Conference Papers* (p. 8). Queensland, Australia: Queensland University of Technology.

Fainholc, B. (1999). *La interactividad en la educación a distancia*. Bs.Aires: Edit Paidos. Bs.Aires.

Fainholc, B. (2002). *La tecnología educativa apropiada y critica*. Retrieved from www.cediproe.org.ar

Fainholc, B. (2007). *Usos y abusos de la Web 2.0*. Retrieved from www.cediproe.org.ar/Historial

Fainholc, B. (2008, July 3-9). Nuevas tecnologías de la información y la comunicación-TICs-y la mujer en la sociedad de la información. Mundos de Mujeres 2008, Perspectivas sobre Tecnología y la Construcción de Género, Madrid, España.

Fink, L. D. (2003). *Creating significant learning experiences*. San Francisco: Jossey-Bass.

Fountain, R. (2005). *Wiki pedagogy*. Retrieved from http://www.profetic.org/dossiers/rubrique.php3?id_rubrique=110

Frydenberg, M. (2008). Wikis as a tool for collaborative course management. *MERLOT Journal of Online Learning and Teaching, 4*(2). Retrieved from http://jolt.merlot.org/vol4no2/frydenberg0608.htm

Guth, S. (2007). Wikis in education: Is public better? *Proceedings of the 2007 International Symposium on Wikis* (pp. 61-68). Montreal: Association for Computing Machinery.

Heidegger, G. (1992). Machines, computers, dialectics: A new look at human intelligence. *6*(1), 27-40. DOI 10.1007/BF02472767. Germany: Bremen University, Institute of Technology and Education.

Holmes, Tangney, FitzGibbon, Savage, & Mehan. (2001). *Communal constructivism: Students constructing learning for as well as with others*. Retrieved from http://www.cs.tcd.ie/publications/t.2004

Hura, G. (2008). Teaching behavioral-based skills online. *MERLOT Journal of Online Learning and Teaching, 4*(3).

Lamb, B. (2002). Open spaces: Wikis, ready or not. *EDUCAUSE Review, 39*(5), 36-48. Retrieved from http://www.educause.edu/pub/er/erm0

Lamb, B. (2004). Wide open spaces wikis ready or not. *EDUCAUSE*. Retrieved from net.educause.edu/ir/library/pdf/ERM0452.pdf

Moneo Rodriguez, M. (1999). *Conocimiento previo y cambio conceptual*. Buenos Aires: Editorial Aique.

Novak, & Gowlan. (1984). The concept map as an advance organizer. *Journal of Research in Science Teaching, 28*(8), 705-711.

Perkins, D. (1995). *La escuela inteligente*. Bs Aires: Edit Paidós.

Russel, T. (2001). *IDECC*. Retrieved from www.nosignificantdifference.org

Siemmens, G. (2004). *A connectivist digital learning*. Creative Commons. Retrieved from http://d.scribd.com/docs/1yhhhthpoaervbohwzkc.pdf

Stone Wiske, M. (2006). *La enseñanza para la comprensión con nuevas tecnologias*. Buenos Aires: Paidós.

ENDNOTES

1 Web 2.0 o Social Web and their implications in Education: http://cent.uji.es/octeto/node/2055, http://www.youtube.com/watch?v=YeoCu3yGvew&feature=related, http://www.youtube.com/watch?v=nsa5ZTRJQ5w&feature=related.

2 Schwartz, et al. (2004) looked at 24 wiki uses in Western universities; wikis tend to particular topics rather than on a campus-wide basis. Their use in scheduling, faculty use, learning support materials, and course management seems to be rare. Wikis have been used for a variety of co-Web purposes in educational contexts, such as co-creating and co-monitoring projects (writing, design) over time. Specific uses include collaborative concept elaboration, as the case we are presenting. See Fountain, R. (2005). *Technoscientific literacies via wiki research collectives: A project overview.* http://wikipedagogy.tuxcafe.org

3 E-portfolios: Also known as an e-portfolio or digital portfolio, is a collection of electronic evidence assembled and managed by a user on the Web or Webfolio. Such electronic evidence may include inputted text, electronic files, images, mulitmedia, blogs, and hyperlinks. E-portfolios are both demonstrations of the user's abilities and platforms for self-expression, and if they are online, they can be maintained dynamically over time. Some e-portfolio applications permit varying degrees of audience Access, so the same portfolio might be used for multiple purposes. http://en.wikipedia.org/wiki/EPortfolio

4 Metacognition: self-regulation that they were able to deploy during their learning.

5 Socio-emotional and communicative skills and competences have been inferred in the exchange of discourses, dialogues, discussions, and analysis of the process and the content which allows them to direct their performance and evaluate it. It is important to include the critical reading of Internet, too. Fainholc, B. (2004). *Lectura critica en Internet.* Rosario, Argentina: Edit Homo Sapiens.

6 Inspired on Elliott (1996), its steps are: 1-Identification of a problem area of the educational practice; 2-Organization of the work team; 3-Exploration of the initial situation by research procedures and a theoretical study of the field topic; 4-Writing of objectives after a common oral discussion; 5-Identification of factors to be modified and formulation of the hypothesis in action; 6-Planning of strategies to be developed; 7-Implementation, and so on.

7 Cognitive Residue: Effects of technology obtained during intellectual partnership with it and effects of it in terms of the transferable cognitive residue that this partnership leaves behind in the form of better mastery of skills and strategies (Salomon, Perkins & Globerson, 1991 in Pea, 1997).

8 Knowledge and self-control are included, as well as the control of the process, which involves commitment, flexible attitudes, and attention, to recognize their own emotions and reactions in order to prepare the self monitoring process, the (possible) transformation and improvement.

9 This is not new: at the end of the seventeenth century, the empiricist John Locke expressed: "It seems to me, that the limits of our mind are, without a doubt, simple ideas which we received through our sensations and reflection, limits beyond which the mind, no matter how hard it tries, cannot advance an inch; nor discover anything whern it tries to introduce itself in nature and the hidden causes of those ideas" (Essays on human understanding, 1960).

Chapter 14
Budding Researchers in the Humanities: An Intercultural Online Project

Vander Viana
Queen's University Belfast, UK

Anna Chesnokova
Kyiv National Linguistic University, Ukraine

Sonia Zyngier
Federal University of Rio de Janeiro, Brazil

Willie van Peer
Ludwig Maximilian University Munich, Germany

EXECUTIVE SUMMARY

This chapter aims at describing the networks within the Research and Development in Empirical Studies (REDES) Project, an intercultural enterprise aimed at promoting empirical studies of culture. Probably the first and only international project in the Humanities which prepares new researchers through on-line communication, the experience is believed to be scalable to other areas as it enables students to acquire the methodological techniques needed to undertake and carry out research in an environment totally different from the traditional university classroom. The chapter traces the history of the project from its foundation in 2002, explains in detail how it was set up, and evaluates the contributions of this joint effort. The problems met along the way are also pointed out. The chapter concludes with the challenges still to be faced. This case report stands as a proof of the impact of technology in preparing human resources for the Humanities.

INTRODUCTION

Perhaps one of the most distinguishing aspects of human beings, one which has never changed and has enabled us to survive as a species, is our need to communicate and live in social groups. What has actually changed, however, is the way in which communication has been carried out. As we enter the 21st century, a new group of human beings, also known as the "Net Generation" (Baron & Maier,

DOI: 10.4018/978-1-60566-870-3.ch014

2005), is on the make. These youngsters, born in many parts of the world after 1980s, have grown up with technology, which they have used for communication. Baron & Maier (2005, p. 57) describe them as "digitally literate; connected; social; prefer to working in groups; achievement oriented; require structure and guidelines; crave interactivity; have short attention spans; are experiential, visual, kinaesthetic learners; and prefer working on things that matter". They also state that the Net Generation is representative of the population in all the universities around the world. Still, they are a large group, who, to some extent, are trying to show us that a new way of learning must find its way into the curriculum. Indeed, if one of the major goals of the university is to produce new knowledge, working online in a research community which goes beyond geographical boundaries is one of the most effective ways to prepare students to become budding researchers and thus effectively participate in the scientific knowledge production. It may be argued that the idea of research communities is not new and hence presents no innovation. However, roughly until the turn of this century, research groups depended on the physical presence of their members and were generally affiliated to a single university or to those within a similar geographical area. Nowadays, technology has opened the access to exchange which hardly knows space or time limitations. The Internet has made some aspects of research quicker, more economic and more accessible to all, especially to students. As Geer (2001) rightly notes, there are still limitations but they may be of a cultural nature. She writes that:

Although as human beings, we communicate with others in many ways and across many mediums, communication is not always easy, even when we feel we know the other person. Cultural influences are often at the root of the communication challenge where misunderstandings and misinterpretations occur. Interaction and collaboration become much harder when communicating with total strangers in the online environment (p. 557).

It must be stressed, however, that before the Net Generation was born, scholars were already discussing the implementation of projects aimed at providing researchers in the Humanities with personal computers (Andersen, 1984) so as to facilitate communication among research groups. Fortunately, the world has moved beyond this stage. According to De Smedt (1999), "[t]he fact that students and staff have a computer on their desks instead of a pile of books is one visible change affecting the learning and teaching situation, even if it is a superficial one" (p. 2). Still, the way computers and their distance-learning advantages influence the process of producing and disseminating knowledge could be much different – from simply helping literature searches and storing information to full-scale research collaboration crossing the boundaries of classrooms, laboratories, libraries, universities, and countries.

This chapter aims at presenting a case study showing how to prepare new researchers in the Humanities through online communication. Relying on the Internet for about seven years, it has been possible to create a research group that works simultaneously in four different countries. Additionally, it has allowed the development of cross-cultural studies that would be impossible to carry out if students had to travel to all of the countries involved (Fisher & McGeveran, 2006). The present chapter indicates in a sense that the New Generation may be growing in several different places around the world.

BACKGROUND

The project to be focused in this chapter is the Research and Development in Empirical Studies[1] ('REDES'), which was founded on September 10,

Figure 1. REDES and its national centers

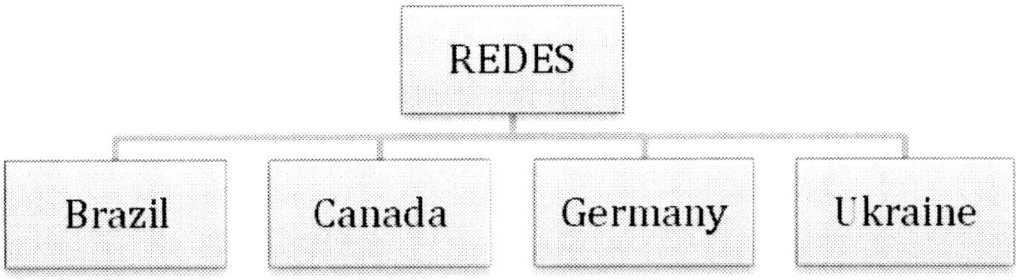

2002. At the moment this chapter is being written, the group is active in four different countries, as illustrated in Figure 1.

The major concept within REDES is to bring together both senior and junior researchers[2] in order to promote the development of empirical studies in the area of the Humanities. Members are appreciated for the kind of work they do in spite of the titles they may hold. This practice entails that B.A., M.A. and Ph.D. students equally participate in the group provided that they want to do so. In other words, the same opportunities are offered to all those who meet the minimum requirements to be a member.

One of the most important aims of REDES is to engage students in carrying out research as soon as possible in their career. It is believed that they will profit the most from it if they are immersed in the culture of inquiry (Brew, 2006; Bender, 2006) at the earlier stages of their studies. According to Freire (1970/1987, p. 58), traditional forms of education tend to see teaching as making 'deposits' of knowledge into learners' minds. Within REDES, however, the idea that knowledge should be produced by junior researchers during (and also after) their academic lives is crucial to the project's operation. These researchers are expected to develop high-quality work and to present it as papers at conferences whether local, national, or international. To this end, encouragement, support and guidance are offered by both senior researchers and by more experienced junior colleagues, which may range from designing a study to submitting it to a journal.

One of the most essential tenets of REDES is maintaining ethical standards. Especially when working collaboratively, an ethical approach is of the highest importance. At the simplest level, it entails giving credit to those who have helped one in the course of an investigation. Such standards are no less vital when international collaboration or the use of information technology is at play.

SETTING THE STAGE

REDES had been developing mainly by means of e-mail since its foundation in 2002, but after the seminar on 'Intercultural Reading' in the summer of that year, some of the Munich students with an interest in the possibilities for communication over the Internet started setting up a website. Junior members of the project from Ludwig Maximilian University Munich began constructing a portal for international communication between the various groups participating. This was an initiative deployed first and only by the students, without any interference from senior researchers in the group. It was done out of the need to acquire the necessary tools for communication. The structure of the project, whose members are dispersed all over the world (and in different time zones), demanded a simple and cheap form of communication that could be used by all without extra costs or infrastructure. The World Wide Web offered itself as a viable way to allow this communication and the means to achieve it was an Internet forum. Thus, every member was able to read and distribute news, ideas and private greetings (which certainly in the beginning made out a good deal of the communication). Initially, the local REDES groups in Brazil, Germany and Ukraine, showing a high continuity of members, were especially active in the forum.[3]

Shortly afterwards new requirements to the forum arose: it soon emerged that exchanges of the form 1:*n* (one participant writes to *n* potential readers) led to problems in the asymmetric cooperation between local groups and subgroups (see Auracher, Schwaiger & Tiourikov, 2007). On the one hand, it was not possible to respond only to parts of a message in the forum, which meant that many messages soon contained several different 'threads', which in turn were replied to by different people, leading – after a few exchanges – to serious problems maintaining an overview of the ongoing discussion. Although contributions in a forum may be ordered thematically, participants often deviated from the theme or brought up new ones. However, because of the chronological order of contributions, this often led to the situation where it was no longer possible (even for participants in a particular thread) to realize to which aspect individual comments responded. Such problems could be solved by direct communication between some members, for instance, through direct email contact, which meant that other participants could feel excluded.

An attempt was made to circumvent this problem by extending the forum (toward the end of 2004) with a content management system (henceforth CMS). This had two advantages. First, it created a dynamic Internet performance, in which one could optionally add further content and modules. This functionally modular structure allowed each individual participant a free choice between various communication styles adapted to diverse types of messages and their expected reactions by different participants. In this way, a highly flexible system in which a wide range of options of a truly dynamic Internet lay just a few mouse clicks away. Second, this allowed a choice of all possible combinations of direct or relayed communication and of 1:1 or 1:*n* messages. The forum was furthermore extended with a Wiki, a picture gallery and a download facility, where members could download minutes of meetings, questionnaires to be tried out, or other such documents. While chat rooms (which had been available from the beginning) allow a simultaneous exchange – preventing that an idea is frayed out in time – a Wiki provides every member the possibility to contribute to one specific topic together with other participants producing after some time a common text to which diverse authors have contributed. Direct exchange with individual participants also remained possible via private messages.

Because the Internet forum also presented the REDES project to the public at large, the dynamic aspects were complemented with static pages, where an interested visitor might learn about the history, aims and contents of the project.

This initial portal ran on the public German university server as CMS, in order to optimally increase the use of Internet resources and also allow static information. Structurally, only open source software was used (which was changed in 2008) as a functional modular construction. It was construed in an open-ended fashion, so that it could be extended as the need arose. A number of facilities were offered in this user-friendly tool. Registered members could post stories, add/see pictures, download/upload files, add/check relevant web links, send private messages, use chat rooms and describe themselves and their interests in the profile section. However, the most important aspect of the portal was the possibility of enabling researchers to communicate with one another. This tool both guaranteed the networking within REDES and helped enhance a sense of community. The portal contained a forum in which members could discuss their research projects, ask for help and try to find international partners, but also exchange general information, for instance reviews of recent publications, critical discussions and the like. In addition, members attending courses offered by the different area coordinators posted projects, comments or questions. This way, the portal also functioned as a site where discussions initiated in the classroom were extended and further developed online within the REDES community.

Participation in the forum required a posture which promoted criticism of their peers' work, and thus helped them develop their research. This openness entailed a change in attitude. The portal was a real revolution within REDES. The first reactions in the group were almost euphoric: all of a sudden individual participants disposed of technologies that made it possible to link up people separated geographically by large distances. Part of the success was certainly due to this enthusiasm for these new technological possibilities. As soon as it was operational, it began to be visited by the Brazilian, German and Ukrainian groups to such an extent that soon subcategories of communication had to be created, like 'General', 'Suggestions for reading', 'Technical questions', 'Research Projects', which themselves developed into further subdivisions. A lively interaction over the portal ensued for several years. To give an example: the rubric 'Research Projects' generated 432 posts to 34 threads, being viewed more than 5,000 times.

In addition to the forum, international meetings[4] were held at major international events.[5] During these meetings, administrative issues were dealt with, prospects for future developments discussed and, most importantly, members had their sense of community enhanced once they got to know their colleagues in person and were able to socialize with them.

However, the world is not perfect, and a breaking point occurred when the original generation that had built the portal left the university. The old portal can still be read at http://www.redes.lmu.de/portal/modules.php?name=Forums. Yet, it is no longer functional and only works as an archive. The main problem here was that the portal came under attack from hackers, posting pornography on the site, and later (when the previous damage was repaired) from ultra-rightist groups who posted their materials. This made the site unclear for a long time, since repairing it cost a good deal of time and money. These problems raised the question of how to secure the site from attacks. Still believing that the portal was needed, the new forum was launched and can now be found at www.redes.lmu.de/forum. This was again taken care of by junior members of the REDES group in Germany.

Until the middle of 2008 the old portal used PHP-Nuke for the simple posting of content on the website. For communication, phpBB was used, a software that, like PHP-Nuke, is freely available over the Internet. Extensions and improvements of these programs are produced worldwide by volunteers, which are then made available free to the audience at large. This means that they enjoy wide dissemination. But exactly this extensive diffusion of open source software, which in itself is to be greatly appreciated, also harbors a downside: security. The more a system is used, the more also it will be attempted to bypass its security mechanisms. Windows and Linux are good examples: the by large greater amount of harmful software is programmed especially for computers using Windows, since these computers are much more widespread than computers using Linux. The same befalls, unfortunately, both PHP-Nuke and phpBB, which are victims to hacker attacks time and again. This happened to the REDES portal twice, as mentioned above. Although extra security updates were installed several times, the maintenance of such is in the long run hardly possible, also because of costs.

Hence it was decided to further use a form of commercial software, in this case the well established system vBulletin. It contained all functions that used to be installed complementarily before. Besides, it is relatively simple to keep the system updated, as one is regularly informed about latest developments in the administrator's part. Everything is managed from one central point, while before all changes had to be carried through from various areas. Security in this software is not only checked regularly by users: the firm Jelsoft, in addition, constantly ferrets out and closes security holes. What is more, the use of the program reCAPTCHA averts that spam-bots (programs that register automatically in forums and install advertising) can access the forum.

The contributions and contributors from the old portal, however, could not be transferred to the new one. In the former, no new users could register. Additionally, one of the greatest security gaps, the upload of documents, was blocked. Finally, no new themes or replies could be posted. Finally, with the changeover to a real forum software an important separation was reintroduced, namely, that between communication and information platforms. Information purposes are served by the REDES main page, which is managed for a limited number of users through a self-developed CMS. For interaction, the communication platform is used.

CASE DESCRIPTION

The REDES group has been carrying out collective research for about seven years now. One of the most relevant contributions of joint work was the writing of *Muses and Measures: Empirical Research Methods for the Humanities* (van Peer, Hakemulder & Zyngier, 2007), totally conceived and written online by three REDES senior members over a period of five years. In the meantime, collaboration between two of the chapters, Brazil (REDES-BRA) and Ukraine (REDES-UA), also stands as an example of activity based on purely technologically supported long-distance contacts. In these countries where financial constraints are an issue, such projects would never have taken off if senior and especially junior members had had to travel worldwide to meet their colleagues.

The REDES research group brought a real change to the way students of Literature and Language, especially undergraduates, had studied so far: it opened up the road to conducting their own independent research for the first time. Studies in the Humanities had long relied on a hermeneutic tradition (Wiseman, 2002, p. 13), and imaginative texts have normally been viewed from a 'literary' perspective, which usually means paraphrasing parts of a text, providing comments to it, reviewing previous scholarship on the passage, or proposing ways to interpret it. As it will be clear from the examples, such methods can hardly be called 'research' from a scientific point of view. In this sense, researchers in literary studies tend to work independently from one another, and the very idea of combining the traditional literary view with technology may be quite alien to them. It should be emphasized here that the kind of research carried out in the framework of REDES is still genuinely 'literary' or 'linguistic' or 'cultural' since it deals with such objects of study like texts, readers, authors, etc., but that the approach differs from the more traditional one. The range of questions asked is considerably stretched: whereas previously the focus was mainly on the texts themselves, what they mean, how to interpret them, the range was now broadened to include questions to do with the functioning of these texts in the lives of individuals or groups, the various kinds of effects the reading of such texts may have, or why such texts are there in the first place, following the tenets of the Empirical Science of Literature (Schmidt, 1982). In a way, REDES has moved from a text-centered approach to encompass a larger scope which involves the sociology and psychology of literature (Schram & Steen, 2001).

This is a significant widening and deepening of perspective, which does not do away with the older and more philological methods. However, it now embeds these in a social framework where such traditional questions and methods may be critically reviewed and, in some cases, left for other ones. After all, literature is not created for commentators at universities, but to fulfill real functions in people's lives. Neglecting to investigate these has greatly impoverished the study of literature and culture over the past decades. This is in itself strange, since the beginnings of literary studies in the West owe their existence precisely to such social perspectives. Both the literary theories emanating from the writings of Plato and Aristotle, the foundations of Western poetics and literary studies, start from the question about what literature contributes to society, and only later ask how the poets do this textually. Without ignoring traditional analytical strategies, REDES members set out to enrich them with questions and methods better adapted to analyze the psychological, sociological and anthropological aspects of literature. All this has been possible and has been made faster through the easy, fast and efficient communication of present-day information technology.

For REDES researchers, observation, theory, and application are equally important, and technology is part of the methodological tools. Consequently, actively using the REDES portal and establishing day-to-day on-line contacts between junior and senior researchers acquired extreme importance in supporting and encouraging participants' efforts. New topics appeared, and the number of posts both in general and regional sections grew every day.

This kind of trust in long-distance technology was especially relevant to Ukrainian and Brazilian junior and senior researchers, who tend to remain on the margins of mainstream Euro-centered exchanges. The cooperation resulted in a number of full-scale projects presented at various international forums and a history of joint publications, as mentioned above.

One specific product of cooperation between REDES-BRA and REDES-UA was testing strategies for Literary Awareness (Zyngier, 1994a) in the Ukrainian context. The methodology to increase readers' sensitivity to imaginative texts had been applied in Brazil for several years before the idea was expanded to the Ukrainian context. It would have been hardly possible to have such development without the information technology as all discussions were carried out on-line through the REDES portal under the topic 'A new idea about literary research'. All in all, the topic attracted 1,993 views and 190 comments in 27 months. The first post was submitted on January 14, 2005 by a Brazilian member, who invited other groups to join her in applying the method in different contexts. Three days later, a reply was posted by a Ukrainian REDES member, and the discussion started. The participants shared their ideas about the content and structure of the classes, about the material to be used during practical sessions, etc.

Many of the resources themselves were sent to the Ukrainian group by their Brazilian colleagues as DOC or PDF files; simultaneously, local meetings in Brazil and Ukraine were held on a regular basis. The Brazilian senior member in charge of REDES in this country, who actually conceived the program of Literary Awareness, then joined the discussion on March 10, 2005 to ensure the project would get off the ground and cooperation would start between junior members. The Ukrainian REDES chapter coordinator also participated in shaping up the project. PowerPoint presentations of the research in progress were discussed as well and presented at a conference in Brazil, at the REDES seminar in Germany, and at the UNESCO conference in Ukraine.

As a result of the on-line cooperation, in 2005, the course on Literary Awareness (Zyngier, 1994a, 1994b) was launched at Kyiv National Linguistic University and then expanded to yet another Ukrainian university, Horlivka State Pedagogical Institute of Foreign Languages. Both Ukrainian groups of teachers who were delivering the course as well as their students communicated long-distance – not to mention their frequent consultations with the author of the book, who was thousands of miles away from both schools. In October 2005, she was invited to Kyiv to conduct a workshop on Literary Awareness, which materialized in April 2006. The preparation for the visit, including planning the workshop, was also carried out long-distance.

The pedagogical component of introducing Literary Awareness in Ukraine by way of on-line cooperation is crucial to Ukrainian education, in the Humanities in particular. In many countries today, university education faces a serious threat, as Barreto & Fialho (2005) put it: "the students do not experience *flow* while reading a literary text [...], do not have a positive response towards literary texts and tend to build an emotional and critical distance towards them" (p. 95), though it is well known that "[r]eading, in any age, also means reading the world, attempting to interpret, to come to terms with, to assimilate, perhaps even one day to understand what surrounds us, in order that we be better equipped to live in the world" (McRae, 1997, p. 19).

The cooperation reported above resulted in a publication (Fedorova et al., 2006) and in a paper (Zyngier, Chesnokova & Viana, 2007) which was also written collaboratively on-line. The Literary Awareness course is to the present moment still being taught at KNLU as an optional subject for 20 undergraduates each academic year.

Continuing with the cooperation, another research project was carried out in 2007–2008 to examine how Brazilian and Ukrainian readers responded to Poe's "Annabel Lee" in Portuguese and Ukrainian. As the project was being developed, the Ukrainian coordinator traveled to the USA and expanded the scope by involving North American participants. The results of this joint project have been published (Chesnokova et al., 2009).

Other research projects also resulted from the cooperation between the Brazilian and the German groups (Fausto, 2006; Teles, 2006) as regards reactions to anti-smoking advertising campaigns, and movie subtitling and dubbing. More recently, Coachman & Weigert (2008) looked at the way Brazilian, German and Japanese respondents reacted to vulgar language in a comparable short text.

Over seven years of stable cooperation (2002–2009) have inspired many other research projects and enabled the creation of a global research community where distance communication tools help cross the boundaries of cultures and continents. This is especially valuable for students as they become more and more skilled to operate various computer programs, work with statistics, present their projects before international audiences and work in a team (often under severe time pressure) to successfully progress in their careers – in academia but also beyond it.

All these examples stand as illustrations of the REDES philosophy, itself a product of cooperative work, which states that in this academic global online community "we share experience and knowledge, doing as much teamwork as possible, and using the Internet as a meeting ground" (see www.letras.ufrj.br/redes/philosophy.htm).

CURRENT CHALLENGES

The experience of budding Humanities researchers over the Internet does not go without problems. Nowadays the research group faces different challenges, which may be grouped into three main areas – namely administrative, personal and technological issues.

As regards administrative matters, the first issue to be considered is that of academic calendars. As the project involves at present four different educational systems (Brazilian, Canadian, German, and Ukrainian), it is not always easy to carry out research as part of a relevant course taught and this problem is enhanced further by the non-synchronization of university calendars around the world. For instance, in Brazil, the academic year is divided into two semesters: from February through July and from August through December. In Ukraine, the first term or semester starts in September and ends in late December with January being an examination period; the second semester comprises February to May, and exams are usually taken in June.

Another administrative problem is related to where each branch coordinator is located. In the Brazilian context, the senior researcher is part of the Postgraduate Program in Applied Linguistics, which allows her to offer her courses as optional subjects. In the Ukrainian situation, the area coordinator is currently taking her full-time post-doctoral course and teaching part-time, which only includes term papers and the supervision of Ph.D.'s as well as graduation exams. The German coordinator has, at the moment, taken upon himself more administrative load and, although still involved in research, has been kept away from tutoring.

The status each coordinator has at his/her university also influences the project because freedom at taking decisions is subject to different university constraints. The Ukrainian area coordinator, for example, in spite of being the Head of the Department for several years, could not fully influence the academic program, as curricula are usually predefined by the Ministry of Education, and introducing new courses involves years of changes and paperwork. In Brazil and Germany, however, the coordinators have had more freedom of action as regards their courses and what to focus on.

Research traditions in each of the four countries should also be considered. Whereas in all of them literary studies are not geared toward empirical investigation, the Brazilian group is stimulated to work in an interdisciplinary way. In Canada, the rigid disciplinary boundaries are aggravated by the bureaucracy with getting permissions to do empirical research. In Ukraine, literary studies and language research are conducted within two different majors and their intersection, together with extensive reference to psychology, philosophy and sociology is not welcome. In Germany, the problem now arising is trying to get students who need to pay for their studies opt for research instead of a seminar they can get away with quickly.

As for personal matters, the most serious problem is that of motivation. More recently, junior researchers have shown a decrease in motivation, especially due to lack of funds. Participation in the group depends on the willingness to see it as an opportunity for personal and academic growth. However, when junior researchers do not get any grants, they also have to finance themselves through their university life. This competition the research group meets with the job market is unfair, since the job opportunities students find outside academia are often more appealing. When having to choose between carrying out independent studies or accepting a job which will provide them with the means to survive – junior researchers have to opt for the latter even though they would rather become full-fledged researchers. In Germany and Canada, the panorama is even more somber. Participants mainly get involved in the project because they get credits in return. As one may anticipate, this situation threatens the group's cohesion and makes long-term projects nearly impossible. In the German and the Canadian settings, voluntary involvement does not seem to work as it does in the Brazilian and the Ukrainian context. In both latter settings, the successful functioning of the group mainly depends on the individual efforts of highly interested students who find enough motivation to see the project as a fruitful ground for their intellectual and personal growth while also financing themselves so as to attend national and international conferences. Finding such individuals in everyday academic environment is always hard, but not impossible.

These problems have raised complex situations. Junior researchers need to understand that carrying out intercultural research implies first and foremost strong commitment with a colleague in another country. They cannot, for instance, decide to leave the research study they may be carrying out unfinished, since this will implicate a student from another group who started working with them. Even if they do need to leave the group, they must help their partners complete the study, as this is what is ethically expected of them.

There are also linguistic issues involved in interpersonal relationships. As five different languages are involved in project (Brazilian Portuguese, English, German, Ukrainian and Russian), there is the need to use English as a language of integration. Most Brazilian and Ukrainian participants spontaneously use English in the forum in spite of some linguistic inadequacies. German participants, however, have often to be reminded that they must not communicate in their mother tongue unless they risk losing interest and international cooperation.

Over the past seven years work in REDES has shown that junior researchers need to understand that their participation in the research group is a great opportunity to be tutored by experienced senior researchers in their field of study and to share this experience with colleagues from different parts of the world, people they might never have a chance to meet. Moreover, in REDES, they are introduced to empirical methods of research in the Humanities, which is also unusual. Finally, they will be encouraged to carry out their studies autonomously from the very beginning, again not a common practice. Autonomy does not mean that they will work alone. Instead, they get the needed guidance on how to plan and carry out their study, prepare oral presentations, write papers, and so on.

The use of an online discussion forum already solves some problems such as that of time differences. The asynchronous mode of interaction enables participants to access it at their convenience. However, if participants do not visit the forum on a regular and frequent basis, this may cause some problems. In fact, a vicious circle may be created: if the forum is not accessed, there are no new posts; and if participants do not post, they do not feel compelled to visit the forum. With time, this setting may lead to collapse, bringing the forum to a halt. As with several online communication platforms, it is of utmost importance that participants promptly reply to posts. This is why sometimes members need to be reminded to do so as one of the area coordinators did in the following excerpt from the forum: "[Name] has been waiting for a reply as of OCTOBER 9th last year!!! Please make sure you reply to her and get your project off the ground!" If there is no immediate feedback, members may experience frustration. In turn, this may affect one's motivation and engagement in the activity.

Another problem is avoiding the forum and communicating through parallel means. Once participants establish contact with their research partners, cooperation should take place in the forum itself. On some occasions, junior research members have used their private e-mails to exchange messages on a research project. When this occurred, other REDES members were excluded from the exchange and could not comment on the study. In the same post quoted above the area coordinator also draws a junior researcher's attention to this issue: "And please do not use your personal e-mail to discuss the project. Use the Portal... otherwise we won't be able to comment on it." One of the tenets of this community is that advice and/or positive criticism must be offered by all members. If interaction is carried out by e-mail, this becomes a private exchange which does not benefit from the group as a whole.

At present, the group is looking for new ways to make the project thrive. In order to do so, answer to three main questions must be found:

(a) How to reshape the research group?
(b) How to draw/raise active participants' interest?
(c) How to expand research projects to other countries?

The results obtained so far have shown the project is worth the effort and should be continued.

CONCLUSION

Although there are a fair number of problems, they do not prevent the group from producing and preparing future researchers in the field of Humanities as early as possible in their academic lives. The experience so far has shown how to do away with individualism in the university and stress the importance of collaborative work. In fact, the case described above is applicable to many different disciplines in higher education which aim at preparing budding researchers. REDES has shown that it is a new way of doing research and of preparing independent, creative, and autonomous researchers who abide by the following tenets of its philosophy.

(a) The purpose of university education lies first and foremost in the training of critical abilities which are fostered through the production of new knowledge. The members of REDES are of the opinion that traditional ways of handling knowledge in Humanities departments are not conducive toward such aims, but remain largely confined to the dissemination of existing knowledge.

(b) Even where some allowance is made that students gather research experience, this is usually offered to them at the very end of the study period. We are of the opinion that doing research at the end of their studies is not only far too late, it is also counter-productive, in that it often produces frustration (because students are ill-prepared) rather than interest, confidence and enthusiasm. REDES attaches great value to instruction into modes of knowledge-production from the very beginning of university training.

(c) Members of REDES are dedicated to the insight that the aims of the group can only be achieved through active participation in the research process of all concerned, including beginning students.

(d) Methodological training is underdeveloped in the Humanities; this is paradoxical, because the level of complexity in culture is many times higher than that in natural phenomena. Nevertheless, it is students of the natural sciences who are initiated in methodological matters at a deeper level. We are of the opinion that this situation should be redressed, and, therefore, attach great importance to serious education in methodology.

(e) The REDES project puts emphasis on empirical methods of investigation: we believe that there is a too one-sided emphasis on hermeneutic-interpretative and subjectivist methods in the Humanities. Such methods may have merits, but should be complemented by more scientific investigations of culture.

(f) Those who take part in REDES see research not as an individual enterprise, but as a form of team work that is carried out by groups of people who actively communicate about their plans and progress.

(g) Such groups do not stop at national borders, but should be essentially transnational and cross-cultural forms of cooperation.

(h) Members of this project actively engage with cultural products and propagate the value that cultural products bring to individuals' lives and to society as a whole. Emphasis on empirical ways of investigation is in no way a depreciation of culture, high or low. On the contrary, participants are deeply respectful of cultural artifacts, traditions and preferences.

REDES was born out of the need to offer students a more intense and involved journey towards knowledge. It implies commitment, self-discipline, academic generosity, mutual respect, and, in many cases, volunteer work. In this online research community, we want to prepare qualified thinkers by developing and promoting the exchange of knowledge and ideas, and thus to collaborate for a better world.

In fact, years of cooperation have shown that this community works and functions as a source of cultural education and international exchange, where friendships develop and where differences are recognized. Members understand they are part of a scientific group as individual beings and thus help each other on several different levels. Belonging to this online project has proven to be a long-term commitment to working together and supporting others' efforts in their research and lives, a dedication which requires more than a minimal engagement during normal office hours. As a consequence, the goals of working towards an ideal academic setting which aims at scientific advancement and the feeling of personal growth, of learning and of sharing has been achieved, mediated by the Internet as a meeting ground.

Besides offering the conditions where learning outside the classroom and the curriculum takes place, the Internet has enabled REDES to become more than just a formal research group. It has turned out to be a place where interpersonal relationships are established and friendships develop. It is this bond of affection between researchers that guarantees that knowledge and experience become long-lasting and memorable. And all this has only been possible because of the advent of information technology. It is true that this work requires a higher than usual level of energy, enthusiasm, openness and dedication. But to those who have made this possible, research has been a thrilling, rewarding and unusually enriching experience.

REFERENCES

Andersen, S. (1984). Computers in the Humanities: Providing faculty with new tools. In R. W. Lutz, E. Jacobson & B. Rader (Eds.), *Proceedings of the 12th Annual ACM SIGUCCS Conference on User Services* (pp. 143-148). NY: ACM.

Auracher, J., Schwaiger, M., & Tiourikov, A. (2007). Say it again: Ideas regarding the correct usage of Internet communication. In S. Zyngier, A. Chesnokova & V. Viana (Eds.), *Acting and connecting: Empirical approaches to language and literature* (pp. 203-218). Munster: LIT Verlag.

Baron, J., & Maier, H. (2005, December). *The challenge of maintaining the momentum.* Paper presented at the Conference of the Australasian Society for Computers in Learning in Tertiary Education (ASCILITE), Brisbane. Retrieved on October 20, 2008, from http://www.ascilite.org.au/conferences/brisbane05/blogs/proceedings/07_Baron.pdf

Barreto, J., & Fialho, O. (2005). Flow+foregrounding: A possible relation? In S. Zyngier, V. Viana & F. Fausto (Eds.), *Venturas & desventuras. Coletânea dos trabalhos do V ECEL* (pp. 87-96). Rio de Janeiro: Setor de Publicações da Faculdade de Letras da UFRJ.

Bender, S. L. (2006). *Quasitative inquiry/research/approach: Leading a new culture of inquiry?* Retrieved on December 28, 2008, from http://www.sharonbender.com/quasitative_inquiry_research_approach_culture_of_inquiry.html

Brew, A. (2006). *Research and teaching: Beyond the divide.* London: Palgrave Macmillan.

Chesnokova, A., Zyngier, S., Viana, V., Jandre, J., & Nero, S. (2009). Universal poe(try)? Reacting to "Annabel Lee" in English, Portuguese and Ukrainian. In S. Zyngier, V. Viana & J. Jandre (Eds.), *Linguagem, criatividade & ensino: Abordagens empíricas e interdisciplinares* (pp. 193-211). Rio de Janeiro: Publit.

Coachman, E., & Weigert, T. (2008, July). *Readers' response to vulgar language: A comparative study.* Paper presented at XI Congress of the International Society for the Empirical Study of Literature (IGEL), Memphis, TN.

De Smedt, K. (1999). Introduction. In K. De Smedt, H. Gardiner, E. Ore, T. Orlandi, H. Short, J. Souillot & W. Vaughan (Eds.), *Computing in Humanities education: A European perspective* (pp. 1-12). Bergen: University of Bergen. Retrieved on December 27, 2007, from http://gandalf.aksis.uib.no/AcoHum/book/intro.html

Fausto, F. (2006). O impacto das advertências contra o cigarro: Um estudo intercultural. In S. Zyngier, V. Viana & A. M. Spallanzani (Eds.), *Linguagens e tecnologias: Estudos empíricos* (pp. 67-81). Rio de Janeiro: Publit.

Fedorova, Y., Ivanyuk, L., Korolchuk, V., & Yemets, N. (2006). *The catchers in the rhyme.* Kyiv: Lenvit.

Fisher, W. W., & McGeveran, W. (2006). *The digital learning challenge: Obstacles to educational uses of copyrighted material in the digital age.* Harvard: The Berkman Center for Internet & Society. Retrieved on December 31, 2007, from http://cyber.law.harvard.edu/sites/cyber.law.harvard.edu/files/BerkmanWhitePaper_08-10-2006.pdf

Freire, P. (1970/1987). *Pedagogia do oprimido.* São Paulo: Editora Paz e Terra.

Geer, R. (2001). The necessity of considering cultural influences in online collaborative learning. Retrieved on October 29, 2007, from http://gre-guns2.gre.ac.uk/pcet/PROGRAMMES/CeLTT/CeLTRes.nsf/0/720dbac82511cadd802569b100460f93/$FILE/greer.pdf

McRae, J. (1997). *Literature with a small "l."* London: MacMillan Publishers.

Schmidt, S. (1982). *Foundation for the empirical study of literature: The components of a basic theory.* Hamburg: Helmut Buske.

Schram, D., & Steen, G. (Eds.). (2001). *The psychology and sociology of literature: In honor of Elrud Ibsch.* Amsterdam & Philadelphia: John Benjamins.

Teles, C. (2006). Identificação, envolvimento, e afeto: A influência das legendas na apreciação de filmes estrangeiros. In S. Zyngier, V. Viana & A. M. Spallanzani (Eds), *Linguagens e tecnologias: Estudos empíricos* (pp. 223-242). Rio de Janeiro: Publit.

Van Peer, W., Hakemulder, J., & Zyngier, S. (2007). *Muses and measures: Empirical research methods for the Humanities.* Newcastle: Cambridge Scholars Publishing.

Wiseman, M. (2002). The empirical study of literature in the 21st century. In S. Zyngier & A. C. F. Valente (Eds.), *Fatos & ficções: Estudos empíricos de literatura* (pp. 11-16). Rio de Janeiro: Setor de Publicações da Faculdade de Letras da UFRJ.

Zyngier, S. (1994a). *At the crossroads of language and literature: Literary awareness, stylistics, and the teaching of EFLit.* Unpublished doctoral dissertation, University of Birmingham.

Zyngier, S. (1994b). Introducing literary awareness. *Language Awareness, 3*, 95–108.

Zyngier, S., Chesnokova, A., & Viana, V. (2007). Literary awareness: Connecting Brazil and Ukraine. *UNESCO Chair Messenger: Philology, Pedagogics . Psychology (Savannah, Ga.), 15*, 125–129.

ENDNOTES

1 Previously, the group has been referred to as 'Research to the Development of Empirical Studies' and 'Research Development in Empirical Studies'. Although the final version of its name has been used for a couple of years by now, references to these previous versions may still be found.

2 We opt to refer to mature scholars and students alike using the same term, namely, 'researchers'.

[3] A former Dutch group contributed very little to the discussions presumably because of the high turnover of its members. The Canadian group had not joined the project at that time.

[4] Due to financial constraints, not every REDES member can take part in these meetings every year or two. Efforts are made so that at least one representative per national center is present.

[5] There have also been several REDES meetings at the events organized by its national centers.

Chapter 15
Emerging Trends of Student Support Services in Indian Distance Education

Ramesh C. Sharma
Indira Gandhi National Open University, India

EXECUTIVE SUMMARY

Even though the concept of distance education emerged in the West/Europe, it is evident that the impact of this concept in terms of emergence of distance education institutions, student enrolment, and utilization of a variety of media are quite predominant in the context of Indian subcontinent especially India. The chapter traces briefly the history and growth of distance education in India. Various facets of student support are discussed followed by how various modes of ICT are employed as a part of student support. This chapter also examines the best practices adopted and emerging trends of students support by open universities in India. It also examines the role ICT has played and the future trends the ICT would enable effective student support.

INTRODUCTION

As India has entered into the 11th Plan (2007-2012) period, she is witnessing new developments in education, particularly post-school higher education having set a target of increasing the higher education access rate to 15% by 2012 from the current 10%. The 11th Plan has proposed a record 10-fold increase in outlay for higher education (Singh, 2008). Other areas of focus in the 11th Plan are capacity building, training, employability, life-long education and continuing education, which can be achieved through Open and Distance Learning (ODL). In developing countries like India, ODL has been main instrument towards social and economic development. A paradigm shift in terms of convergence of traditional and ODL is leading to the creation of a knowledge-based society. Heavy component of Information and Communication Technology (ICT) tools is providing new mechanisms of support to the learners. Education is now available to individuals and masses through various means like TV, radio, Internet, and Satellite. Owing to the success of ODL in the country, now there is one

DOI: 10.4018/978-1-60566-870-3.ch015

Table 1.

S No	Name of Open University	Year of Establishment
1	Dr Bhim Rao Ambedkar Open University, Hyderabad	1982
2	Indira Gandhi National Open University, Delhi	1985
3	Vardhman Mahaveer Open University, Kota, Rajasthan	1987
4	Nalanda Open University, Patna, Bihar	1987
5	Yashwantrao Chavan Maharastra Open University, Nasik, Maharastra	1989
6	Madhya Pradesh Bhoj Open University, Bhopal, Madhya Pradesh	1991
7	Dr Baba Saheb Ambedkar Open University, Ahmedabad, Gujarat	1994
8	Karnataka State Open University, Mysore, Karnataka	1996
9	Netaji Subhash Open University, Kolkata, West Bengal	1997
10	Uttar Pradesh RajRishi Tandon Open University, Allahabad, Uttar Pradesh	1998
11	Tamil Nadu Open University, Chennai, Tamil Nadu	2002
12	Pandit Sundarlal Sharma Open University, Raipur,	2005
13	The Global Open University, Nagaland	2006
14	K K Handique Open University, Guwahati, Assam	2007
15	Uttranchal Open University, Dehradun, UttraKhand	2007

National Open University, 14 State Open Universities, nearly 150 Distance Education Institutes in conventional universities, the National Institute of Open Schooling with a nation-wide network of open school support centers.

BACKGROUND

At the advent of the independence era, just when the British left, the Indian Higher Education had 20 Universities and 590 colleges for its 350 million people. Seventy percent among the populace were illiterates and lived in villages. As the leaders of India found it to be a major challenge to provide education to the masses, several commissions and committees were set up from time to time to opine and resort to a suitable measure for providing education to the populace spread out over a wide geographical terrain of the sub-continent.

On the recommendations of an Expert Committee headed by Dr D S Kothari constituted by Central Advisory Board of Education, a Directorate of Correspondence Courses and Continuing Education at University of Delhi in 1962 which paved the way for the first correspondence course.

On acclaiming this experiment in University of Delhi to be successful, the Punjabi University and the University of Rajasthan also followed suit in the year 1968. In 1969 Meerut and Mysore Universities also followed the foot prints.

In 1982, the first ever open university was established in India as Hyderabad as Andhra Pradesh Open University, which was renamed as Dr Bhim Rao Ambedkar Open University. This University had jurisdiction all over the State of Andhra Pradesh. Enjoying the success reaped by this Distance Education University, Indira Gandhi National Open University (IGNOU) was established in 1985 with national jurisdiction. In order to provide quality higher education other states in India also established their own open universities. The establishment of open universities in India is shown in Table 1.

The expansion of distance education system in the country is a clear indication for catering to the societal requirements by the open universities. Indira Gandhi National Open University (IGNOU), during its 24 years of existence, has played a vital role in this qualitative and quantitative expansion of the higher education system.

SETTING THE STAGE

The Indira Gandhi National Open University was established by an Act of Parliament in 1985. As on the end of the year 2008, IGNOU serves the educational aspirations of about 2 million students in India (10% of who get into Higher Education) and 33 countries abroad through 21 Schools of Studies and an elaborate network of 59 Regional centres, 2300 study centres, and 52 overseas centres. The University is expanding its efforts to take higher education to the doorsteps of the hitherto unreached. Teaching, research, extension and training form the mainstay of its academic activities. It also acts as a national resource centre; and more importantly, functions as an apex body to promote and maintain standards of distance education. IGNOU has been recognised as one of its centres of excellence by The Commonwealth of Learning. The Secretariats of SAARC Consortium of Open Distance Learning (SACODiL) approved by Heads of Govts. of SAARC Nations and Global Mega Universities Network (GMU-NET) initially promoted by UNESCO are being hosted by IGNOU.

Indira Gandhi National Open University (IGNOU) caters to learners from all possible strata of the society, spreading beyond geographic territories like rural, urban and tribal areas, going beyond the employer and the employed to the physically challenged and prison inmates, from rehabilitation houses, and governments and non-government sectors, parents and home-makers, personnel of armed and paramilitary forces. The University has put special focus on the women, minority communities, socially and economically disadvantaged groups, besides the north-east region of India, and other tribal and low literacy areas of the country by opening Special study centres exclusively for each of these categories.

The University has an extensive network of Regional Centres, the Study centres and Partner Institutions through which it reaches out to its learners located in different parts of the country and overseas. At these Centres, services in respect of subject specific academic counseling, listening. viewing of A/V programmes, library facilities, teleconferencing, video conferencing, CD-ROMs, services related to rules, regulations, counseling and examination schedules of the University, assignment submission, evaluation and feedback, term-end examination, computer access, laboratory work and other practical work are provided to learners. Tele-learning Centres were established at its Regional Centres for online programmes by the University. In addition, keeping in view with the requirement, the University is collaborating with private entrepreneurs to provide tele-learning facilities to its learners. Interactive Radio Counselling is provided from the University studio as well as from FM radio stations called Gyan Vani.

Mechanisms of Student Support

Programme delivery is one of the most crucial components of the teaching –learning process. It involves face-to-face academic counseling, assignment evaluation, information sharing, hands-on training and skill development, radio lectures, TV educational programmes and tele-counselling.

Technology is heavily used as a part of student support. At the study centres, the facilities extended to learners generally include the following:

1. Subject specific academic counseling
2. Audio, video and teleconferencing
3. Library consultation
4. Submission and evaluation of assignments
5. Information services

6. Provision for submitting for the application form for the term-end examinations.
7. Touch screen dialogue systems

CASE DESCRIPTION

Applications of ICT in Indian Distance Education Scenario

For the Government of India, ICT has been a good resource for mass education to bridge the digital divide. Starting from the use of radio, television and then satellite in 1970s to the launching of virtual education experiments, India has definitely struggled hard on the way to take advantage of IT and to join the league of IT superpowers. In spite of having more than 20000 colleges and nearly 400 university level institutions, the vital issue of concern remains how to take education to more than 400 million illiterates along with large number of school dropouts. The constitution of a National Task Force in 1998 for human resource development in IT related areas and use of ICT in education was aimed at achieving hundred per cent literacy at secondary and senior secondary school level (GOI, 1999).

Technological Support for Student Services

The major technological support provided to the learners by the University is through the production and dissemination of educational audio-video programmes. Curriculum-based audio-video programmes are developed by highly trained and skilled in house staff using the state-of-the-art facilities, to supplement the courseware given as ready to reading materials. This specific function is being overseen by a separate division in IGNOU namely Electronic Media Production Centre (EMPC). In addition, freelance filmmakers have been empanelled and suitable programmes are also drawn from agencies such as UKOU, Transtel DW TV, RCI etc. Both broadcast and non-broadcast modes of delivery are adopted by IGNOU. The cassettes/CDs are sent to regional centers and study centres and also sold at an affordable rates through the Marketing Unit in Exhibitions.

The various mechanisms of providing technological support in IGNOU are:

- Production of audio and video academic programmes and their dissemination to the IGNOU regional centers and study centres via non- broadcast mode.
- Transmission of video and audio programmes through public broadcasters – DoorDarshan-the official Telecast Station of the Government of India (DD1) and All India Radio (AIR) Stations at Hyderabad and Mumbai.
- Transmission of video programmes through dedicated satellite based educational TV channels – Gyan Darshan channel-GD1, GD3 & GD4.
- Transmission of live teleconferencing, telecounselling sessions through dedicated satellite based interactive educational TV channels – Gyan Darshan- GD2.
- Transmission of educational and enrichment audio programmes through FM Radio network of 26 Gyan Vani Stations located at Ahmedabad, Allahabad, Aurangabad, Bangalore, Bhopal, Coimbatore, Chennai, Delhi, Guwahati, Jabalpur, Kolkata, Mumbai, Mysore, Rajkot, Raipur, Varanasi, Vishakapatnam, Lucknow, Shillong, Jaipur, Panaji, Indore, Kanpur, Patna Hyderabad and Nagpur.
- Transmission of live Interactive Radio Counseling (IRC) sessions for IGNOU students through AIR Stations for a payment of the license fee to the AIR.
- Technical support and turnkey consultancy expert services to State Open Universities and DE institutions.

Educational Radio and Television Channels

Radio and television have been innovatively introduced as the support mechanisms in Indian Distance Education. In IGNOU, the use of Educational radio and television channels is tailor-made to suit the various facets of the target group. They are being discussed in the following paragraphs.

Gyan Darshan

It is a satellite-based fully digital 24 hour exclusive national educational and development channel, by the Ministry of Human Resource Development and Prasar Bharati with IGNOU as the nodal agency. It beams programmes produced by IGNOU, University Grants Commission (UGC), National Council for Educational Research and Training (NCERT), Central Institute for Educational Technology (CIET), Indian Institute of Technology (IIT) and certain specific Ministries of Govt. of India. Gyan Darshan expanded into a digital bouquet of four channels namely GD-1, GD-2, GD-3 and GD-4, being uplinked from earth station at EMPC.

- GD 1 is 24 hr channel having the programmes produced by IGNOU and best educational programmes from other educational channels of Gyan Darshan channels.
- GD 2 is devoted entirely for interactive distance education, which is a unique feature of the bouquet of channel.
- GD 3 (Eklavya Channel) is 24 hr Channel which brings quality education to the student pursuing engineering throughout the country.
- GD 4 (Vyas Channel) is 24 hr Channel which provide quality education to the students pursuing higher education throughout the country.

Gyan Vani

This is a unique low cost, interactive medium for enhancing and supplementing the teaching-learning process by reaching out to the widespread learners. Gyan Vani stations operate as media cooperative with the day-to-day programmes being contributed by various educational institutions, NGOs, government and semi-government organizations, UN agencies, Ministries such as Agriculture, Environment, Health, Women and Child Welfare, Science and Technology, etc. besides national level institutions such as NCERT, NIOS and State Open Universities.

Each Gyan Vani Station has a range of about 60 kms and covers an entire city/town and the adjoining rural area. Gyan Vani serves as an ideal medium for niche audiences to address the local educational, development and socio-cultural requirements. The flavour of the channel is by and large local, and the medium is English, Hindi or language of the region. The broadcast duration varies from station to station but the overall content pertains to the following aspects and levels of education:

- Sarva Shiksha Abhiyan (Education for All)
- Primary and Secondary Education
- Adult Education
- Technical and Vocational Education
- Higher Education
- Distance Education
- Extension Education

Teleconferencing

In order to infuse interactivity in distance learning, one-way video and two-way audio teleconferencing facility is being offered through GD2. Important nation-wide programmes for learners, lectures by eminent experts/dignitaries, discussions with RC Staff, Induction for new students and convocations for graduating students are conducted live through teleconferencing every year. Besides, tele-counseling sessions are conducted for selected application oriented programmes such as nursing, Clinical Cardiology, Hospital management, information technology, MBA etc. in which the students get ample opportunities to interact with the faculty. Besides IGNOU, many other institutions also use this facility for reaching out cost-effectively to their learners scattered all over the country.

Interactive Radio Counseling (IRC)

IGNOU offers one hour of live phone-in counseling programme weekly through the National Network of AIR and the Gyan Vani stations where studio invited experts clarify student queries put across to them from their homes via telephone. This value added services is presently being offered through nearly 188 All India Radio stations every Sunday from 4-5 p.m. A toll free conferencing facility is available to the learners in 80 cities to interact freely with the experts. The Interactive Radio Counseling (IRC) sessions on Gyan Vani Delhi are held from 5 to 6 p.m. on every Sundays (various topics) and Tuesdays (topics related to IGNOU programmes). Currently, this has been held at a dormant stage as the renewal of the License Fee is being discussed between the Government and the University. Other Gyan Vani Stations offers this facility at least one hour per week at the convenience of the students.

Edusat

The launch of EDUSAT (a dedicated satellite for education) has ushered in an era where both internet and intranet can be used for transmission, interaction, dialogue, digital repositories, digital multimedia content, and for virtual education and research. EMPC-IGNOU has already established 134 interactive terminals across the country with the two way interactive facility. The teleconferencing sessions on GD-2 are simulcasted on the EDUSAT network. The facility is expected to enhance the capacity of the learners to access the resources as per the schedule. Networking of all Gyan Vani FM Radio Stations is also under consideration.

The use of Edusat with the launch of the Rajiv Gandhi Project moved the use of technology in reaching to the populace in the Schools. The Rajiv Gandhi Project for Edusat Supported Elementary Education was inaugurated by Hon'ble Minister Sh. Arjun Singh in December, 2005. This is a collaborative project of MHRD-ISRO-IGNOU and State Governments of Madhya Pradesh, Bihar, Chhattisgarh and Uttar Pradesh. The network has a provision of one teaching end with sub-hub at Jabalpur with connectivity to 850 schools in the Sidhi District and 50 schools each in Sonbhadra in U.P., Koria in Chhattisgarh and Vaishali in Bihar.

Educational Radio in India

- Radio Technology came into popular usage during the early twentieth century. Although television offers visuals, radio in its own capacity can reach a wide geographic audience at a low production cost with proven educational results (Couch, 1997). In India, the first radio programme was broadcast by the Radio Club of Bombay in June 1923 (Sharma, 2002a). Vyas, Sharma & Kumar (2002a) have provided a brief of some of the major Educational Radio Projects in India which are as follows:
- **School Broadcast Project** commissioned in 1937 and the target group was School students.
- **Adult education and community development project** agriculture-based project, commenced in 1956.
- **University broadcast project** launched for University students in 1965, with an aim to expand higher education as widely

as possible among the different strata of society.

- **IGNOU-AIR Broadcast** started in January 1992 with a target group of students of Open / Conventional Universities.
- **IGNOU-AIR Interactive Radio Counselling** started in 1998 for students of Open / Conventional Universities, this project is also very successful. In order to bridge the gap between Institutions and learners by instantly responding to their queries and also to provide Academic Counselling in subject area (Sharma, 2002a).
- **Gyan-Vani** Educational FM Radio Channel of India, launched in 2001 for the target group of students of Open / Conventional Universities. It has been conceived as a unique decentralised concept of extending mass media for education and empowerment, suiting to the educational needs of the local community (Sharma, 2002b). Gyan Vani is not only meant for the conventional educational system but is also a main tool in making available the dream of education for all come true. The main intention of Gyan Vani is to take education to the doorsteps of the people.

Educational Television in India

Television first came to India [named as 'Doordarshan' (DD)] on Sept 15, 1959 as the National Television Network of India. Some of the major educational projects in India (Vyas, Sharma & Kumar (2002b) are given below:

- **Secondary School television project** started in 1961 to improve the standard of teaching in view of shortage of laboratories, space, equipment and dearth of qualified teachers in Delhi.
- **Delhi Agriculture Television** (DATV) Project (Krishi Darshan) launched in 1966 for providing agricultural information to the farmers.
- **Satellite Instructional Television Experiment** (SITE) conducted in 1975 for the villagers and their Primary School going children of selected 2330 villages in six states of India.
- **Indian National Satellite project** (INSAT) started in 1982 and was aimed at making the rural masses aware of the latest developments in the areas of agricultural productivity, health and hygiene.
- **UGC-Higher Education Television Project** (HETV) launched in 1984 for university students.
- **IGNOU-Doordarshan Telecast** started in 1991 for Distance learners
- **Gyan-Darshan Educational Channel** launched jointly by the Ministry of Human Resource Development, Information and Broadcasting, the Prasar Bharti and IGNOU on 26th January 2000 as the exclusive Educational TV Channel of India.

Teleconference

Teleconferencing was started in 1983 to interact with learners at different locations. It is one-way-video-two-way-audio teleconferencing. It enables the experts (teaching end) to address and interact with large audience placed at different locations (learning end) through live and spontaneous interactions. This has been reported Dash (1997) and Mishra (1999) to be quite effective for content presentation, learner-expert interactions and learner supported activities. In IGNOU, this facility has been exploited to reach the masses of the newly enrolled learners through Induction Meeting, Train the Academic counselors to the distance education system through Orientation Programme and to virtually meet the Regional and Study Centre Staff during convocation, inauguration and launch of new programmes.

Education through Satellite in India

EDUSAT is the first Indian satellite designed and developed exclusively to meet the demand for an interactive satellite based distance education system for the country (Sharma, 2006). It is a collaborative project of Ministry of Human Resource Development (MHRD), and Department of Space, Indian Space Research Organization (ISRO). The IGNOU (Indira Gandhi National Open University) acts as the Nodal Institution supported by AICTE (All India Council for Technical Education), ICAR (Indian Council of Agricultural Research), NCERT (National Council of Educational Research and Training), and UGC (University Grants Commission) towards content generation for the Programmes to be telecast.

E-Learning Initiatives

Owing to some barriers like low telephone density, high cost of Internet services, non-availability of skilled manpower etc, the developments in the field of E-Learning were slow till early 2000. It was then through the initiatives of Government of India and the private sector to the development of IT sector, the access of masses to technology (like mobile telephony and internet) has been increased. As a consequence of recommendations of the National Task Force on IT, with sudden growth of telephony (both landline and mobile) and certain private ISPs entering the market place, more and more institutions (educational and commercial) have either adopted online technologies for their services or gone online. Mishra and Sharma (2005) suggested establishment of a "e-Learning Consortium" including member educational institutions to offer e-learning programmes without duplicating efforts both in infrastructure and intellectual inputs. A brief overview of virtual educational institutions in India like Netvarsity (www.netvarsity.com), virtual campus initiative of Indira Gandhi National Open University (www.ignou.ac.in), Tamil Virtual University (www.tamilvu.org), Punjab Technical University (www.ptuonline.com), and IIT Delhi (www.iitd.ac.in) etc. have been provided by these two authors.

IGNOU has started virtual education in different programmes (Figure 1):

- VIRTUAL EDUCATION IN LAW (VEL): http://www.ignouonline.ac.in/vel/
  - PG Certificate in Cyber Law (PGCCL)
  - PG Diploma in Legal Process Outsourcing (PGDLPO)
- VIRTUAL EDUCATION IN LIBRARY & INFORMATION SCIENCE (LIVE): http://www.ignouonline.ac.in/live/
  - Master Degree in Library and Information Science (MLIS)
- VIRTUAL EDUCATION IN HEALTH SCIENCES (VEHS): http://www.ignouonline.ac.in/acupuncture/
  - PG Diploma in Acupuncture (PGDACP)
- VIRTUAL EDUCATION IN AGRICULTURE (SOA): http://www.ignouonline.ac.in/soa/
  - PG Diploma in Food Safety and Quality Management (PGDFSQM)
  - PG Certificate in Agriculture Policy(PGCAP)
- VIRTUAL EDUCATION IN FOREIGN LANGUAGES (SOFL): http://www.ignouonline.ac.in/cpsl/
  - Certificate Programme in Spanish Langugae(CPSL)
- VIRTUAL EDUCATION IN SUSTAINABLE DEVELOPMENT (SAVE): http://www.ignouonline.ac.in/save/
  - Appreciation Programme on Sustainability Science (APSS)

These Online Programmes have the following features:

Figure 1. http://www.ignouonline.ac.in/VirtualClass.htm

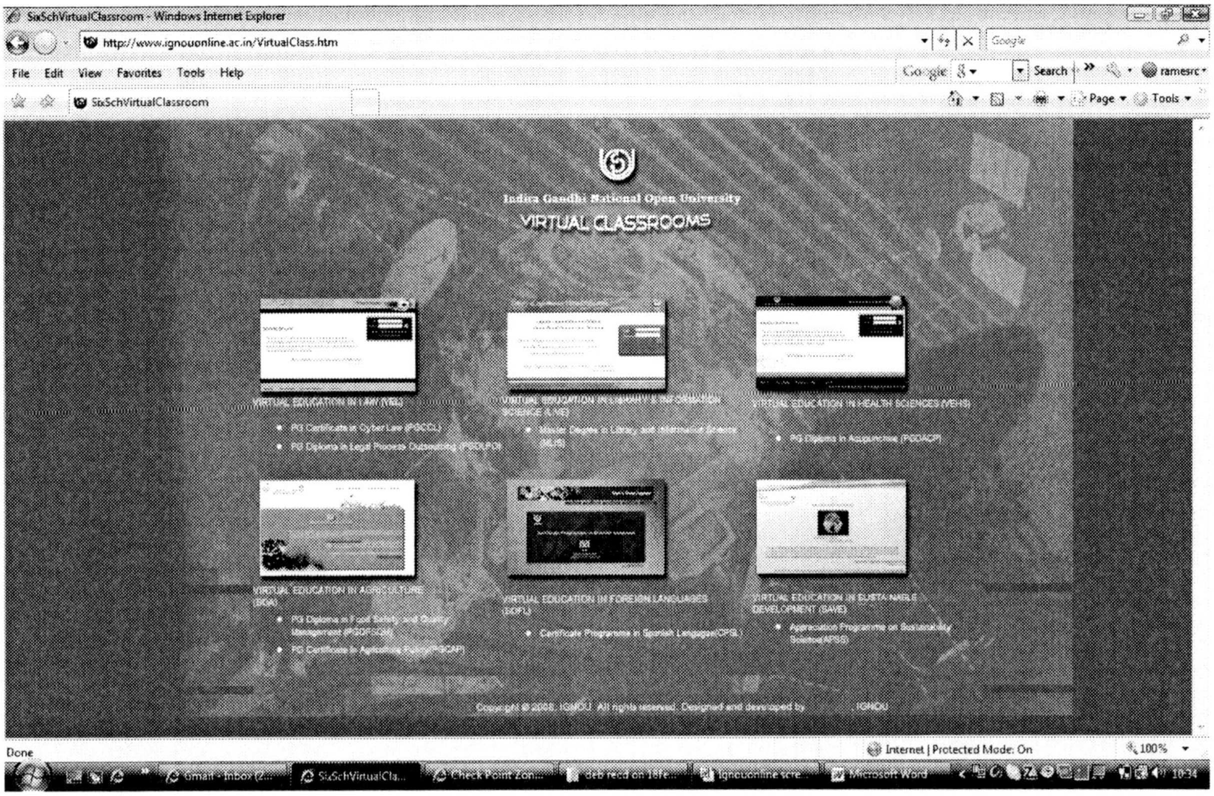

- *Walk in admission*: Admission will be available throughout the year. There shall be four cycles in an academic year i.e. January, April, July and October. The two months period between the two cycles shall be used for advertisement and registration of the students for the next cycle.
- *Integrated multimedia courseware*: once registered, learners will have access to personalized learning space (My Page). This includes self instructional material, related audio/ video, slides, self check exercises, etc. weaved in one platform.
- *Online counselling and mentoring*: web cast based counselling integrated with text based chatting facility shall be used for counselling purposes.
- *24×7 learner support*: Asynchronous and synchronous modes of interaction will be used to provide just in time support to learners.
- *Group based online Seminar*: Blogs, chat rooms and discussion forums and web conferencing facility shall be used for group based seminars.
- *Online Project platform:* Templates shall be made available for synopsis submission, project uploading and evaluation. Viva voce will be conducted through web based video conferencing tool.
- *Online Term End Examination*- Online examinations will be conducted in specified centres for in proctored environment.

Sakshat

Sakshat (www.sakshat.ac.in and www.sakshat.edu.in) is a landmark initiative of the Ministry

Figure 2.

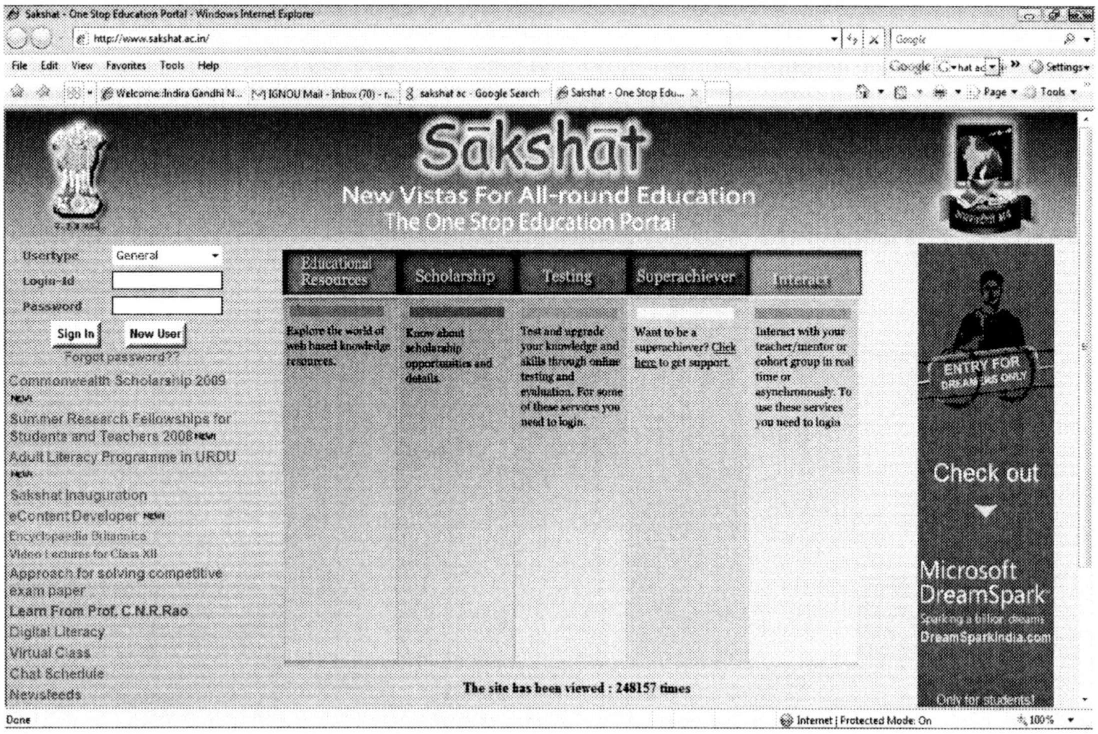

of Human Resource Development to develop a One Stop Education Portal to cater to the educational and learning needs of students, teachers, scholars and lifelong learners (see Figure 2). This portal has been developed with the objective of providing one stop solution to learners ranging from K- level to 20 study different fields of study including vocational education and learning for life skills.

India is a big country where the cultural diversity and geographical terrain have a great relation with vast disparity of educational facilities available across the country. Sakshat portal would bridge this gap by providing just-in-time quality educational resources and teachers 24x7 to learners irrespective of their social, economic and educational status. To develop world class content for this portal, premier institutions of India like IGNOU, National Council of Educational Research and Training, Kendriya Vidyalaya Sangthan, University Grants Commission, Indian Institutes of Technology, Indian Institute of Science, and Central Board of Secondary Education etc are anchoring the efforts. The objectives of the Sakshat portal are to:

- provide barrier-free web based learning resources
- develop a repository of world class interactive multimedia content
- nurture learners by providing web based guidance and encouragement through synchronous and asynchronous modes of interaction
- facilitate learners to make informed decision on choice of profession through online career counselling facility

Wireless Mobile Internet Model for Open and Distance Education

Jasola and Sharma (2005) proposed a wireless mobile Internet model for open and distance education to address two basic barriers namely

Figure 3. http://www.egyankosh.ac.in/

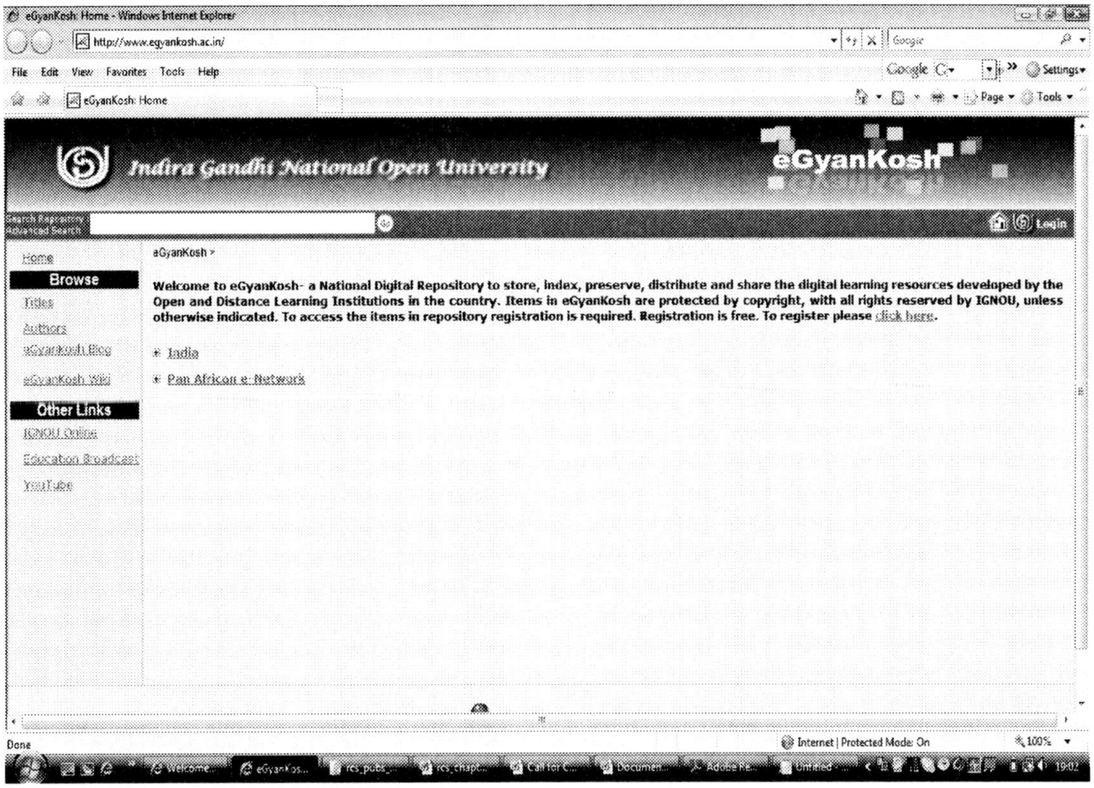

the learner's feeling of isolation and remoteness; and the time it takes to complete an instructional transaction. This model theorizes that educational content must not be driven by technology, instead be based on the learners needs and the perceived outcomes.

eGyanKosh

The eGyanKosh is a National Digital Repository to store, index, preserve, distribute and share the digital learning resources developed by the Open and Distance Learning Institutions in the country (Figure 3).

SMS Alerts Service

The Indira Gandhi National Open University (IGNOU), has launched SMS services to its students across the country as well as for the staff of the university. It is a Pilot in the run up for disseminating study material and conducting of limited objective tests on mobile phones. This service will enable the IGNOU to send SMS alerts to students from time to time about various developments. Currently, the IGNOU is having 1.85 million students. The IGNOU also plans to declare results of its term end examinations through SMS. It wants to harness the power of SMS to support students in their endeavours as well as to solve their problems which are time consuming through other existing modes. Future SMS services will enable students to check their marks, information in their profile on demand. Further, the IGNOU also would like to enable students to update their profiles through SMS. Latest profiles of students is very important for the IGNOU to support students. For example, availability of latest address of a student enables the IGNOU to send material to the student in time.

Current Challenges

The growth of information technology across various sectors like business, education, and health etc has led to new opportunities to tap the potential of ICT. The educational institutions be them regular or distance mode, are offering their services with a heavy input of technology. The educational institutions thus face the challenge of how to leverage the digital divide. Another challenge being faced by institutions is to enhance the learner-tutor interaction. The changing pedagogies, new assessment criteria, unique modes of offering learning modules are some of the areas needing immediate attention of the service providers.

Kawachi (2005) reflected on the challenge of enhancing learning effectiveness. He observed that due to increasing access to internet, availability of educational softwares in local languages and thus the access to content over Internet is getting easier. Kawachi recommends a shift from student-to-content interaction to student-to-teacher and student-to-student interaction, although with a note of caution that it may prove to expensive in doing that. This transformation in reference to Asia to happen "would depend on whether students move from individual and co-operative mode towards more critical thinking and collaborative learning mode. Where this is indeed occurring in the centres of excellence in Hong Kong, Japan, Korea and Singapore, the digital divide in education is widening between these centres and the rest of Asia" (Kawachi 2005, p. 112).

Adoption of technology to a great extent by the formal education system is also creating a challenge, which has so far been the domain of distance education. Thus observing the role the formal face-to-face, open and distance learning, online learning and satellite instructions are playing in teaching learning process and thereby extending the outreach to unreached section of society, convergence of ODL systems with the conventional patterns of education in the traditional universities has been initiated since 2008. What impact this convergence would have on the managerial or organizational processes or learning outcomes is yet to be seen.

Govt of India launched a *National Education Mission through ICT* on Tuesday, February 03, 2009 (GOI, 2009). Under this Mission, about 400 university level institutions & 20000 colleges would be connected through broadband connectivity and suitable e-learning material would be made available to them. State Governments would share 25% of the cost of providing broadband connectivity to the institutions under their control. Private institutions will have to share 25% of the cost of providing the broadband connectivity. Thus the educational institutions (be at University or college level) would face the challenge of identification and nurturing of talent and lifelong learning, development of knowledge modules to address to the personalized needs of learners, having right content to take care of their aspirations, and their delivery through ICT.

Certification of competencies of the human resources acquired through formal or non-formal means as also to develop and maintain the database having profile of human resources. In order to deliver the benefits of ICT enabled Learning including academic, vocational and life skills, the Mission would also focus attention on six challenge areas with a view to achieving technological breakthrough in the field of:

(a) very low cost, low power consuming access for every Indian,
(b) research in pedagogical techniques and development of pedagogically efficient learning modules for the entire spectrum of learners,
(c) standardization and quality assurance of contents to make them world class

(d) research in the field of virtual reality laboratories,
(e) techniques and strategies for imparting digital literacy to every Indian, and
(f) creation of virtual technological university.

The Mission would also have to attempt to build knowledge network between and within institutions of higher learning in the country with a view to achieving critical mass of researchers in any given field.

Most crucial challenge in the context of Indian Open and Distance Education system is to strengthen the learner support systems. As shown in the table in the beginning of this chapter, the network of student support has grown too fast. With 15 Open Universities, around 150 Distance Education institutions, more than 5000 learners support centers of all these open universities, a need arises to ensure the quality of instruction and provision of minimum necessary infrastructure, in addition to proper upkeep of equipments and materials. The Govt of India has planned to provide Internet connection and computer facilities to educational institutions at all levels. The challenge to keep these things running needs to be taken very seriously by administrators.

CONCLUSION

Education systems all over the world are evolving to better serve the students. This evolution is clearly reflected in the ways that IGNOU continuously researches and develops its academic courses. The University offers certificate programmes and a variety of courses that address the wants and needs of society and all adult students throughout the sub-continent. These courses span a wide range of content, structure, designs and delivery methods to suit the diversity that is India. The variety of courses reflects the local and global concerns of contemporary social development. These cater to each sector of society providing academic, professional, continuing education, extension and training programmes - with an appropriate use of delivery and learning technologies to excite the imagination of the students and support their learning throughout their lives.

REFERENCES

Couch, L. (1997). *Digital and analog communication systems*. Upper Saddle River, NJ: Prentice-Hall.

Dash, N. K. (1997). Reactions of primary school teachers towards training through interactive television. *Indian Journal of Open Learning, 6*(1&2), 77–90.

GOI. (1999). National task force on information technology and software development: IT action plan part III–long term national IT policy. New Delhi: Government of India. Retrieved from http://it-taskforce.nic.in/actplan3/

GOI. (2009). *National mission on education through information and communication technology launched*. Press release. Press Information Bureau. Retrieved on February 25, 2009, from http://pib.nic.in/release/release.asp?relid=47221

Jasola, S., & Sharma, R. (2005). Open and distance education through wireless mobile Internet: A learning model. *International Journal of Instructional Technology and Distance Learning, 2*(9), 35-47. Retrieved from http://www.itdl.org/Journal/Sep_05/article04.htm

Kawachi, P. (2005). Computers, multimedia, and e-learning. In U. V. Reddi & S. Mishra (Eds.), *Perspectives on distance education: Educational media in Asia* (pp. 97–122). Vancouver: Commonwealth of Learning.

Mishra, S. (1999). An empirical analysis of interactivity in teleconference. *Indian Journal of Open Learning, 8*(3), 243–253.

Mishra, S., & Sharma, R. C. (2005, March 14-20). Development of e-learning in India. *University News, 43*(11), 9–15.

Sharma, R. (2006). EDUSAT: Taking Indian distance education to new frontiers. *Distance Learning Link*, January-June, 6-8. USA: Information Science Publishing.

Sharma, R. C. (2002a). Interactive radio counseling in distance education. *University News, 40*(10), 8–11.

Sharma, R. C. (2002b). Gyan Vani: The educational FM radio network of India. *Field Notes, 2*(2). News from Asia. Retrieved from http://www.irrodl.org/content/v2.2/field.html

Singh, M. (2008). *Foreword to eleventh five year plan 2007-12, volume I: Inclusive growth* (p. iv). Retrieved on December 9, 2008, from http://planningcommission.nic.in/plans/planrel/fiveyr/11th/11_v1/11v1_foreward.pdf

Vyas, R. V., Sharma, R. C., & Kumar, A. (2002a). Educational radio in India. *Turkish Online Journal of Distance Education, 3*(3). Retrieved from http://tojde.anadolu.edu.tr/tojde7/articles/educationalradio.htm

Vyas, R. V., Sharma, R. C., & Kumar, A. (2002b). Educational television in India. *Turkish Online Journal of Distance Education, 3*(4). Retrieved from http://tojde.anadolu.edu.tr/tojde8/articles/educationaltv.htm

Compilation of References

Agarwal, R., Edwards, S. H., & Pérez-Quiñones, M. A. (2006). Designing an adaptive learning module to teach software testing. [Houston, TX: ACM.]. *SIGCSE, 06*, 259–263. doi:10.1145/1124706.1121420

Almeda, M. B. (1985). Upgrading the image. *Continuum, 49*(3), 198–202.

Almeda, M. B. (1988). Speaking personally with Gayle B. Childs. *American Journal of Distance Education, 2*(2), 68–74. doi:10.1080/08923648809526626

American Association for Collegiate Independent Study. (2008). *Awards.* Retrieved in November 2008, from http://www.aacis.org/main/main.htm?page=awards

American Council on Education & Adult Higher Education Alliance. (2000). *Principles of good practice for alternative and external degree programs for adults.* Washington, D.C.: ACE & Alliance.

American Council on Education Task Force. (2000). *Principles of good practice for alternative and external degree programs for adults.* Washington, D.C.: American Council on Education and The Alliance.

Andersen, S. (1984). Computers in the Humanities: Providing faculty with new tools. In R. W. Lutz, E. Jacobson & B. Rader (Eds.), *Proceedings of the 12th Annual ACM SIGUCCS Conference on User Services* (pp. 143-148). NY: ACM.

Anderson, D., Moxley, V., Maes, S., & Reinert, D. (2008). Multi-institution academic programs: Dealmakers and dealbreakers. *Continuing Higher . Educational Review, 72*, 103–119.

Anderson, L. W., & Krathwohl, D. R. (Eds.). (2001). *A taxonomy for learning, teaching, and assessing: A revision of Bloom's taxonomy of educational objectives.* New York: Longman.

Anderson, T. (2002). *An updated and theoretical rationale for interaction* (Athabasca University's IT Forum Paper No. 63). Athabasca University, Albuquerque, Canada. Retrieved on November 5, 2008, from http://it.coe.uga.edu/itforum/paper63/paper63.htm

Auracher, J., Schwaiger, M., & Tiourikov, A. (2007). Say it again: Ideas regarding the correct usage of Internet communication. In S. Zyngier, A. Chesnokova & V. Viana (Eds.), *Acting and connecting: Empirical approaches to language and literature* (pp. 203-218). Munster: LIT Verlag.

Aycock, A. Garnham, C., & Kaleta, R. (2002). Lessons learned from the hybrid course project. *Teaching with Technology Today, 8*(6). Retrieved on October 22, 2008, from http://www.uwsa.edu/ttt/articles/garnham2.htm

Balci, B., & Inceoglu, M. (2007). A Web-based learning content design platform and metadata editing. *37th ASEE/IEEE Frontiers in Education Conference* (pp. F1H-13-F1H-18). Milwaukee, WI.

Barbosa, E. F. (2006). Towards the establishment of a standard process for developing educational modules. *36th ASEE/IEEE Frontiers in Education Conference* (pp. T2J-5-T2J-10). San Diego, CA: IEEE.

Baron, J., & Maier, H. (2005, December). *The challenge of maintaining the momentum.* Paper presented at the Conference of the Australasian Society for Computers in Learning in Tertiary Education (ASCILITE), Brisbane. Retrieved on October 20, 2008, from http://www.ascilite.org.au/conferences/brisbane05/blogs/proceedings/07_Baron.pdf

Barr, R., Pandy, M., Petrosino, A., & Svihla, V. (2005). Challenge-based instruction: The VaNTH biomechanics learning modules. 35th ASEE/IEEE Frontiers in Education Conference (pp. SIG–8-IG–13). Indianapolis, IN.

Barreto, J., & Fialho, O. (2005). Flow+foregrounding: A possible relation? In S. Zyngier, V. Viana & F. Fausto (Eds.), *Venturas & desventuras. Coletânea dos trabalhos do V ECEL* (pp. 87-96). Rio de Janeiro: Setor de Publicações da Faculdade de Letras da UFRJ.

Bartolic-Zlomislic, & Bates. (2002). The costs and benefits of investing in online learning. *Distance Education Report, 6*(9), 3.

Bender, S. L. (2006). *Quasitative inquiry/research/approach: Leading a new culture of inquiry?* Retrieved on December 28, 2008, from http://www.sharonbender.com/quasitative_inquiry_research_approach_culture_of_inquiry.html

Berg, G. A. (2001). Distance learning best practices debate. *WebNet Journal,* (5-6), 17. Retrieved on November 26, 2008, from http://www.aace.org/pubs/webnet/v3no2/3=2DistanceLearning.pdf

Berge, Z. L. (1995). Facilitating computer conferencing: Recommendations from the field. *Educational Technology, 35*(1), 22–30.

Berge, Z., & Mrozowski, S. (2001). Review of research in distance education, 1990 to 1999. *American Journal of Distance Education, 15*(3), 5–19. doi:10.1080/08923640109527090

Bergmann, H. F. (2001). The silent university: The society to encourage studies at home, 1873-1897. *The New England Quarterly, 74*(3), 447–477. doi:10.2307/3185427

Bershin, J. (2004). *The blended book of learning.* San Francisco, CA: Pfeiffer.

Bloom, B., Englehart, M. D., Furst, E. J., Hill, W. H., & Krathwohl, D. R. (1956). *Taxonomy of educational objectives: The classification of educational goals. Handbook I: Cognitive domain.* New York, Longman.

Blumenstyk, G. (1994). Networks to the rescue? *The Chronicle of Higher Education,* A21.

Blumenstyk, G. (1995). Campuses in cyberspace. *The Chronicle of Higher Education,* A18.

Bonk, C. J., & Graham, C. R. (2006). *The handbook of blended learning: Global perspectives, local designs.* San Francisco: Pfeiffer.

Bourne, J., & Moore, J. (Eds.). (2002). *Elements of Quality Online Education.* Sloan Center for Online Education.

Braumoeller, B. F., & Gaines, B. J. (2001). Actions *do* speak louder than words: Deterring plagiarism with the use of plagiarism-detection software. *PS: Political Science and Politics, 34*(4), 835–839. doi:10.1017/S1049096501000786

Brent, B. (1999). Distance education: Implications for equity and cost-effectiveness in the allocation and use of educational resources. *Journal of Education Finance, 25*(2), 229–254.

Brew, A. (2006). *Research and teaching: Beyond the divide.* London: Palgrave Macmillan.

Brigham, D. (2001). Converting student support services to online delivery. *International Review of Research in Open and Distance Learning, 1*(2), 1-15. Retrieved on September 5, 2008, from http://www.irrodl.org/index.php/irrodl/article/view/23/365

Burcaw, S. S. (1981). *Standards for independent study: The American experience in higher education.* Paper presented at the Conference of the International Council for Distance Education, June, Vancouver, British Columbia, Canada.

Burgess, S. (2003). Course modularization for improved delivery and assessment. *Consortium for Computing Sciences in Colleges, Northwestern Conference* (pp. 128–134).

Burton, G. (2008, November). *Establishing best practices in distance and independent study.* Paper presented at the

Sixteenth Annual American Association for Collegiate Independent Study Conference, St. Louis, MO. Retrieved from http://www.aacis.org/pages/conferences/2008conf.htm

Carey, J. (2001). Effective student outcomes: A comparison of online and face-to-face delivery modes. *DEOSNEWS, 11*(9). Retrieved from http://www.ed.psu.edu/ascde/deos/deosnews/deosarchives.asp

CAT. (2005). *Center for Academic Transformation at the Rensselaer Polytechnic Institute*. Retrieved in March 2005, from http://www.center.rpi.edu

CAT. (2005). *Center for Academic Transformation, Program in Course Redesign*. Retrieved in March 2005, from http://www.center.rpi.edu/PewGrant.html

Chatterji, M. (2003). Designing and using tools for educational assessment. Boston, MA: Allyn and Bacon.

Chattratichart, J. (2007). A theory-based approach to designing student learning context. [San Jose, CA: ACM.]. *CHI, 2007*, 1721–1728.

Chesnokova, A., Zyngier, S., Viana, V., Jandre, J., & Nero, S. (2009). Universal poe(try)? Reacting to "Annabel Lee" in English, Portuguese and Ukrainian. In S. Zyngier, V. Viana & J. Jandre (Eds.), *Linguagem, criatividade & ensino: Abordagens empíricas e interdisciplinares* (pp. 193-211). Rio de Janeiro: Publit.

Chickering, A., & Ehrmann, S. (1996). Implementing the seven principles: Technology as lever. *American Association for Higher Education Bulletin,* 3-6.

Chickering, A., & Gamson, Z. (1987). Seven principles of good practice in undergraduate education. *AAHE Bulletin, March*, 3-7.

Childs, G. (1973). "Correspondence Study: Concepts and Comments." Speech given at the NUEA Conference: Omaha, NE. 17 April, 1973.

Churcher, N., Irwin, W., & Kriz, R. (2003). Visualizing class cohesion with virtual worlds. *Conferences in Research and Practice in Information Technology* (pp. 24). Australian Computer Society.

Clark, R. C., & Mayer, R. E. (2008). *E-learning and the science of instruction*. San Francisco: Pfeiffer.

Coachman, E., & Weigert, T. (2008, July). *Readers' response to vulgar language: A comparative study*. Paper presented at XI Congress of the International Society for the Empirical Study of Literature (IGEL), Memphis, TN.

Committee on Developments in the Science of Learning. (1999). Technology to support learning. In J. Brandsford, A. Brown & R. Cocking (Eds.), *How people learn: Brain, mind, experience, and school*. Washington, D.C.: National Research Council. Retrieved on May 30, 2000, from http://books.nap.edu/html/howpeople1/ch9.html

Couch, L. (1997). *Digital and analog communication systems*. Upper Saddle River, NJ: Prentice-Hall.

Crynes, B. L. (1996). Curriculum granularity. *IEEE FIE '96 Proceedings* (pp. 740–744).

Curran, C. (2008). Online learning and the university. In W. Bramble & S. Panda (Eds.), *Economics of distance and online learning: Theory, practice, and research* (pp. 26-51). New York: Rutledge.

Dansberry, B. E., & Cates, C. L. (2004). A social responsibility learning module for use in cooperative education. *34th ASEE/IEEE Frontiers in Education Conference* (pp. F3F-3-F3F-6). Savannah, GA: IEEE.

Darcy, S., & Nagel, T. (2003, October). *Fait accompli: The integration of CDIS with the University of Missouri*. Paper presented at the Annual Meeting of the American Association for Collegiate Independent Study, Sioux Falls, SD. Retrieved on March 23, 2009, from http://www.aacis.org/pages/conferences/2003conf.htm

Dash, N. K. (1997). Reactions of primary school teachers towards training through interactive television. *Indian Journal of Open Learning, 6*(1&2), 77–90.

De Smedt, K. (1999). Introduction. In K. De Smedt, H. Gardiner, E. Ore, T. Orlandi, H. Short, J. Souillot & W. Vaughan (Eds.), *Computing in Humanities education: A European perspective* (pp. 1-12). Bergen: University of Bergen. Retrieved on December 27, 2007, from http://gandalf.aksis.uib.no/AcoHum/book/intro.html

DeFleur, M. H., & Adams, J. (2004). Acceptability of online degree programs as criteria for admission to graduate programs. *Journal of Computing in Higher Education, 16*(1), 150–163. doi:10.1007/BF02960287

Desire2Learn, Inc. (2008). *Competencies and rubrics user guide (8.3)*. Retrieved on November 28, 2008, from http://community.desire2learn.com

Dewey, J. (1933). *How we think.* New York: Heath.

Duff, A., & Quinn, D. (2004). *Benefits of online learning.* Online Learning at UniSA (University of South Australia). Retrieved in November 2008, from http://www.unisanet.unisa.edu.au/learn/learnOnline/?PATH=/Resources/Workshop-ready4online/Learning+Online/&default=Welcome.htm

Duffy, P., & Axel, B. (2006). The use of blogs, wikis, and RSS in education: A conversation of possibilities. *QUT Online Learning & Teaching Conference Papers* (p. 8). Queensland, Australia: Queensland University of Technology.

Duning, B. (1985). The independent study image dilemma. *Continuum, 49*(3), 209–212.

Edelson, P. J. (1991). Codification and exclusion: An analysis of the early years of the national university extension association (UCEA) 1915-1923. *Continuing Higher Education Review, 55*(3), 176–189.

Edelstein, S., & Edwards, J. (2002). If you build it, they will come: Building learning communities through threaded discussion. *Online Journal of Distance Learning Administration, 5*(1). Retrieved from http://www.westga.edu/~distance/ojdla/spring51/edelstein51.html

Eisen, M. J. (2005). Shifts in the landscape of learning: New challenges, new opportunities. In M. A. Wolf (Ed.), *Adulthood: New terrain. New directions for adult and continuing education* (No. 108, pp. 15-26). San Francisco, CA: Jossey-Bass.

Eterovich, F. H. (1980). *Aristotle's Nicomachean ethics: Commentary and analysis.* Washington, D.C.: University Press of America.

Fainholc, B. (1999). *La interactividad en la educación a distancia.* Bs.Aires: Edit Paidos.

Fainholc, B. (2002). *La tecnología educativa apropiada y crítica.* Retrieved from www.cediproe.org.ar

Fainholc, B. (2007). *Usos y abusos de la Web 2.0.* Retrieved from www.cediproe.org.ar/Historial

Fainholc, B. (2008, July 3-9). Nuevas tecnologías de la información y la comunicación-TICs-y la mujer en la sociedad de la información. Mundos de Mujeres 2008, Perspectivas sobre Tecnología y la Construcción de Género, Madrid, España.

Fausto, F. (2006). O impacto das advertências contra o cigarro: Um estudo intercultural. In S. Zyngier, V. Viana & A. M. Spallanzani (Eds.), *Linguagens e tecnologias: Estudos empíricos* (pp. 67-81). Rio de Janeiro: Publit.

Fedorova, Y., Ivanyuk, L., Korolchuk, V., & Yemets, N. (2006). *The catchers in the rhyme.* Kyiv: Lenvit.

Fink, L. D. (2003). *Creating significant learning experiences.* San Francisco: Jossey-Bass.

Fisher, W. W., & McGeveran, W. (2006). *The digital learning challenge: Obstacles to educational uses of copyrighted material in the digital age.* Harvard: The Berkman Center for Internet & Society. Retrieved on December 31, 2007, from http://cyber.law.harvard.edu/sites/cyber.law.harvard.edu/files/BerkmanWhitePaper_08-10-2006.pdf

Fountain, R. (2005). *Wiki pedagogy.* Retrieved from http://www.profetic.org/dossiers/rubrique.php3?id_rubrique=110

Freire, P. (1970/1987). *Pedagogia do oprimido.* São Paulo: Editora Paz e Terra.

Freire, P. (1974). *Pedagogy of the oppressed* (M. B. Ramos, Trans.). NY: Seabury Press.

Frydenberg, M. (2008). Wikis as a tool for collaborative course management. *MERLOT Journal of Online Learning and Teaching, 4*(2). Retrieved from http://jolt.merlot.org/vol4no2/frydenberg0608.htm

Gabelnick, F., MacGregor, J., Matthews, R. S., & Smith, B. L. (1990). Learning community models. San Francisco, CA: Jossey-Bass.*New Directions for Teaching and Learning*, (41): 19–37.

Garnham, C., & Kaleta, R. (2002). Introduction to hybrid courses. *Teaching with technology today, 8*(6). Retrieved on October 22, 2008, from http://www.uwsa.edu/ttt/articles/garnham.htm

Garrison, D. R., & Anderson, T. (2003). *E-learning in the 21st century: A framework for research and practice.* London: Routledge Falmer.

Garrison, D. R., & Cleveland-Innes, M. (2005). Facilitating cognitive presence in online learning: Interaction is not enough. *American Journal of Distance Education, 19*(3), 133–148. doi:10.1207/s15389286ajde1903_2

Geer, R. (2001). The necessity of considering cultural influences in online collaborative learning. Retrieved on October 29, 2007, from http://gre-guns2.gre.ac.uk/pcet/PROGRAMMES/CeLTT/CeLTRes.nsf/0/720dbac82511cadd802569b100460f93/$FILE/greer.pdf

Gerrity, T. (1976). *College sponsored correspondence instruction in the United States: A comparative history of its origins (1873-1915) and its recent developments (1960-1975).* Unpublished doctoral dissertation, Teachers College, Columbia University, NY.

Gibson, C. C. (1998). The distance learner in context. In C. Gibson (Ed.), *Distance learners in higher education: Institutional responses for quality outcomes* (pp. 113-125). Madison, WI: Atwood Publishing.

Gimbert, B., Moore, M., & Sahin, F. (2008). *A comparison of nontraditionally and traditionally prepared teacher performance on a pedagogical content test in Ohio: The impact of blended preparation.* Paper presented at the Annual Meeting of the American Educational Research Association, New York.

Giordano, R., & Bell, D. (2000). Participant stakeholder evaluation as a design process. [Arlington, VA: ACM.]. *CUU, 00*, 53–60. doi:10.1145/355460.355472

GOI. (1999). National task force on information technology and software development: IT action plan part III–long term national IT policy. New Delhi: Government of India. Retrieved from http://it-taskforce.nic.in/actplan3/

GOI. (2009). *National mission on education through information and communication technology launched.* Press release. Press Information Bureau. Retrieved on February 25, 2009, from http://pib.nic.in/release/release.asp?relid=47221

Graham, C. R. (2006). Blended learning systems: Definition, current trends, and future directions. In C. J. Bonk & C. R. Graham (Eds.), *The handbook of blended learning: Global perspectives, local designs* (p. 5). San Francisco, CA: Pfeiffer.

Great Plains IDEA policy and procedure manual. (2003). Kansas State University on behalf of the Great Plains IDEA. Retrieved on November 25, 2008, from http://www.gpidea.org/alliance/ResourceCenter/P&PManual.pdf

Greenwald, S. (2008, September 3). *University of Missouri—Columbia campus visit, slide #32.* PowerPoint presented in 203 Whitten Hall on the University of Missouri, Columbia Campus.

Grijalva, T. C., Nowell, C., & Kirkvliet, J. (2006). Academic honesty and online courses. *College Student Journal, 40*(1), 180–185.

Grossman, P. (2005). Research on pedagogical approaches in teacher education. In M. Cochran-Smith & K. M. Zeichner (Eds.), *Studying teacher education: The report of the AERA panel on research and teacher education.* Mahwah, NJ: Lawrence Erlbaum Associates, Inc.

Grützner, I., Angkasaputra, N., & Pfahl, D. (2002). A systemic approach to produce small courseware modules for combined learning and knowledge management environments. [Ischia, Italy: ACM.]. *SEKE, 02*, 533–539. doi:10.1145/568760.568855

Guth, S. (2007). Wikis in education: Is public better? *Proceedings of the 2007 International Symposium on Wikis* (pp. 61-68). Montreal: Association for Computing Machinery.

Harris, B. A. (2003). Application of curriculum outcomes from an adult baccalaureate degree-completion program. *The Journal of Continuing Higher Education, 51*(2), 33–45.

Haughey, M., & Muirhead, B. (2003). *Evaluating learning objects for schools.* Retrieved on November 1, 2008, from http://www.usq.edu.au/electpub/e-jist/docs/vol8_no1/fullpapers/Haughey_Muirhead.pdf

Hawkes, M. (2000). The cost factor: When is interactive distance technology justifiable? *T.H.E. Journal, 28*(1), 26–32.

Hawkins, B., & Barone, C. (2003). Assessing information technology: Changing the conceptual framework. In P. A. McClure (Ed.), *Organizing and managing information resources on your campus* (pp. 129-145). San Francisco: Jossey-Bass.

Hawkins, B., & Rudy, J. (2007). Educause core data service 2007 summary report. *Educause Connect.* Retrieved on October 21, 2008, from http://connect.educause.edu/Library/Abstract/EDUCAUSECore-DataService20/47414

Hawryszkiewycz, I. T. (2005). *A framework for integrating learning into business processes.* Australian Computer Society, Inc.

Heidegger, G. (1992). Machines, computers, dialectics: A new look at human intelligence. *6*(1), 27-40. DOI 10.1007/BF02472767. Germany: Bremen University, Institute of Technology and Education.

Heinze, A., & Procter, C. (2004). Reflections on the use of blended learning. *Education in a Changing Environment Conference Proceedings.* Salford, UK: University of Salford. Retrieved from http://www.ece.salford.ac.uk/proceedings/papers/ah_04.rtf

Helps, C. R. G., & Renshaw, S. A. (2004). Design of a flexible case-study instructional module for operating systems for information technology. [Salt Lake City, UT: ACM.]. *SIGITE, 04,* 56–59.

Helps, R. (2007). Dancing on quicksand gracefully: Instructional design for rapidly evolving technology courses. [Destin, FL: ACM.]. *SIGITE, 07,* 1–7. doi:10.1145/1324302.1324304

Hermann, N., Popyack, J. L., Char, B., Zoski, P., Cera, C. D., Lass, R. N., & Nanjappa, A. (2003). Redesigning introductory computer programming using multi-level online modules for a mixed audience. [Reno, NV.]. *SIGCSE, 03,* 196–200. doi:10.1145/792548.611967

Higher Education Reauthorization Act. (HR 4137, Title IV, Part H, Sec. 495). (n.d.). Retrieved on November 17, 2008, from http://www.govtrack.us/congress/bill.xpd?bill=h110-4137

Hitchcock, L. (2007). Industry certification and academic degrees: Complementary or poles apart? *SIGMIS–CPR '07.* St. Louis, MO: ACM.

HLC Report. (2003). [*A Classroom of One: How Online Learning Is Changing Our Schools and Colleges.* New York: Palgrave Macmillan.]. *Gene,* I.

Holmes, Tangney, FitzGibbon, Savage, & Mehan. (2001). *Communal constructivism: Students constructing learning for as well as with others.* Retrieved from http://www.cs.tcd.ie/publications/t.2004

Holstein, E. (2003). Iowa's educational legacy and vision: Widening the portal for the baccalaureate degree. *The Journal of Continuing Higher Education, 51*(3), 35–40.

Hord, S. M. (1997). *Professional learning communities: Communities of continuous inquiry and improvement.* Austin, TX: Southwest Educational Development Laboratory.

Houle, C. (1973). *The external degree.* San Francisco: Jossey-Bass.

Howell, S. L., Williams, P. B., & Lindsay, N. K. (2003). Thirty-two trends affecting distance education: an informed foundation for strategic planning. Retrieved 1/5/2009 from the *Online Journal of Distance Learning Administration, V. VI, No. III, Fall 2003.*http://www.westga.edu/~distance/ojdla/fall63/howell63.htm.

Hsu, S. (1999). How to design a virtual classroom: 10 easy steps to follow. *T.H.E. Journal, 27*(2), 96–98.

Hura, G. (2008). Teaching behavioral-based skills online. *MERLOT Journal of Online Learning and Teaching, 4*(3).

Hybrid classes bloom in arts and science. (2006). Retrieved on October 23, 2008, from http://www.unm.edu/~market/cgi-bin/archives/001242.html

Hybrid courses: Maximizing resources and student learning. (2002). Retrieved on October 22, 2008, from http://courses.durhamtech.edu/tlc/www/html/Special_Feature/hybridclasses.htm

Hyde, P., Booth, R., & Wilson, P. (2003). The development of quality online assessment in VET. In H. Guthrie (Ed.), *Online learning: Research readings* (pp. 87-105). Leabrook, Australia: National Centre for Vocational Education Research.

Imel, S. (2002). *Adult learning in cohort groups. Practice application brief.* Columbus, OH: ERIC Clearinghouse on Adult, Career, and Vocational Education, Ohio State University. Retrieved from http://www.calpro-online.org/ERIC/docgen.asp?tbl=pab&ID=114

Institute for Higher Education Policy. (1994). Assessing distance learning from a public policy perspective. *Policy Steps, 3,* 7–9.

Iqbal, R., & Every, P. (2005). Scenario based method for teaching, learning, and assessment. [Newark, NJ: ACM.]. *SIGITE, 05,* 261–266. doi:10.1145/1095714.1095775

Irele, M. (2002). *Institutional mainstreaming of distance education: Guiding principles.* Unpublished doctoral dissertation, Pennsylvania State University.

Irele, M. E. (2005). Can distance education be mainstreamed? Retrieved 9/19/2008 from the *Online Journal of Distance Learning Administration, V. VIII, No. II, Summer 2005.* http://www.westga.edu/~distance/ojdla/summer82/irele82.htm.

J.E. Brindley. (Personal Communication, September 20, 2008).

Jasola, S., & Sharma, R. (2005). Open and distance education through wireless mobile Internet: A learning model. *International Journal of Instructional Technology and Distance Learning, 2*(9), 35-47. Retrieved from http://www.itdl.org/Journal/Sep_05/article04.htm

Jensen, D., Self, B., & Rhymer, D. (2002). A rocky journey toward effective assessment of visualization modules for learning enhancement in engineering mechanics. *Educational Technology & Society, 5*(3). Retrieved on March 11, 2009, from http://www.ifets.info/journals/5_3/jensen.html

Jonassen, D. (1991). Objectivism vs. constructivism: Do we need a new philosophical paradigm? *Educational Technology Research and Development, 39*(3), 5 14. doi:10.1007/BF02296434

Jonassen, D. (1994). *Operationalizing mental models: Strategies for assessing mental models to support meaningful learning and design–supportive learning environments.* Retrieved on March 20, 2009, from http://education.uncc.edu/jagretes/instructional%20design%20articles%20-%20&%20constructivist%20articles/operationalizing%20mental%20models.pdf

Kawachi, P. (2005). Computers, multimedia, and e-learning. In U. V. Reddi & S. Mishra (Eds.), *Perspectives on distance education: Educational media in Asia* (pp. 97–122). Vancouver: Commonwealth of Learning.

Kearns, P., McDonald, R., Candy, P., Knights, S., & Papadopoulos, G. (1999). *VET in the learning age: The challenge of lifelong learning for all.* Canberra, Australia: National Centre for Vocational Education and Training.

Kemp, W. C. (2002). Persistence of adult learners in distance education. *American Journal of Distance Education, 16*(2), 65–81. doi:10.1207/S15389286AJDE1602_2

Kilpatrick, S., Barrett, M., & Jones, T. (2003). *Defining learning communities.* (Discussion Paper D1/2003. CRLRA Discussion Paper Series). Launceston, Australia: University of Tasmania. Retrieved from http://www.aare.edu.au/03pap/jon03441.pdf

King, C. G., Guyette, R. W., & Piotrowski, C. (2009). Online exams and cheating: An empirical analysis of business students' views. *The Journal of Educators Online, 6*(1), 1–11.

Kirschner, P. A., Sweller, J., & Clark, R. E. (2006). Why minimal guidance during instruction does not work: An analysis of the failure of constructivist, discovery, problem-based, experiential, and inquiry based teaching. *Educational Psychologist, 41*(2), 75–86. doi:10.1207/s15326985ep4102_1

Kleiber, P., & Zhu, I. (1995). Planning for curriculum development in independent study. *The Journal of Continuing Higher Education, 43*(3), 25–33.

Knowles, M. S. (1970). *The modern practice of adult education: Andragogy versus pedagogy.* NY: Cambridge Books.

Kock, N., Chatelain-Jardon, R., & Carmona, J. (2008). An experimental study of simulated Web-based threats and their impact on knowledge communication effectiveness. *IEEE Transactions on Professional Communication, 51*(2), 183–197. doi:10.1109/TPC.2008.2000345

Koku, E. F., & Wellman, B. (2002). Scholarly networks as learning communities: The case of the TechNet. In S. Barab & R. Kling (Eds.), *Designing virtual communities in the service of learning* (pp. 299-337). Cambridge: Cambridge University Press.

Kolb, A., & Kolb, D. A. (2001). *Experiential learning theory bibliography 1971-2001.* Boston: McBer and Co.

Koszalka, T. A., & Ganesan, R. (2004). Designing online courses: A taxonomy to guide strategic use of features available in course management systems (CMS) in distance education. *Distance Education, 25*(2), 243–256. doi:10.1080/0158791042000262111

Krathwohl, D. R., & Payne, D. A. (1971). Defining and assessing educational objective. In R. L. Thorndike (Ed.), *Educational measurement* (pp. 17-45). Washington, D.C.: American Council on Education.

Lamb, B. (2002). Open spaces: Wikis, ready or not. *EDUCAUSE Review, 39*(5), 36-48. Retrieved from http://www.educause.edu/pub/er/erm0

Lamb, B. (2004). Wide open spaces wikis ready or not. *EDUCAUSE.* Retrieved from net.educause.edu/ir/library/pdf/ERM0452.pdf

Lanier, M. M. (2006). Academic integrity and distance learning. *Journal of Criminal Justice Education, 17*(2), 244–261. doi:10.1080/10511250600866166

LaPadula, M. (2003). A comprehensive look at online student support services for distance learners. [from the EBSCO database.]. *American Journal of Distance Education, 17*(2), 119–128. Retrieved on September 5, 2008. doi:10.1207/S15389286AJDE1702_4

Leary, J., & Berge, Z. L. (2007). Challenges and strategies for sustaining eLearning in small organizations. Retrieved 9/19/2008 from the *Online Journal of Distance Learning Administration, V. X, No. III, Fall 2007.* http://www.westga.edu/~distance/ojdla/fall103/berge103.htm.

Lee, K., & Lee, J. (2004). Benefits of a virtual learning environment in enabling collaborative and constructivist learning in a community of specialist nursing practice. *Proceedings of the IEEE International Conference on Advanced Learning Technologies (ICALT '04).* IEEE.

Lewis, B. A. (2002). Learning effectiveness: Efficacy of quizzes vs. discussions in online learning. *Dissertation International Abstracts, 63*(3), 911A.

Lightfoot, K. (2006). Using Bloom's taxonomy to develop learning outcomes and assessment tools. In *Step-by-step guide to an ACE evaluation, 2006* (pp. 12-14). Trenton, NJ: American Council on Education College Credit Recommendation Service.

Lightfoot, K. (2007, November). *Course development models and the role of instructional designers and course developers. How do they go about it and who does what?* Paper presented at the Annual Conference of the American Association for Collegiate Independent Study, Ogden, UT.

Little, C. B., Titarenko, L., & Bergelson, M. (2005). Creating a Successful International Distance-Learning Classroom. [from http://www.jstor.org/]. *Teaching Sociology, 33*(4), 355–370. Retrieved October 18, 2008.

Liu, G. Z. (2008). Innovating research topics in learning technology: Where are the new blue oceans? *British*

Journal of Educational Technology, 39(4), 738–747. doi:10.1111/j.1467-8535.2008.00851.x

Lorenzetti, J. (2002). Toward an expanded cost-benefit analysis for distance programs. *Distance Education Report, 6*(24), 1–6.

Lovett, C. (1996). How to start restructuring our colleges. *Planning for Higher Education*, 19.

Macdonald, J., & Twining, P. (2002). Assessing activity-based learning for a networked course. *British Journal of Educational Technology, 33*(5), 603–618. doi:10.1111/1467-8535.00295

Mackenzie, O., & Christensen, E. (1971). *The changing world of correspondence study: International readings.* University Park, PA: Penn State.

Mager, R. (1975). *Preparing instructional objectives, 2nd ed.* Belmont, CA: Fearon-Pitman Publishers.

Magjuka, R. J., Shi, M., & Bonk, C. J. (2005). Critical design and administrative issues in online education. Retrieved 9/19/2008 from the *Online Journal of Distance Learning Administration, V. VIII, No. IV, Winter 2005.* http://www.westga.edu/~distance/ojdla/winter84/magjuka84.htm.

Maimon, E. P. (1997). Teaching across the curriculum. In J. G. Gaff, J. L. Ratcliff & Associates (Eds.), *Handbook of the undergraduate curriculum* (pp. 377-392). San Francisco: Jossey-Bass.

Mallory, H. (1915). Discussion. *Proceedings of the First National University Extension Conference.* Madison: University of Wisconsin Press.

McClelland, R. J. (2002). Evolving Web-based delivery: Managing the transition from VLE's designed by academics to commercial VLE's, the post-graduate experience. *Proceedings of the 28th Euromicro Conference*. IEEE.

McGrath, L. (2006). Developing eLearning policies at the departmental level. *MERLOT Journal of Online Learning and Teaching. 2:3 pp. 177-186.*

McGreal, R. (2004). *Learning object granularity. Learning objects: A practical definition.* Retrieved on November 1, 2008, from http://itdl.org/Journal/Sep_04/article02.htm

McNabb, L. (2008). Chronicle article focuses on distance-ed academic integrity. Retrieved on October 18, 2008, from Listserv: Sloan-C.

McNutt, L., & Brennan, M. (2005). Work in progress—learning styles and e-learning, what is the connection? *35th ASEE/IEEE Frontiers in Education Conference* (pp. F1H-27-F1H-31). IEEE.

McRae, J. (1997). *Literature with a small "l."* London: MacMillan Publishers.

McVay Lynch, M. (1998). *Constructivism in instructional design for distance education.* Retrieved on March 11, 2009, from http://web.pdx.edu/~mmlynch/constructivist.html

Merriam, S. B. (1998). *Qualitative research and case study applications in education.* San Francisco: Jossey-Bass.

Merriam-Webster Inc. (2003). *Merriam-Webster's collegiate dictionary, 11th ed.* Springfield, MA: Britannica.

Miller Brown, S. (2002). Strategies that contribute to nontraditional/adult student development and persistence. *PAACE Journal of Lifelong Learning, 11*, 67–76.

Miller, N. E. (1941). The frustration-aggression hypothesis. *Psychological Review, 48*, 337–342. doi:10.1037/h0055861

Mishra, S. (1999). An empirical analysis of interactivity in teleconference. *Indian Journal of Open Learning, 8*(3), 243–253.

Mishra, S., & Sharma, R. C. (2005, March 14-20). Development of e-learning in India. *University News, 43*(11), 9–15.

Moneo Rodriguez, M. (1999). *Conocimiento previo y cambio conceptual.* Buenos Aires: Editorial Aique.

Monolescu, D., Schifter, C., & Greenwood, L. (2004). *The Distance Education Evolution: Issues and Case Studies.* Hershey, PA: Information Science Publishing.

Moore, M. (2000). *Speaking Personally about Distance Education*. University Park, Pennsylvania: The American Center for the Study of Distance Education.

Moore, M. G. (1972). Learner autonomy: The second dimension of independent learning. *Convergence*, *5*(2), 76–88.

Moore, M. G. (1989). Editorial: Three types of interaction. *American Journal of Distance Education, 3*(2), 1-7. Retrieved from http://www.ajde.com/Contents/vol3_2.htm#editorial

Moore, M. G., & Kearsley, G. (1996). *Distance education: A systems view*. New York: Wadsworth Publishing Company.

Moxley, V. M. (2005, February). *Inter-institutional academic alliances-when, why, who, and how*. Paper presented at the Twenty-Second Annual Conference on Academic Chairpersons: The Chair's Role in Empowering Change, Orlando, FL.

Moxley, V., & Maes, S. (2003). The Great Plains Interactive Distance Education Alliance. *Continuing Higher . Educational Review*, *67*, 141–154.

Moxley, V., & Maes, S. (2007). The Great Plains IDEA story. In S. Johnstone (Ed.), *Advancing campus efficiencies: A companion for campus leaders in the digital era* (pp. 161-165). San Francisco: Jossey-Bass.

Murphy, D., Casey, D., & Fraser, J. (2007). Talk about walkabout: Evaluation of a flexible learning initiative. In P. Tsang, R. Kwan, & R. Fox (Eds.), *Enhancing learning through technology* (pp. 11-26). Hackensack, NJ: World Scientific Publishing Company.

Murphy, K., Mahoney, S., & Harvell, T. (2000). Role of contracts in enhancing community building in Web courses. Educational Technology & Society, 3(3).

Murry, J. W. Jr, & Hall, B. L. (1998). A systematic approach to designing and evaluating effective adult degree-completion programs. *The Journal of Continuing Higher Education*, *46*(2), 19–26.

National Council of Teacher Quality. (2005). *Increasing the odds: How good policies can yield better teachers.* Washington, D.C.: NCTQ. Retrieved from http://www.nctq.org/p/publications/docs/nctq_io_20071129024229.pdf

National Survey of Student Engagement. (2008). *Promoting engagement for all students: The imperative to look within*. Bloomington, IN: Indiana University Center for Postsecondary Research.

NCA Higher Learning Commission. (2000). *Best practices for electronically offered degree and certificate programs.* Retrieved on March 23, 2009, from http://www.ncahlc.org/download/Best_Pract_DEd.pdf

Novak, & Gowlan. (1984). The concept map as an advance organizer. *Journal of Research in Science Teaching, 28*(8), 705-711.

Nulden, U. (1999). Thematic modules in asynchronous learning network: Designing introductory courses. Proceedings of the 32nd Hawaii International Conference on System Sciences (pp. 1-10). IEEE.

Olcott, D. (1991). Bridging the gap: Distance learning and academic policy. *Continuing Higher Education Review*, *55*(1 & 2), 49–60.

Olcott, D. (1992). Policy issues in statewide delivery of university programs by telecommunications. *American Journal of Distance Education*, *6*(1), 14–26. doi:10.1080/08923649209526771

Opper, J. (2002). Funding and cost containment of educational technology: Shifting policy and practices. Retrieved in September 2002, from http://www.wcet.info/projects/tcm/whitepaper2.pdf

Palloff, R. M., & Pratt, K. (1999). *Building learning communities in cyberspace: Effective strategies for the online classroom.* San Francisco: Jossey-Bass.

Palmer, R. R. Colton, Joel & Kramer, Lloyd (2007). *A history of the modern world since 1815*, 10th ed. McGraw-Hill.

Palmer, R. R., Colton, J., & Kramer, L. (2007). *A history of the modern world to 1815, 10th ed.* McGraw-Hill.

Paolucci, R., & Gambescia, S. F. (2007). Current administrative structures used for online degree program offerings in higher education. Retrieved 9/19/2008 from the *Online Journal of Distance Learning Administration, V. X, No. III, Fall 2007.* http://www.westga.edu/~distance/ojdla/fall103/gambescia103.htm.

Pappas, J. P., & Jerman, J. (Eds.). (2004). *Developing and delivering adult degree programs. New directions for adult and continuing education* (No. 103). San Francisco: Jossey-Bass.

Pawan, F., Paulus, T. M., Yalcin, S., & Chang, C. (2003). Online learning: Patterns of engagement and interaction among in-service teachers. *Language Learning & Technology*, *7*(3), 119–140.

Perkins, D. (1995). *La escuela inteligente*. Bs Aires: Edit Paidós.

Petersen, R., & Gearhart, D. (2004, November). *Bridging the digital divide through independent study.* Paper presented at the Annual Meeting of the American Association for Collegiate Independent Study, Bellingham, WA. Retrieved on March 23, 2009, from http://www.aacis.org/pages/conferences/2004conf.htm

Picciano, A. G. (2001). *Distance learning making connections across virtual space and time.* Upper Saddle River, NJ: Merrill Prentice Hall.

Pisel, K. P. (2008). A strategic planning process model for distance education. Retrieved 9/19/2008 from the *Online Journal of Distance Learning Administration, V. XI, No. II, Summer 2008.* http://www.westga.edu/~distance/ojdla/summer112/pisel112.htm

Pittman, V. (1992). Evaluation and independent study. *Adult Assessment Forum*, *11*, 6–13.

Pittman, V. (1998). Low-key leadership: Collegiate correspondence study and "campus equivalence." . *American Journal of Distance Education*, *12*(2), 36–44. doi:10.1080/08923649809526993

Pittman, V. (2001). Distance education's scarlet letters. *The Journal of Continuing Higher Education*, *49*(3), 42–43.

Plaisant, C., Rose, A., Rubloff, G., Salter, R., & Schneiderman, B. (1999). The design of history mechanisms and their use in collaborative educational simulations. [Palo Alto, CA.]. *Proceedings of the Computer Support for Collaborative Learning, CSCL*, *99*, 348–359.

Poe, D., McGowan, K., Hansen, C., & Singh, G. (2008). Refining educational content through a closed-loop FLOW approach. *Inroads— . SIGCSE Bulletin*, *10*(2), 102–106. doi:10.1145/1383602.1383645

Popham, W. J. (2000). *Modern educational measurement: Practical guidelines for educational leaders, 3rd ed.* Boston, MA: Allyn and Bacon.

Popham, W. J. (Ed.). (1971). *Criterion-referenced measurement: An introduction.* Englewood Cliffs, NJ: Educational Technology Publications.

Prensky, M. (2000). *Digital game-based learning.* NY: McGraw-Hill.

Quality Matters. (2008). *Research literature and standards set support for quality matters review standards.* Retrieved on November 28, 2008, from http://www.qualitymatters.org

Reay, J. (2001). Blended learning–a fusion for the future. *Knowledge Management Review*, *4*(3), 6.

Reigle, R. R. (2007). *The online bully in higher education.* Educational Resources Information Center (ERIC). Retrieved on April 10, 2009, from http://www.eric.ed.gov/ERICWebportal/detail?accno=ED495686

Rekkedal, T. (1983). The written assignments in correspondence education. Effects of reducing turn-around time. *Distance Education*, *4*, 231–250. doi:10.1080/0158791830040211

Rohfeld, R. (1996). Access and the search for academic respectability in continuing higher education. *Continuing Higher Education Review*, *6*, 54–66.

Rooney, J. E. (2003). Blending learning opportunities to enhance educational programming and meetings. *Association Management*, *55*(5), 26–32.

Rose, C. (1991). *Accelerated Learning.* Aylesbury, Bucks.: Accelerated Learning Systems Ltd.

Rovai, A. P., & Jordan, H. M. (2004). Blended learning and sense of community: A comparative analysis with traditionally and fully online graduate course. *International Review of Research in Open and Distance learning, 5*(2), 1-12. Retrieved from http://www.irrodl.org/index.php/irrodl/article/view/192/274

Rumble, G. (1999). Cost analysis of distance learning. *Performance Improvement Quarterly, 12*(2), 122–137.

Russel, T. (2001). *IDECC*. Retrieved from www.nosignificantdifference.org

Russell, T. L. (2001). *The no significant difference phenomenon, 5th ed.* Montgomery, AL: International Distance Education Certification Center.

Sands, P. (2002). Inside outside, upside downside: Strategies for connecting online and face-to-face instruction in hybrid courses. *Teaching with Technology Today, 8*(6). Retrieved on October 22, 2008, from http://www.uwsa.edu/ttt/articles/sands2.htm

Schauer, J., Rockwell, S. K., Fritz, S. M., & Marx, D. B. (2005). Implementing distance education: issues impacting administration. Retrieved 9/19/2008 from the *Online Journal of Distance Learning Administration, V. VIII, No. III, Fall 2005.* http://www.westga.edu/~distance/ojdla/fall83/schauer83.htm

Schlossberg, N. K., Waters, E. B., & Goodman, J. (1995). *Counseling adults in transition, 2nd ed.* New York: Springer.

Schmidt, S. (1982). *Foundation for the empirical study of literature: The components of a basic theory.* Hamburg: Helmut Buske.

Schön, D. (1983). *The reflective practitioner.* NY: Basic Books.

Schram, D., & Steen, G. (Eds.). (2001). *The psychology and sociology of literature: In honor of Elrud Ibsch.* Amsterdam & Philadelphia: John Benjamins.

Schrum, L. (1996, March). Teaching at a distance: Strategies for successful planning and development. *Learning and Leading with Technology, 23*(6), 30–33.

Selman, B., Keely, J., Washington, C., & Nagel, T. (2008, November). *myZou makeover at MU: Taking advantage of a student information system change.* Paper presented at the Annual Meeting of the American Association for Collegiate Independent Study, St. Louis, MO. Retrieved on November 17, 2008, from http://www.aacis.org/pages/conferences/2008conf.htm

Sharma, R. (2006). EDUSAT: Taking Indian distance education to new frontiers. *Distance Learning Link*, January-June, 6-8. USA: Information Science Publishing.

Sharma, R. C. (2002). Interactive radio counseling in distance education. *University News, 40*(10), 8–11.

Sharma, R. C. (2002). Gyan Vani: The educational FM radio network of India. *Field Notes, 2*(2). News from Asia. Retrieved from http://www.irrodl.org/content/v2.2/field.html

Shea, P., Swan, K., Fredericksen, E., & Pickett, A. (2002). Student satisfaction and reported learning in the SUNY learning network. In J. Bourne & J. Moore (Eds.), *Elements of quality online education* (pp.145-156). Needham, MA: Sloan Center for OnLine Education.

Sherry, L. (1996). Issues in distance learning. *International Journal of Educational Telecommunications, 1*(4), 337–365.

Siemmens, G. (2004). *A connectivist digital learning.* Creative Commons. Retrieved from http://d.scribd.com/docs/1yhhhthpoaervbohwzkc.pdf

Simonson, M., Smaldiino, S., Albright, M., & Zvacek, S. (2000). *Teaching and learning at a distance: Foundations of distance education.* Upper Saddle River, NJ: Prentice Hall.

Sims, R., Dobbs, G., & Hand, T. (2002). Enhancing quality in online learning: Scaffolding planning and design through proactive evaluation. *Distance Education, 23*(2), 135–148. doi:10.1080/0158791022000009169

Singh, M. (2008). *Foreword to eleventh five year plan 2007-12, volume I: Inclusive growth* (p. iv). Retrieved on December 9, 2008, from http://planningcommission.nic.in/plans/planrel/fiveyr/11th/11_v1/11v1_foreward.pdf

Smith, E. (2004, November). *A difference in kind or a difference in degree? Developing a distance degree-completion program.* Paper presented at the Annual Meeting of the American Association for Collegiate Independent Study, Bellingham, WA. Retrieved on March 23, 2009, from http://www.aacis.org/pages/conferences/2004conf.htm

Sosniak, L. A. (1994). The taxonomy, curriculum, and their relations. In L. W. Anderson & L. A. Sosniak (Eds.), *Bloom's taxonomy: A forty-year retrospective, ninety-third year-book of the national society for the study of education* (pp. 103-125). Chicago: University of Chicago Press.

Soules, M. (n.d.). *Hybrid online courses & strategies for collaboration.* Retrieved on October 22, 2008, from http://records.viu.ca/~soules/hybrid.htm

Standridge, C. R. (2000). Teaching simulation using case studies. *Proceedings of the 2000 Winter Simulation Conference* (pp. 1630–1634).

State of Ohio. (2003, January 1). *Teacher education and licensure standards.* Columbus, OH: State Board of Education, Ohio Department of Education.

Stewart, I., Hong, E., & Strudler, N. (2004). Development and validation of an instrument for student evaluation of the quality of Web-based instruction. *American Journal of Distance Education, 18*(3), 131–150. doi:10.1207/s15389286ajde1803_2

Stone Wiske, M. (2006). *La enseñanza para la comprensión con nuevas tecnologias.* Buenos Aires: Paidós.

Sullivan, E., Stewart, D. W., & Spille, H. A. (1997). *External degrees in the information age: Legitimate choices.* The American Council on Education and the Oryx Press.

Swan, K., & Shea, P. (2005). The development of virtual learning communities. In S. R. Hiltz & R. Goldman (Eds.), *Learning together online: Research on asynchronous learning networks* (pp. 239-260). Mahwah, NJ: Erlbaum.

Tallman, F. (1994). Satisfaction and completion in correspondence study: The influence of instructional and student-support services. *American Journal of Distance Education, 8*(2), 43–57. doi:10.1080/08923649409526854

Taylor, J. A. (2000, June 22). *Adult degree-completion programs: A report to the NCA Board of Trustees from the task force on adult degree-completion programs and the award of credit for prior learning at the baccalaureate level.* ED 449847. Retrieved on March 23, 2009, from http://www.eric.ed.gov/ERICDocs/data/ericdocs2sql/content_storage_01/0000019b/80/16/d5/7d.pdf

Teles, C. (2006). Identificação, envolvimento, e afeto: A influência das legendas na apreciação de filmes estrangeiros. In S. Zyngier, V. Viana & A. M. Spallanzani (Eds), *Linguagens e tecnologias: Estudos empíricos* (pp. 223-242). Rio de Janeiro: Publit.

The Florida Community College Distance Learning Consortium. (2004). Retrieved in October 2004, from http://www.distancelearn.org/mainPage.cfm

The Sloan Consortium. (2004). Entering the mainstream: The quality and extent of online education in the United States, 2003 and 2004. Retrieved in November 2004, from http://www.sloan-c.org/resources/survey.asp

Thurmond, V., Wambach, K., Connors, H., & Frey, B. (2002). Evaluation of student satisfaction: Determining the impact of a Web-based environment by controlling for student characteristics. *American Journal of Distance Education, 16*(3), 169–189. doi:10.1207/S15389286AJDE1603_4

Trigano, P. C., & Pacurar-Giacomino, E. (2004). Toward a Web-based environment for evaluation and design of pedagogical hypermedia. *Educational Technology & Society, 7*(3), 21–37.

Troy University Mission Statement. (2008). Retrieved March 16, 2009 from http://www.troy.edu/mission.htm.

Truluck, J. E. (2003). Questions and turning points when designing and implementing distance education degree programs. *Proceedings of the 23rd Annual Adult Higher Education Alliance Conference* (pp. 325-327), Asheville,

NC, October. Retrieved on March 23, 2009, from http://www.ahea.org/conference/proceedings/2003.pdf

Turner, C. G., & Monk-Turner, E. (2001). The returns to education and degrees. *Educational Research Quarterly, 26*(3), 45–53.

U.S. Census Bureau. (2000). *Americanfactfinder. DP-2. Profile of selected social characteristics: 2000*. Geographic area: Missouri. Retrieved on November 13, 2008, from http://censtats.census.gov/data/MO/04029.pdf

U.S. Department of Education, Office of Innovation and Improvement. (2004). *Alternative routes to teacher certification*. Jessup, MD: Education Publication Center. Retrieved on December 12, 2007, from http://www.ed.gov/admins/tchrqual/recruit/altroutes/report.pdf

Van Peer, W., Hakemulder, J., & Zyngier, S. (2007). *Muses and measures: Empirical research methods for the Humanities*. Newcastle: Cambridge Scholars Publishing.

Verleger, M., Diefes-Dux, H., Rickus, J., & Schaffer, S. (2005). An online bioengineering learning module: Gene regulation. *35th ASEE/IEEE Frontiers in Education Conference* (pp. SIG 20-SIG-25). Indianapolis, IN.

Vyas, R. V., Sharma, R. C., & Kumar, A. (2002a). Educational radio in India. *Turkish Online Journal of Distance Education, 3*(3). Retrieved from http://tojde.anadolu.edu.tr/tojde7/articles/educationalradio.htm

Vyas, R. V., Sharma, R. C., & Kumar, A. (2002b). Educational television in India. *Turkish Online Journal of Distance Education, 3*(4). Retrieved from http://tojde.anadolu.edu.tr/tojde8/articles/educationaltv.htm

Wagner, E. D. (1994). In support of a functional definition of interaction. *American Journal of Distance Education, 8*(2), 6–29. doi:10.1080/08923649409526852

Wallace, Raven McCrory (2004). A Framework for Understanding Teaching with the Internet. *American Educational Research Journal* Vol.41, No.2, 447-488. Retrieved October 18, 2008, from http://www.jstor.org/

Ward, J., & LaBranche, G. A. (2003). Blended learning: The convergence of e-learning and meetings. *Franchising World, 35*(4), 22-23. Retrieved from http://www.accessmylibrary.com/coms2/summary_0286-23358935_ITM

Warnock, S. (2006). Awesome job!–or was it? The 'many eyes' of asynchronous writing environments and the implications of plagiarism. *Plagiary, 1*, 178–190.

Watkins, B. (1991). A quite radical idea: The invention and elaboration of collegiate correspondence study. In B. Watkins & S. Wright (Eds.), *Foundations of American distance education: A century of collegiate correspondence study*. Dubuque, IA: Kendall/Hunt.

Wedemeyer, C. (1963). *Brandenburg memorial essays on correspondence study, vol. I*. Madison, WI: University of Wisconsin-Extension.

Wedemeyer, C. (1966). *Brandenburg memorial essays on correspondence study, vol. II*. Madison, WI: University of Wisconsin-Extension.

Wedemeyer, C. (1977). Independent study. In A. S. Knowles (Ed.), *The international encyclopedia of higher education* (Vol. 5, pp. 2114-2132). San Francisco: Jossey-Bass Publication.

Wedemeyer, C. (1981). *Learning at the back door: Reflections on non-traditional learning in the lifespan*. Madison, WI: University of Wisconsin Press.

Wedemeyer, C., & Childs, G. (1961). *New perspectives in university correspondence study*. Chicago, IL: The Center for the Study of Liberal Education for Adults.

Wedemeyer, C., & Najem, R. (1969). *AIM: From concept to reality: The articulated instructional media program at Wisconsin*. NY: Syracuse University Press.

Wenger, E. (1997). *Communities of practice: Learning, meaning, and identity*. NY: Cambridge University Press.

Wheeler, B. C., Valacich, J. S., Alavi, M., & Vogel, D. (1995). A framework for technology-mediated inter-institutional telelearning relationships. *Journal of Computer-Mediated Education, 1*(1). Retrieved on December 2, 2008, from http://jcmc.indiana.edu/vol1/issue1/wheeler/essay.html

Wills, B. (1993). *Distance education: A practical guide*. Educational Technology Publications.

Wiseman, M. (2002). The empirical study of literature in the 21st century. In S. Zyngier & A. C. F. Valente (Eds.), *Fatos & ficções: Estudos empíricos de literatura* (pp. 11-16). Rio de Janeiro: Setor de Publicações da Faculdade de Letras da UFRJ.

Wolf, M. A. (Ed.). (2005). *Adulthood: New terrain. New directions for adult and continuing education* (No. 108). San Francisco: Jossey-Bass.

Woolfolk, A. (2005). *Educational psychology: Active learning edition, 9th ed.* NY: Allyn Bacon.

Wright, B. D. (1997). Evaluating learning in individual courses. In J. G. Gaff, J. L. Ratcliff & Associates (Eds.), *Handbook of the undergraduate curriculum* (pp. 571-590). San Francisco: Jossey-Bass.

Yarnit, M. (2000). *Towns, cities, and regions in the learning age: A survey of learning communities*. London: LGA Publications.

Young, J. R. (2002). Hybrid teaching seeks to end the divide between traditional and online instruction. *The Chronicle of Higher Education*. Retrieved on October 22, 2008, from http://chronicle.com/free/v48/i28/28a03301.htm

Youngblood, P., Trede, F., & DeCorpo, S. (2001). Facilitating online learning: A descriptive study. *Distance Education, 22*(2), 264–284. doi:10.1080/0158791010220206

Zalenski, A. (2000, November). *Graduation, persistence, and attrition from distance learning programs: An analysis of selected variables.* Paper presented at the Annual Meeting of the American Association for Collegiate Independent Study, Austin, TX.

Zhu, L., Wong, J., Sturz, B., Shi, Y., & Tang, C. (2008). An integrated evaluation method for module-based undergraduate information retrieval education. Consortium for computing sciences in colleges. *JCSC, 23*(6), 148–156.

Zyngier, S. (1994a). *At the crossroads of language and literature: Literary awareness, stylistics, and the teaching of EFLit*. Unpublished doctoral dissertation, University of Birmingham.

Zyngier, S. (1994b). Introducing literary awareness. *Language Awareness, 3*, 95–108.

Zyngier, S., Chesnokova, A., & Viana, V. (2007). Literary awareness: Connecting Brazil and Ukraine. *UNESCO Chair Messenger: Philology, Pedagogics . Psychology (Savannah, Ga.), 15*, 125–129.

About the Contributors

Deb Gearhart is the director of eCampus for Troy University. Twenty one of Troy University's degree programs are offered online. Dr. Gearhart has worked in the field of distance education for 23 years. Previously Deb served as the founding Director of E-Education Services at Dakota State University in Madison, South Dakota and was there for the 11 years. Before joining Dakota State she spent 10 years with the Department of Distance Education at Penn State. Deb was an associate professor for educational technology at Dakota State University teaching at both the undergraduate and graduate levels. She has co-authored at textbook entitled Designing and Developing Web-Based Instruction and is editing another publication titled Cases on Distance Delivery and Learning Outcomes: Emerging Trends and Programs, due out in 2009. Dr. Gearhart has earned a BA in Sociology from Indian University of Pennsylvania. She earned a M.Ed. in Adult Education with a distance education emphasis and an M.P.A. in Public Administration, both from Penn State. Deb completed her Ph.D. program in Education, with a certificate in distance education, from Capella University.

* * *

Dawn Anderson is the Senior Coordinator of the Kansas State University Institute for Academic Alliances. She specializes in collaborative academic program development and implementation, project management, consortia management, grant writing and academic market research. Anderson has managed several U.S. Department of Education Fund for the Improvement of Postsecondary Education (FIPSE) projects. She graduated from Iowa State University with a B.S. in Family Services and Consumer Sciences, and an M.S. in Family Studies (Gerontology).

Leah Blakey, Ph.D. is an Assistant Professor of History and Political Science at Drury University. She has pursued an academic career involving international business, government work and pedagogy, culminating in a doctorate from Saint Louis University. Her interest in teaching methodology continued at Drury through online and blended courses. She was named Educator of the Year in 2007, and now oversees training and assessment of the University's blended courses as Blended Courses Coordinator. In addition, Leah continues to teach in both traditional and non-traditional formats. Through her ongoing interest in distance education and pedagogy, she has spoken at conferences around the United States.

Gera Burton is a graduate of both the National University of Ireland, Dublin, and the University of Missouri, where she earned a Ph.D in Latin American and Afro-Hispanic Literature. As Associate Director for the University of Missouri's Center for Distance and Independent Study, she oversees curriculum development for more than 20,000 students from 44 countries. She is a former President of the American Association for Collegiate Independent Study (AACIS) and serves on the Executive Boards of the International Society for Educational Biography (ISEB) and the Society for Irish Latin American Studies (SILAS). She is a Peer Evaluator for the Higher Learning Commission's North Central Accrediting Association. An experienced researcher, she serves on UCEA's Wedemeyer and Powell Committees.

Beatriz Fainholc

- Degree of Profesor in Educational Sciences by Nacional University of Buenos Aires in 1996
- Master in Social Sciences by University of Sao Paulo, Brasil, 1970
- Doctorate in Education Sciences en La Plata University Argentina
- University Professor (face to face and virtual) in many Argentinean and foreigner Universities.
- Asesor of the Ministery of Education, Argentina from 1989 to 1992.
- General Director of CEDIPROE Foundation- Design, Production, Evaluation and Research of Multimedia Materials Cnter, Buenos Aires, Argentina.
- She won academic awards from the Canadian Government for her proposals in "Approppriate Educational Technology" framework.
- Consultant of nacional and international organizations and universities.

Belinda Gimbert is an assistant professor, Educational Administration in the School of Educational Policy and Leadership, Educational Administration, The Ohio State University, Columbus, Ohio. Prior to attaining an M.S in Education at Mansfield University, Pennsylvania, Dr Gimbert taught mathematics and computing science for fifteen years in secondary schools (Grades 7-12) in rural New South Wales, Australia. She is the Principal Investigator for Project KNOTtT (2007-2012), a consortium of colleges/ universities and school districts in Kansas, Nevada, Ohio, and Texas that will transition 545 participants to teaching, while addressing the national issue of teacher quality. She also was Co-Director for OhioTtT (2007-08) and Project Director for the Newport News Public Schools and Old Dominion University Transition To Teaching Project (2002-2008). Dr Gimbert's scholarship and research interest include: recruitment and retention of teachers and administrators in high need schools, teacher development; teachers' professional development and work lives, specifically the impact of pay for performance on teacher's sense of efficacy and on student achievement; school-university partnership work; school leadership (with an emphasis on urban and rural contexts).

Shalin Hai-Jew works as an instructional designer at Kansas State University (K-State). She has co-designed courses and trainings in numerous fields. She writes as Eruditio Loginquitas for the Instructional Design Open Studio (IDOS) blog and the E-Learning and Teaching Exchange (ELATE) wiki. She has worked for WashingtonOnline as an instructor for many years. She was a tenured faculty member (in English and Communications) at Shoreline Community College in Washington State. She works as a reviewer for *Educause Quarterly* and *MERLOT's Journal of Online Learning and Teaching (JOLT)*, both by invitation. She is interested in immersive learning, digital imagery in e-learning, simulations, socio-technical systems, and all facets of instructional design.

David Kendrick is the Director for the Center for the Enhancement of Teaching and Learning at the University of Northern Colorado. He has served as faculty, administrator, and consultant in higher education, k-12, and industry technology learning environments. His current interest is in cost benefits in traditional to e-learning transformations. He received his B.A. and M.A. in English at Colorado State University and his M.S. and Ed.D. in Educational Technology and Educational Leadership - Higher Education Administration at Texas A&M University-Corpus Christi.

Kenneth Lightfoot is an Instructional Designer at Thomas Edison State College. He has several years experience in the field of education in the UK, the USA, and Germany. He was awarded the Postgraduate Certificate in Education (PGCE) from the University of Wales, an MA in history from the University of Connecticut, and a PhD in history from the University of Wales. For the past twelve years Dr. Lightfoot has been an active member of the American Association for Collegiate Independent Study (AACIS). He is a frequent presenter at AACIS conferences and is currently serving as the chair for both the Archives Committee and Awards Committee.

Deborah Lundberg Windes earned her PhD in Organizational Behavior at the University of Illinois at Urbana-Champaign. She has developed and taught online and distance courses for 21 years. In her current position as Visiting Program Director for Distance Education in the Office of Continuing Education at the University of Illinois, she has assisted in developing and coordinating a number of online programs that serve both undergraduate and graduate audiences. Her current research interests include faculty perspectives on online education, an examination of who the innovators are as business schools go online, and retention initiatives in online programs.

Sue Mae, Ph.D., is the Dean of the Division of Continuing Education at Kansas State University and Co-director of the Institute for Academic Alliances, established to encourage and assist higher education institutions in developing and delivering collaborative online academic programs. Maes served as President of the University Continuing Education Association. She was a founding member of two national organizations: the Rural Clearinghouse for Lifelong Education and Development, and the Association of Community Based Organizations. In 2000 she was elected into the International Adult and Continuing Education Hall of Fame and later served as the Hall's president.

Maria Hruby Moore, Ph.D. is currently an Assistant Professor at Ohio Dominican University, Columbus, Ohio in Human Resource Management. Dr. Moore was a Senior Research Specialist with The Ohio State University, Center on Education and Training for Employment (CETE), where she served as a Co-principal Investigator for the Ohio Transition to Teaching program from 2006-2008 and was a member of the grant and production team since the start of OhioTtT in 2002. In 2007, she became Lead for Strand 1: Online Learning Community and e-Tutorial Development for Project KNOTtT (2007-2012). Maria Moore received her Ph.D. from OSU in Workforce Development and Lifelong Learning, M.A. in Adult Education and Human Resource Development, and M.B.A. from Case Western Reserve University in Human Resource Management. Dr. Moore's scholarship and research interests include distance education, adult learning, training, professional and career development.

Virginia Moxley, Ph.D., is a professor at Kansas State University where she is Dean of the College of Human Ecology and Co-director of the Institute for Academic Alliances. Moxley was a founding member of the Great Plains Interactive Distance Education Alliance and served as the first Board Chairman. Moxley's scholarship focuses on policies and practices that sustain multi-institutional academic alliances. Moxley is a founder of the Undergraduate Research Community for the Human Sciences and currently serves as a director on the Board on Human Sciences.

Terrie Nagel is the Assistant Director for Student Services at the Center for Distance and Independent Study at the University of Missouri. She has published articles on the impact of goal orientation and motivation on course completion and has won national student services and research awards. Nagel epitomizes the non-traditional learner, returning to graduate school nearly two decades after completing her undergraduate studies. As part of the requirements for her doctorate in educational psychology, Nagel is currently completing her dissertation, a hierarchical linear model of independent study. Her interest in lifelong learning was fostered by her parents, a teacher and an engineer, and strengthened by her own circuitous route.

David Schwager has more than sixteen years experience working in test development and the assessment of "non-traditional" and adult learners. Initially employed by Educational Testing Service (ETS) and The Chauncey Group International (a wholly-owned subsidiary of ETS), he served in diverse roles within the Statistical Analysis area, supporting test development activities on various types of assessment programs, including CLEP and DANTES. For the past ten years, Mr. Schwager has worked in the Office of Learning Assessment at Thomas Edison State College, where he has had the opportunity to expand beyond statistical analysis and take an active role in all phases of test development.

Mingming Shao received her first master's degree in Education from Northern Arizona University. In 2001, she completed her second master's degree in Computer Information Systems in Dakota State University. Since 2001, she has been working as an instructional technologist, providing support services in instructional design and technology with the Extended Programs at Dakota State University.

Evan S. Smith grew up in Kansas, the son of a professor and an artist. An interest in languages and culture led him into graduate work in linguistics at the University of Kansas (M.A., 1975) and Indiana University (Ph.D., 1982). He taught linguistics for 5 years, also becoming interested in editing and producing academic works. All this led to independent study at Indiana (1985-1989) and the University of Missouri (1989-present). Smith has presented at many conferences, particularly AACIS and UCEA, and has published articles in print or online journals between 1994 and the present. Personal interests include movies, history, Anglo-Celtic folk music, and dioramas.

Haomin Wang has a master's degree in applied linguistics and an Ed D in instructional technology. He has several years of experience in designing and developing online courses, and supporting faculty in offering online and hybrid courses. He is currently the manager of instructional technology, and also an associate professor teaching courses in instructional technology at Dakota State University.

Peg Wherry is Director of Online and Distance Learning at Montana State University in Bozeman. She has worked in distance learning for over twenty years and has led the development and implementation of online programs since 1995, when she joined the team developing Weber State University's award-winning WSU Online. Her experience with academic integrity issues goes back to her days as a graduate teaching assistant in the English Composition program at Kansas State University in the 1970s. During the time period covered by this case study, she was Assistant Head of Academic Outreach at the University of Illinois at Urbana-Champaign.

Index

D

E

F

G

H

I